MEDICAL RADIOLOGY
Diagnostic Imaging

Editors:
A. L. Baert, Leuven
K. Sartor, Heidelberg

Springer

Berlin
Heidelberg
New York
Hong Kong
London
Milan
Paris
Tokyo

H.-U. Kauczor (Ed.)

Functional Imaging of the Chest

With Contributions by

A. A. Bankier · C. Beigelmann-Aubry · A. Capderou · P. Cluzel · S. R. Desai · B. Eberle
C. Fetita · C. Fink · P. A. Gevenois · P. A. Grenier · D. M. Hansell · H. Hatabu · P. Herzog
H.-U. Kauczor · K. Markstaller · M. U. Niethammer · F. Preteux · S. Schaller · U. J. Schoepf
C. Straus · A. Swift · H. Uematsu · E. J. R. van Beek · J. Verschakelen · J. M. Wild
J. E. Wildberger · M. Zelter

Foreword by

A. L. Baert

With 124 Figures in 219 Separate Illustrations, 48 in Color and 5 Tables

Springer

HANS-ULRICH KAUCZOR, MD
Professor of Radiology
Innovative Krebsdiagnostik und Therapie
Deutsches Krebsforschungszentrum (DKFZ)
Im Neuenheimer Feld 280
69120 Heidelberg
Germany

MEDICAL RADIOLOGY · Diagnostic Imaging and Radiation Oncology
Series Editors: A. L. Baert · L. W. Brady · H.-P. Heilmann · M. Molls · K. Sartor

Continuation of Handbuch der medizinischen Radiologie
Encyclopedia of Medical Radiology

ISBN 978-3-642-62202-1   ISBN 978-3-642-18621-9 (eBook)

DOI 10.1007/978-3-642-18621-9

Library of Congress Cataloging-in-Publication Data

Functional imaging of the chest / H.-U. Kauczor (ed.) ; with contributions by A.A. Bankier
  ... [et al.] ; foreword by A. L. Baert.
      p. ; cm. -- (Medical radiology)
  Includes bibliographical references and index.
  ISBN 978-3-642-62202-1 (alk. paper)
    1. Chest--Imaging. I. Kauczor, H.-U. (Hans-Ulrich), 1962- II. Series.
    [DNLM: 1. Lung--physiology. 2. Magnetic Resonance Imaging. 3. Tomography, X-Ray
  Computed. WF 600 F979 2004]
  RC941.F86 2004
  617.5'40754--dc21   2003050386

http//www. springeronline.com
© Springer-Verlag Berlin Heidelberg 2004
Originally published by Springer-Verlag Berlin Heidelberg New York in 2004
Softcover reprint of the hardcover 1st edition 2004
The use of general descriptive names, trademarks, etc. in this publication does not imply, even in the absence of a specific statement, that such names are exempt from the relevant protective laws and regulations and therefore free for general use.

Product liability: The publishers cannot guarantee the accuracy of any information about dosage and application contained in this book. In every case the user must check such information by consulting the relevant literature.

Cover-Design and Typesetting: Verlagsservice Teichmann, 69256 Mauer

21/3150xq – 5 4 3 2 1 0 – Printed on acid-free paper

Foreword

For many years the role of diagnostic radiology of the lung was based on morphological signs and features. Recent progress in CT and MRI techniques has, however, opened exciting new perspectives for the study of lung function and another new frontier for the discipline of radiology has to be conquered.

This outstanding book covers exhaustively all novel CT and MRI techniques for the investigation of lung ventilation, perfusion, gas exchange and respiratory chest mechanics.

I am deeply indebted to Prof. H.-U. Kauczor, a leading international expert in functional CT and MRI of the lungs, for accepting the challenging task of editing this volume. I would like to congratulate him and the group of internationally renowned specialists who have contributed chapters to the book for their up-to-date coverage of the topic with comprehensive references and numerous well-chosen illustrations.

This superb work will certainly meet with great interest not only from radiologists but also from pneumologists and nuclear medicine specialists involved in research on lung physiology and pathophysiology. I am confident it will encounter the same success with readers as previous volumes published in this series.

Leuven

ALBERT L. BAERT

Preface

Radiological imaging in respiratory medicine, such as chest radiography and computed tomography (CT), was long restricted to the visualisation of morphology and structure, whereas the different aspects of pulmonary function were assessed by a wide battery of pulmonary function tests. Functional imaging of the lung could be achieved only by using nuclear medicine techniques, such as ventilation and perfusion scintigraphy. With the introduction of novel CT technologies, such as high-resolution CT, helical and multislice CT, the capabilities of radiological imaging of structure exceeded general expectations and even surpassed the level of macroscopic inspection.

Chest radiologists, however, soon realised that the mere visualisation of the normal structure of airways and lung parenchyma as well as their pathological changes might not be sufficient to affect therapy decisions. Instead of treating changes of lung structure visible on a CT scan, clinicians will rely on clinical symptoms and measurements of functional impairment, such as hypoxia in respiratory insufficiency, to initiate or change treatment. Thus, chest radiologists investigated structure-function relationships with the idea that the severity of structural pathology will relate to functional compromise. Most of these studies have been carried out using CT, which obviously provides the best representation of structure. However, there are profound methodological differences between global testing of pulmonary function, using a spirometer at the mouth or a whole-body plethysmograph, and high-resolution cross-sectional imaging demonstrating macroscopic cuts of lung tissue without significant superimposition. Nevertheless, the investigation of structure-function relationships in chest radiology has provided significant new insights into lung physiology and respiratory diseases. With the recent improvements in scanner technologies, such as dynamic CT with continuous rotation, multislice CT and magnetic resonance imaging (MRI) with intravenous or inhaled contrast media, both modalities, CT and MRI, have gained a direct approach to real functional imaging as reviewed in this book. The dedicated development and application of such techniques now enable radiologists to provide detailed and regional information about pulmonary perfusion, ventilation, gas exchange and respiratory mechanics. MRI has been the biggest driver in functional imaging of the chest in recent years. With more and more different developments becoming available a lot of interest has been attracted to MRI of the lung, a hitherto widely neglected application. Today, the first steps have been successfully accomplished. Current developments and collaborations of functional imaging strategies strive for technical improvements, especially volume coverage with simultaneous high spatial and dynamic temporal resolution, as well as advances in the interpretation of results, such as the definition of thresholds between physiological variation and pathological changes. At the same time quantitation is a major requirement to achieve acceptance as a surrogate measurement of treatment effects.

The prediction that the pulmonary function tests of the new millennium will be derived from imaging techniques appears to be becoming a reality. A great advantage of the new imaging approaches is the ability to obtain regional function information which can be localised to particular lung zones. The general need for volumetric data acquisition will further promote widespread acceptance. At the same time as pulmonary physicians are learning about the potential and the benefits of image-derived functional parameters, radiologists would be wise to refamiliarise themselves with pulmonary physiology and the basic principles of lung function analysis. Qualitative and quantitative image-based assessments of lung function generate novel parameters which require a new understanding and appropriate appraisal by the respiratory physicians before these results will be accepted and gain clinical and therapeutic impact. This is a major task for ongoing interdisciplinary collaborations.

For me as the editor of this volume it was an enormously invigorating experience bringing together renowned experts, with support from Professor Albert Baert and the publisher, Springer, to provide the first well-illustrated and comprehensive textbook of these novel approaches to "Functional Imaging of the Chest". I hope the book will be of great assistance to all who are establishing functional image-based strategies in the research and clinical arenas for diagnosis and follow-up of patients with lung disease.

Heidelberg   HANS-ULRICH KAUCZOR

Contents

1 General Role of Imaging in the Evaluation of Diffuse Infiltrative and Airways Diseases

DAVID M. HANSELL

CONTENTS

1.1
Definition and Purpose of "Functional Imaging"

Patients with lung disease are usually investigated by a combination of imaging and physiological tests. Historically, there has been only modest expectation that there should be obvious concordance between the results of these two different types of test (which assess structure and function respectively). For example, some individuals with functional evidence of advanced emphysema have a normal chest

D. M. HANSELL, MD, FRCP, FRCR
Professor of Thoracic Imaging, Department of Radiology, Royal Brompton Hospital, Sydney Street, London SW3 6NP, UK

radiograph; conversely, some patients with obvious radiographic features of pulmonary sarcoidosis have pulmonary function tests within the normal predicted range (Fig. 1.1). The failure to successfully reconcile structural abnormalities with functional disturbances led to a somewhat nihilistic view, typified by a remark in the 1970s of an anonymous physiologist: "what you see, or think you see (on a chest radiograph), is a pale reflection of physiological reality." Although insights into the pathophysiology of lung disease were gleaned from chest radiography by meticulous analysis, for example in the evaluation of pulmonary oedema (MINIATI et al. 1988; PISTOLESI et al. 1986), there were, and are, several constraints on the extent and level of information that could be extracted from a chest radiograph.

Striving to understand structure-function relationships may be considered an unfashionable pursuit given the current narrow imperative for diagnostic efficiency. Nevertheless, there are compelling reasons, over and above the desire to increase the sum of

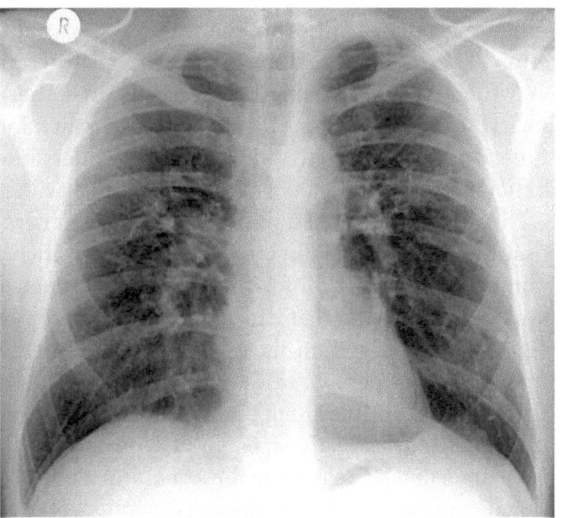

Fig. 1.1. Chest radiograph of a patient with sarcoidosis. The patient was dyspnoeic on exertion but pulmonary function tests of lung volume, airflow and gas diffusing capacity were within normal predicted values

knowledge about the pathophysiology of pulmonary disease, to pursue what may sometimes be regarded as blue-sky research: the discipline of structure-function investigations lends insight into why certain tests may, in diagnostic terms, be insensitive (if not actually misleading) and, over time lead to their abandonment. In a sense there may be leap-frogging between tests; for example, the insensitivity of chest radiography in the face of functional and biopsy evidence of interstitial lung disease (EPLER et al. 1978) was followed by the realisation that abnormalities shown on high resolution computed tomography (HRCT) are not invariably reflected by deranged pulmonary function tests (GURNEY et al. 1992).

The term *functional imaging* is open to two broad interpretations: one is the idea that an understanding of functional abnormalities can be reached by correlation with objective quantification of morphological abnormalities shown on detailed images of the lungs. The second, resulting from recent technological developments, is the ideal of quantifying specific functional characteristics of the lungs (for example, pulmonary ventilation) and depicting the data pictorially (for example as a map of regional ventilation). Many of these techniques have relatively low spatial resolution; the optimal functional investigation of the lungs would combine high spatial resolution and provide precise regional information about functional capacity; in this respect, pulmonary function testing supplies only global estimates of function, hence the attraction of the latest techniques that can provide regional functional imaging.

In most chronic diffuse lung diseases, imaging and functional abnormalities are central to clinical evaluation. The intimate relationship between structure and function has recently been brought into sharp focus by high resolution CT. Studies of structure and function have enabled radiologists to understand better the functional information contained within CT images. Future possibilities include the generation of integrated CT/functional reports that incorporate the information-for-free, such as volumetric lung measurements that can now be made more or less automatically from the latest generation of CT scanners. Considerable work has already been done on the "purer" diffuse lung diseases, for example the meticulous investigations of GEVENOIS et al. that have done much to elucidate the structural and functional consequences of emphysema (GEVENOIS et al. 1996a,b; GEVENOIS and ESTENNE 2001). Using data from the latest generation of CT scanners coupled with increasingly sophisticated post-processing, the way is now open to explore other diffuse lung

diseases that are characterised by combinations of interstitial, airway-centred and other pathology. It is possible to apportion the degree of functional impairment resulting from a mix of pathological processes; an example being asbestos-exposed individuals with coexisting uniform pleural thickening, interstitial fibrosis, and centrilobular emphysema (Fig. 1.2). Many of the considerations that apply to investigation of structure-function relationships using computed tomography apply to more recently developed techniques.

1.2
Imaging Techniques Historically and Currently Used for Structure-Function Studies

1.2.1
Chest Radiography

The projectional chest radiograph contains a wealth of diagnostic information, but by its 2-D and static nature it cannot provide accurate volumetric or temporal data. Nevertheless, the very earliest radiographic examinations, which were performed fluoroscopically, allowed crude functional assessment – for example, Oudin and Barthélemy reported, less than one year after Röentgen's discovery, that it was possible to assess cardiac and aortic pulsa-

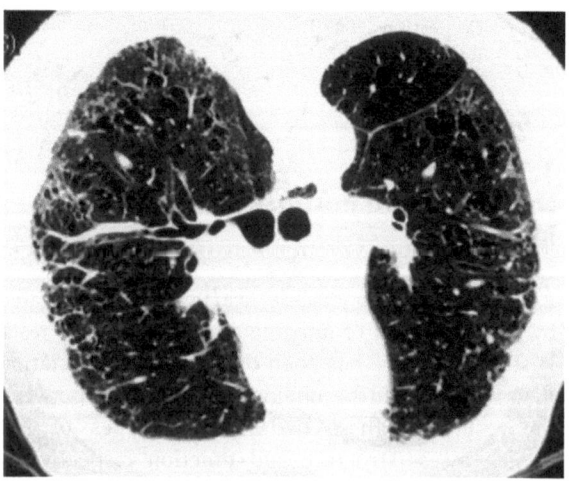

Fig. 1.2. Prone CT of an individual with asbestos-related pleural and parenchymal disease: both the pulmonary fibrosis and extensive visceral pleural thickening contribute to a restrictive functional deficit; this is counteracted to some extent (in plethysmographic measurement) by the centrilobular emphysema which is interspersed between the reticular pattern

tion and the excursion of the diaphragm (WALSHAM and HARRISON ORTON 1906). The vogue for careful measurements of the various dimensions of the chest and mediastinal outline was followed by meticulous radiographic-anatomic correlations (as an aside, it remains a matter of surprise that the true nature of the branching structures shown within the lungs on a plain chest radiograph represented blood-filled vessels, rather than bronchi, occurred as late as 1922; LODGE 1946). Many subtle inferences were made about the nature of density changes seen on plain radiographs in patients with both emphysema and pulmonary fibrosis, and their functional effects on the excursion of the diaphragm were explored in the very early days of chest radiography (STENDER and OESTMANN 1997). Paradoxically, advances in radiographic technique during the first half of the last century, primarily aimed at reducing exposure times, meant that real-time fluoroscopic examinations fell into abeyance; whilst such refinements improved the detection, if not the accurate quantification, of diseases of the lung parenchyma, the momentum to investigate the temporal aspects of pulmonary and cardiac function were lost.

Objective means of quantifying lung disease from a plain chest radiograph became possible with the advent of digital chest radiography. Early work employed digitisation of conventional radiographs, and this was rapidly followed by the development of digital acquisition devices, most recently solid-state thin-film transistor flat panel x-ray detectors (CHOTAS et al. 1999). To approach the spatial resolution of a conventional analogue film radiograph, a picture element (pixel) size of approximately 0.2 mm or smaller, is required with a grey scale of upwards of 1024 levels; this basic requirement predicates up to 8 megabytes of digital data per chest radiograph. The fundamental advantage of digital chest radiography is the ability to subject the data to post-processing to enhance or extract specific features (KIDO et al. 1997).

Despite the obvious attractions of interrogating digital data, the subjective (visual) quantitation of disease has several attractions and, for many years the International Labour Organisation (ILO) classification for the characterisation and quantification of occupational lung disease has been successfully employed, despite the known interobserver variation and other problems (EPSTEIN et al. 1984). The assumption that the functional effects of lung or pleural disease are more accurately evaluated by digital rather than visual quantification does not always hold true: in a study that investigated various methodologies for estimating the extent, and functional consequences,

of pleural disease, a relatively crude visual scoring of the extent of pleural disease on chest radiography was as effective, in functional terms, as that derived from a sophisticated computer-generated mapping of pleural disease (COPLEY et al. 2001).

It is easy to overlook the fact that the projectional chest radiograph can, in specific instances, provide more readily accessible information – most notably the zonal distribution of disease (GURNEY and SCHROEDER 1988) and state of inflation of the lungs – than more sophisticated imaging, such as CT; as an example, the radiographic assessment of hyperinflation of the lungs, in ventilated patients in an intensive care setting, has been shown to correlate very well with higher tidal volume ventilation (JOHNSON et al. 1998). Furthermore, redistribution of zonal blood flow in the lung assessed by visual inspection of chest radiographs alone approaches the performance of sophisticated computer-aided identification of redistribution of blood flow (as judged by the calculation of physical measures of the linear and branching structures in different lung zones) (KIDO et al. 2000).

1.2.2
Computed Tomography

Over the last thirty years computed tomography (CT) has developed from a relatively crude cross-sectional morphological imaging technique to a tool with the potential for many different types of functional imaging. In terms of fine anatomical detail of the lung parenchyma, computed tomography reached its apogee ten years ago, in the form of high-resolution computed tomography (HRCT); there have been no recent technological developments to further the spatial resolution of high-resolution CT using conventional commercially available CT scanners (the minimum resolvable object size being approximately 200 microns). One of the main benefits of the fine detail available from HRCT, in terms of functional imaging, is the ability to distinguish between areas of lung parenchyma showing different disease patterns – for example, abnormalities corresponding to emphysema and interstitial fibrosis respectively; the ability to quantify individual disease processes on a projectional chest radiograph is, by comparison, extremely limited.

As the speed of anode rotation around the patient (and therefore data acquisition) has increased with each succeeding generation of CT scanner, there has been some interest in the use of this increased temporal resolution in acquiring repeated images at a single anatomic level. This intermittent data acquisition has been used to quantify (in terms of Hounsfield Unit changes of lung density) the ingress and egress of intravenously administered contrast medium in any region of interest of the lung (JONES et al. 2001); post-processing of the data can give a highly detailed map of regional variations of perfusion (Fig. 1.3). This kind of time/density analysis can be usefully combined with the better morphological definition obtainable from high-resolution CT (HOFFMAN et al. 1995). Another technique which has relies on the demonstration of serial CT density changes is ventilation mapping with inhaled non-radioactive xenon (HERBERT et al. 1982; SNYDER et al. 1984), but the inherently poor signal-to-noise ratio of the technique has precluded its further development. With the advent of the electron beam CT technology, near-continuous data acquisition is possible such that rapid physiological events (for example, density changes within the lungs on a forced expiratory manoeuvre) can be captured. With the latest generation of mechanical CT scanners, which incorporate slip-ring technology, continuous data acquisition – at a single anatomical level – is now a reality.

The combination of continuous anode rotation with continuous table feed and multi-detector arrays means that a large anatomical area (e.g. the entire lungs) can be rapidly covered at a given state of lung inflation – in clinical practice, most usually

at full inspiration. It has become a relatively simple matter to make precise volumetric measurements of the lung parenchyma by combining "lung extraction" software techniques (ZAGERS et al. 1995) with spirometrically controlled CT acquisition, such that the exact physiologic state of lung inflation is known at the time of the CT volume acquisition (BEINERT et al. 1995; ROBINSON et al. 1999). The accuracy of measurement of lung volume, as judged by volumetric CT, compares very well with plethysmographic measures (KAUCZOR et al. 1998; MERGO et al. 1998) (Fig. 1.4).

It might seem that the highest morphological, temporal and volumetric resolution are desirable simultaneously, but they are, in some respects, mutually exclusive. Nevertheless, it is now a straightforward matter to acquire a volumetric high-resolution CT examination (ENGELER et al. 1994), but the radiation dose to the subject is considerable, and unwarranted in routine clinical practice; in this respect, given that functional investigations using CT tend to the acquisition of large data sets, every attempt to minimise radiation dose should be made (GOTWAY et al. 2000).

1.2.3
Nuclear Medicine and Positron Emission Tomography

Of all the imaging techniques, nuclear medicine is the one most often thought of as providing "true" functional, or physiological, insights. The ubiquitous

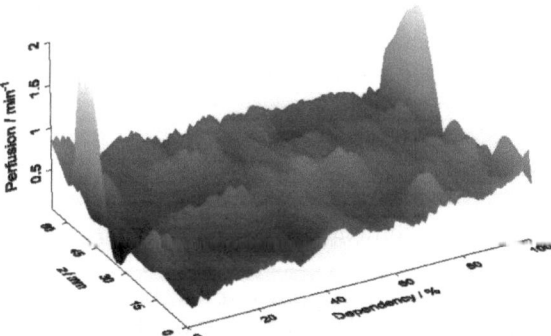

Fig. 1.3. Three-dimensional rendition of regional pulmonary perfusion (vertical axis) in terms of spatial position in a supine individual. Perfusion in the dependent to non-dependent direction is shown on the x axis (100%=the most dependent lung). The cranio-caudal location is provided by the z axis; the total distance of the lung examined in the z axis was 7 cm (centred on the pulmonary artery bifurcation). The data is derived from analysis of time-density analysis of the wash-in and wash-out of contrast through the lungs. The two outlying "peaks" are artefactual (courtesy of Drs J. Dakin, Royal Brompton Hospital and E. Hoffman, University of Iowa)

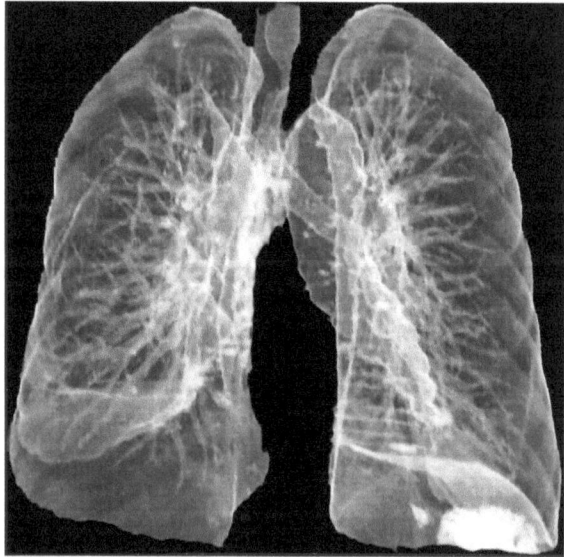

Fig. 1.4. Surface rendered display of the lungs and bronchial tree – precise volumetric data can be readily extracted which correlates well with functional measures of lung volume

ventilation-perfusion scan has served well over the years as a test for pulmonary embolism but, equally importantly, many physiological insights have been gained from studies of regional distribution of both ventilation and perfusion in many different disease states. Furthermore, much of the basic work on the mathematical concepts used to derive tissue perfusion and blood volume measurements in nuclear medicine studies have been transferred to other techniques (for example, the examination of regional pulmonary perfusion using time-density curve for the analysis of a bolus of contrast washing in and out of the lungs on CT) (MILES 1991). Nevertheless, the relatively low spatial resolution, and in the case of perfusion imaging, indirect, nature of evaluating perfusion with nuclear medicine, means that the technique is being supplanted by newer techniques, for example regional assessment of ventilation with hyperpolarized gas on MRI.

The range of radiopharmaceutical agents that can be mapped and tracked is extremely wide and in addition to the classical ventilation and perfusion agents, the dynamics of tissue inflammation and fibrosis, or the trafficking of labelled white cells to sites of infection, can be investigated. There is the potential to radio-label the basic building blocks of the inflammatory process, for example collagen, and in this way "image" the evolution of interstitial pulmonary fibrosis. These considerations apply particularly to PET scanning, given the huge number of molecules entailed in organic processes that can, theoretically at least, be labelled and imaged.

determinants of signal intensity, as displayed in an MR image, flowing blood with or without gadolinium contrast enhancement, lend a further degree of complexity to what is depicted by MR. In terms of functional imaging, the simple use of contrast agents and imaging during the first pass is in many ways analogous to the use of time/density information derived from contrast-enhanced CT. With the introduction of blood pool contrast agents perfusion measurements can be performed during first pass and steady state and quantitation will be easier and more accurate.

Of particular current interest is the application of MR to mapping regional pulmonary ventilation using a variety of gases. Two inert noble gases (3-helium and 129-xenon) have been most intensively investigated to date (KAUCZOR et al. 1996; MCADAMS et al. 2000; SAAM et al. 2000). The attraction of imaging the airspaces of the lung with hyperpolarized noble gases, over and above a conventional radionuclide ventilation scan, are several. In particular, an apparent diffusion coefficient – which reflects alveolar size (and therefore is modified if there is alveolar disruption, for instance in the earliest stages of emphysema) can be derived and displayed as a regional map (KAUCZOR et al. 2001; SAAM et al. 2000). A further recent development has been the exploitation of the weak paramagnetic properties of oxygen, such that oxygen-enhanced MR imaging provides some information about oxygen diffusion from the airspaces into the capillary bed of the lungs (EDELMAN et al. 1996; OHNO et al. 2001; MULLER et al. 2002).

1.2.4
Magnetic Resonance Imaging

Because of the recent developments in volumetric acquisition of the latest generation of CT scanners, there has been a tendency to emphasise the similarities in data available from CT and MR examination; whilst the near-isotropic capabilities of both MRI and CT mean that volumetric renditions (for example, the segmented lungs) are equally feasible with either technique, the fundamental differences in acquisition, and what finally is represented in the data sets, can easily be overlooked. The simple relationship between the CT voxel signal intensity and the attenuation coefficient of different tissues is in marked contrast to the complex interplay between up to ten parameters that determine the final content of the picture elements of an MR image. As well as the myriad of pulse sequences and pathophysiological

1.3
Post-Processing of Image Data and Aspects of Disease Quantification

The most widely cited technique for quantifying diffuse lung disease on CT is subjective visual estimation expressed as the proportion of lung involved, derived from summed lobar scores or selected sections; these methodologies can be applied to any other volumetric imaging technique. The main disadvantage of a subjective approach is variation between observers, but this is balanced by the speed and simplicity of the technique which allows large numbers of subjects to be studied, and so lends statistical power. The instances in which computer-aided recognition and quantification of disease is faster than human evaluation is mainly limited to pure conditions, characterised by relatively simple

CT patterns, such as areas of abnormally decreased attenuation in the case of emphysema (ZAGERS et al. 1995). Subjective visual evaluation of disease extent on CT has been used in a variety of diffuse interstitial and airway diseases (ABERLE et al. 1990; GURNEY et al. 1992; REMY-JARDIN et al. 1994; GAMSU et al. 1995; WELLS et al. 1997a; NG et al. 1999).

Defining the precise features that allow an observer to discriminate visually between areas of normal and diffusely abnormal lung on CT is not always straightforward. The basic "cues" that alert an observer to the presence of disease include: 1) texture changes, usually given simple descriptions such as "reticular pattern"; 2) disturbances of the normal lung architecture – seen as distortion or straightening of the normal branching pattern of the pulmonary vessels, reflecting underlying fibrosis or emphysema, and 3) density changes of the lung parenchyma, typically patchy inhomogeneity of the density of the lungs (the so-called *mosaic attenuation pattern*). In some instances pathophysiological inferences can be drawn from the anatomical location and pattern of diseased lung. For example, minor permeative destruction of the lung parenchyma confined to the upper lobes in a cigarette smoker can be confidently interpreted as representing functionally silent centrilobular emphysema. Nevertheless, over-interpretation of the exact pathological nature of the macroscopic patterns seen on HRCT images is a hazard (see later section on Specific Problems in studies of structure-function). There are numerous permutations in the way in which a "total score" of lung abnormality, whatever its pattern, can be arrived at from the visual analysis of a set of (transaxial) HRCT images. The extent of disease may be graded according to a coarse categorical scale or a fine near-continuous (e.g. to the nearest 5%) scale. Grading may be undertaken at a lobar (or segmental) level or confined to selected (for example, five) levels. Sampling error is inherent in any evaluation of a limited number of sections, however the differences in the degree of error between, for example, three interspaced sections and 10 mm interspacing is less than might be anticipated (KAZEROONI et al. 1997). A weighting factor may be applied to the scores/estimates at each level, to correct for the differing contribution made by each level (an upper zone section has a smaller cross-sectional area, and therefore contributes to the total lung volume, less than a lower zone section) (GURNEY et al. 1997).

A variety of simple visual systems have been used to quantify mosaic attenuation (HANSELL et al. 1997; ARAKAWA and WEBB 1998; LUCIDARME et al. 1998).

In chronic airways disease, air-trapping has been measured on expiratory CT using a grid counting system (ARAKAWA et al. 1998). However, this type of system is laborious. Subjective visual systems, sometimes semi-automated, are often favoured for quantitative studies of small airways disease simply because there is admixture of interstitial disease resulting in highly heterogeneous lung density (the so-called head cheese sign) which confounds automated quantification (Fig. 1.5). The "coarseness" of visual scoring systems may have a marked effect on observer agreement and thus their effectiveness: using a fine grading system, i.e. the extent of the decreased attenuation component of the mosaic pattern to the nearest 5%, quantitation is unsatisfactory as judged by the levels of inter- and intra-observer variation; however, a semi-quantitative (coarser) system yields better levels of agreement, at the cost of some loss in discriminatory power (NG et al. 1999).

Areas of decreased attenuation due to abnormal air-trapping are quantified more easily with expiratory than inspiratory CT images, due to a generalized increase in the attenuation of normal lung parenchyma on expiratory scans, enhancing the contrast between normal and affected lung. In conditions characterised by small airways dysfunction, expiratory CT demonstrates areas of decreased attenuation more frequently than inspiratory CT. The areas of decreased attenuation are greater in extent (NG et al. 1999), due, in part, to recruitment of new areas of air-trapping on expiration and their greater conspicuity, but also to

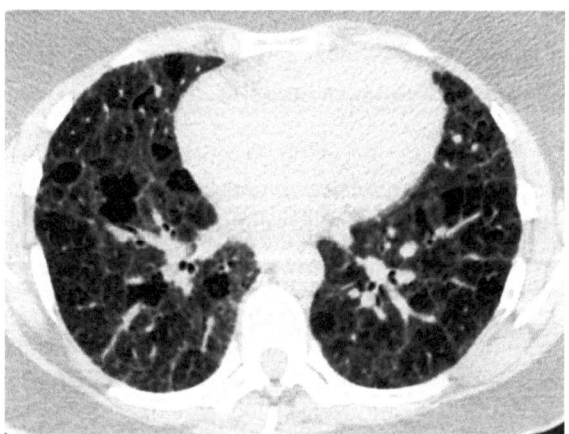

Fig. 1.5. High-resolution CT through the mid zones of a patient with hypersensitivity pneumonitis: there is a mixture of ground glass opacification and a reticular pattern. In addition, there are secondary pulmonary lobules of decreased attenuation which reflect a bronchiolitic component. Such a complex combination of patterns does not readily lend itself to automated quantification

pre-existing areas of air-trapping seen on inspiratory scans, which necessarily occupy a greater proportion of lung on expiratory scans. Furthermore, observer confidence and agreement is substantially higher on expiratory CT than inspiratory CT in conditions such as hypersensitivity pneumonitis, sarcoidosis and asthma (NG et al. 1999).

There are obvious advantages in the use of objective quantitative methods. Objective techniques have been most widely applied to the mapping of emphysema (ADAMS et al. 1991; KINSELLA et al. 1990; KEYZER and GEVENOIS 1999), most commonly the use of a "density mask" to highlight pixels below a certain critical value (KINSELLA et al. 1990). However, the automated identification of areas of decreased attenuation in isolation (without reference to the supplementary signs of the disposition and calibre of the pulmonary vessels) may be misleading: artefactual causes of decreased density of the lung parenchyma include kilovoltage drift and beam hardening from adjacent ribs.

In emphysema, the paradigm of a disease characterized by areas of low (air) density on CT, the extent of disease has been extensively investigated using both subjective scoring and objective density mask techniques. Using precise morphometric measurements of resected lung tissue, BANKIER et al. have shown that observers, irrespective of experience, tend to overestimate the extent of emphysema on CT, whereas CT densitometry correlate better with the morphometric reference (BANKIER et al. 1999). There are attractions in using an objective technique for the quantification of diseased lung (KALENDER et al. 1991), but there is no clear cut difference in density of the lung parenchyma on CT in areas of lung affected by some small airways diseases, such as constrictive obliterative bronchiolitis. Objective CT measurements of diffuse lung disease are best suited to conditions in which there is a definite dichotomy between the density of normal and abnormal lung (such as in emphysema). With the increasing refinement of volumetric acquisition protocols it is now possible to derive, simply and accurately, lung volumes (at any state of inflation if monitored by spirometry) (KAUCZOR et al. 1998; BROWN et al. 1999). Furthermore, it is feasible to apply density thresholding techniques to volumetric data (ZAGERS et al. 1996; MERGO et al. 1998). The lack of studies that have used objective CT measures of disease extent reflect the fact that most diffuse lung diseases have a wide range of lung densities and patterns on CT.

Because the attenuation differences between normal and abnormal areas of lung are sometimes extremely subtle, any method of detecting or enhancing the large area low contrast differences are potentially valuable (YANG and HANSELL 1997; FOTHERINGHAM et al. 1999). Ground glass opacification is considered a "visual sign". However, in some situations, significant disease that ultimately progresses to obvious ground glass opacification may commence as a subliminal change in lung density. Several studies have shown that density perturbations may exist in the absence of visually discernible CT abnormalities. For example, asbestos-exposed individuals may, in the absence of any clinically evident interstitial fibrosis, have a measurable, but invisible, increase in lung density (WOLLMER et al. 1987). As a further example, in normal individuals there is a gravity-dependent density gradient in the lungs on CT, such that parenchyma in the dependent lung is denser than non-dependent lung. In patients with pulmonary arterial hypertension, this density gradient is measurably diminished (CAILES et al. 1996), even though changes in the "grey gradient" may not be obvious to the human eye (Fig. 1.6). However, as with density thresholding for emphysema, densitometric techniques are unlikely to be applicable when there is a combination of pathologic processes, resulting in a complex mixture of CT densities and textures.

More sophisticated approaches to the quantification of interstitial and obstructive lung diseases include analysis of the CT density histogram and evaluation of texture characteristics of the lung parenchyma (RIENMULLER et al. 1991; DELORME et al. 1997; UPPALURI et al. 1997). In patients with pulmonary fibrosis, the density histogram is peaked (kurtotic) and skewed to the left; measurements of the skewness and kurtosis correlate with functional indices (HARTLEY et al. 1994). An operator-independent technique which relies on fractal analysis of CT sections has been reported to have good accuracy for the quantification of fibrosing alveolitis (RODRIGUEZ et al. 1995). Recent work suggests that complex texture analysis is able to distinguish between various forms of obstructive lung disease, which are characterised by apparently non-specific (to the naked eye) areas of decreased attenuation lung (CHABAT et al. 2003) (Fig. 1.7). Another advanced technique involves the "preinterpretation" by a hierarchic network of multiple neural networks of an HRCT image for the presence of ground glass opacities, with checking for false positive detection by an experienced radiologist (KAUCZOR et al. 2000).

A pragmatic approach is the combination of relatively simple (objective) post-processing for image feature enhancement with (subjective) visual estimation of the extent abnormality (this has the advantage that observer experience can take account

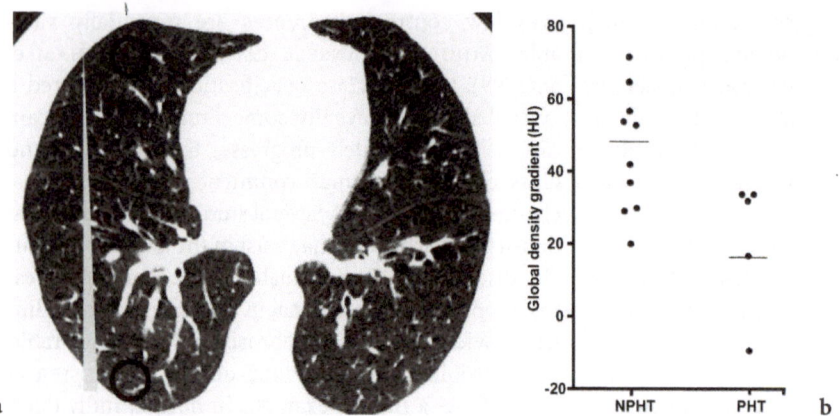

a   b

Fig. 1.6. Patient with systemic sclerosis-related vasculopathy (but no interstitial fibrosis). **a** There is no obvious parenchymal abnormality and in the normal situation. There is a density gradient between the dependent and non-dependent lung (i.e. between the two circular regions of interest). In patients with pulmonary hypertension there is a reduction in the normal gravity-induced density gradient. **b** This trend is shown in the scattergram taken from CAILES et al. (1996) which shows less of a measurable gradient in patients with pulmonary hypertension (PHT) compared to controls (NPHT) – probably reflecting reduced vascular compliance

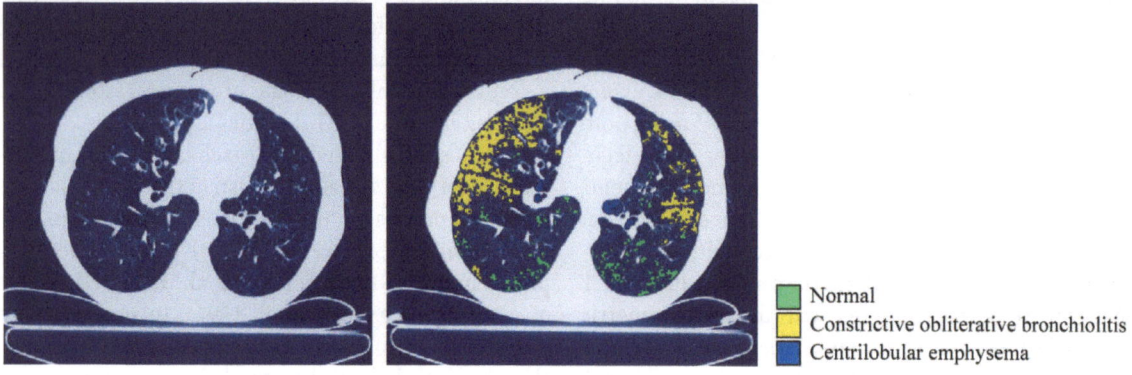

Normal
Constrictive obliterative bronchiolitis
Centrilobular emphysema

Fig. 1.7. Texture classification of lung parenchyma. On the CT section of a patient with constrictive bronchiolitis, the algorithm correctly identifies much of the lung as having a texture consistent with constrictive bronchiolitis (the uncoloured areas of lung have not been assigned a likely texture by the algorithm). The algorithm has misclassified some areas as centrilobular emphysema

of artefacts introduced by the image processing). An example of this approach is the application of Minimum Intensity Projection (MinIP) images and similar post-processing to the quantitation of a mosaic pattern in small airways disease (BHALLA et al. 1996; YANG and HANSELL 1997).

1.4
Physiologic Parameters: Considerations Relevant to Structure-Function Correlative Studies

A detailed discussion of the numerous routinely available and more sophisticated tests of pulmonary func-

tion is not warranted but some generalisations and illustrative points may be helpful for understanding structure-function correlative studies. Pulmonary function tests can be categorised as those that measure the rate of airflow in and out of the lungs, the volume of the lungs at various states of inflation, and the efficiency of gas diffusion and exchange; these tests define the global functional impact of diffuse lung disease, but these tests are modulated by many technical and patient factors. Blood gases are a global measure for overall gas exchange and oxygenation, whereas more detailed data about lung perfusion can only be obtained by invasive measurements or imaging. Diseases that affect the airways will generally be reflected by a reduction in measured airflow and an increase in plethysmographic lung volumes because of air-trapping and hyperinflation. By contrast, interstitial lung diseases will cause a reduction in lung volume and impairment of gas-diffusing capacity. Both disease entities can impair gas exchange and oxygenation significantly. Regional hypoventilation or hypoxia will induce compensatory changes on the perfusion side in order to assure perfusion of well-ventilated lung areas with adequate oxygen uptake and restrict shunt perfusion of hypoventilated and hypoxic areas with insufficient oxygen uptake.

The functional consequences of small airways disease are worth considering as a model of the application of pulmonary function tests in evaluating structure-function relationships.

Tests designed to detect airflow limitation due to diseases affecting the small airways can be divided into robust but relatively insensitive tests performed in most lung function laboratories and more refined techniques which were originally developed to identify disease localised to the small airways, particularly in pre-symptomatic patients. The early promise of these latter techniques has not been fulfilled, largely because of their complexity, poor reproducibility and low specificity (BUIST 1984). The most commonly used tests for the detection of airways disease are forced expiratory volume in one second (FEV_1), forced vital capacity (FVC), residual volume (RV) and flow rates at low lung volumes: the $FEF_{25-75\%}$ or $MEF_{25\%}$ (the maximum expiratory flow rate at 25% above residual volume) are used (McFADDEN and LINDEN 1972). A reduction of $MEF_{25\%}$ is not specific for isolated small airways disease, such as constrictive obliterative bronchiolitis, and may be equally pronounced in patients with pure emphysema in whom the absence of elastic recoil allows collapse and obstruction of the small airways on expiration. Moreover, depression of the maximum mid-expiratory flow rate may indicate either early restriction or early obstruction (FULMER and ROBERTS 1980).

The results of tests of airflow are global measurements of the function of upwards of ten million airways, which contribute unequally to airflow. Thus, expectations about the sensitivity of these tests for diseases centred on the small airways need to be modest. Despite this caveat, a reduction of the FEV_1 (or the FEV_1/FVC ratio) to less than 60% of the predicted value has been widely used as an inclusion criterion in studies of patients with constrictive obliterative bronchiolitis (TURTON et al. 1981). Using this criterion, such studies will only include patients with severe and extensive involvement of the small airways. The FEV_1, and more sophisticated measurements of airflow at low lung volumes, may

be perturbed by coexisting interstitial lung disease (FULMER and ROBERTS 1980). In particular, the FEV_1 is affected by many factors including the size and elastic properties of the lungs, bronchial calibre and collapsibility of the airway walls (PRIDE 1971).

The airflow obstruction of constrictive obliterative bronchiolitis is accompanied by varying degrees of hyperinflation of the lungs and the air-trapping is shown as an increase in residual volume (RV). Total lung capacity (TLC) is usually normal or supranormal. In the presence of coexisting interstitial disease, the RV/TLC ratio is theoretically a more appropriate index of air-trapping as it takes lung restriction into account. In a minority of patients with obliterative bronchiolitis, there may be a pattern of lung restriction and it has been argued that these cases reflect complete (rather than partial) obstruction of the small airways, analogous to removing small units of the lung (MINK et al. 1984).

1.5
Early Structure-Function Correlative Studies

Over the years there have been many investigations of relationships between radiographic abnormalities functional impairment in obstructive lung diseases (SIMON et al. 1973; BURKI and KRUMPLEMAN 1980) and granulomatous and fibrosing lung diseases (McLOUD et al. 1982; NUGENT et al. 1989). However, for the most part, correlations between the estimated extent of disease on chest radiography and physiological measures have been moderate or weak.

There are obvious disadvantages in quantifying disease extent from a projectional (radiographic) image, some of which have already been alluded to. As an example, nearly half of the lungs are obscured by overlying structures such as the heart and diaphragm (CHOTAS and RAVIN 1994). Clinically significant diffuse lung disease may be invisible on the chest radiograph. The texture and density of the lung parenchyma on a chest radiograph is subject to patient-related and technical vagaries so that visual discrimination between normality and early diffuse lung disease may be unclear. Furthermore, it is not surprising that computer-aided detection of digital chest radiographs of patients with and without interstitial lung disease does not seem to outperform the human eye (KIM et al. 1988; MORISHITA et al. 1995). The fundamental problem of the two-dimensional representation of a complex volumetric shape cannot be disregarded: the lungs are approximately pyramidal in shape so that a scoring system that divides the lungs on a frontal radiograph into quadrants takes no account of the considerable differences in volume between the lower and upper quadrants (Fig. 1.1); thus cross-sectional and volumetric imaging techniques have an inherent advantage.

Early CT investigations into structure-function relationship showed only weak correlation between disease extent and functional abnormalities (SIDER et al. 1987). The development of high resolution computed tomography (HRCT) has allowed macroscopic examination of the lung parenchyma with a level of detail only otherwise possible from lung biopsy or post mortem specimens. While much work has relied on CT images for disease quantification, the volumetric acquisition of CT data and, more recently hyperpolarized gas MRI, hold promise for specific structure-function investigations.

1.6
Examples of Structure-Function Correlative Studies in Different Disease States

There is a plethora of examples in the literature of correlative studies that demonstrate either close concordance or sometimes surprising discrepancy between imaging features and functional parameters. In what follows consideration is given to specific examples four conditions with different functional profiles will now be discussed in terms of the relationships between function and structure revealed by CT.

1.6.1
Predominantly Interstitial Disease: Idiopathic Pulmonary Fibrosis

At a functional level, idiopathic pulmonary fibrosis (usual interstitial pneumonitis at histologic examination) is an archetypal restrictive lung disease, and is characterised by reduced lung volume and a decrease in gas diffusing capacity. Several functional indices are routinely used in clinical practice to make an assessment of the extent of the disease and monitor progress. However, the superiority of one PFT measure over another in accurately reflecting the true extent of lung disease is controversial. A combined clinical, radiographic and physiologic (CRP) scoring system was devised in an attempt to improve the accuracy of estimating disease extent (or severity) (WATTERS et al. 1986). However, the choice of pulmonary function parameters used in the CRP system, and their weighting, was based on the strength of previously reported correlations between pulmonary function tests and measures of structural derangement at a histopathologic level. Such correlations take no account of the regional inhomogeneity of fibrosing lung disease and account for the discrepant results between studies that have sought relationships between histopathologic severity of disease and physiologic disturbance (FULMER et al. 1979; CHINET et al. 1990). This is in contrast to more recent work which has shown much stronger correlations between PFT data and global assessment with CT of disease extent (WELLS et al. 1997a,b).

Many patients with idiopathic pulmonary fibrosis are cigarette smokers and thus centrilobular emphysema is a common accompaniment. This combination has long been recognized as being responsible for the additive effect on decreased gas transfer and the spurious preservation of lung volumes (DOHERTY et al. 1997). On HRCT, the morphological distinction between emphysematous and fibrotic lung can usually be readily made; furthermore, smoker's centrilobular emphysema is predominantly upper zone in distribution compared with the basal predilection of fibrosing alveolitis. Nevertheless, there is often a "hinterland" in the mid zones where the distinction between the two processes becomes impossible. The relationship between emphysema and interstitial fibrosis is probably more complex than earlier exclusive definitions allowed (SNIDER et al. 1985); it seems probable that many patients with "pure" emphysema have a fibrotic interstitial component to their smoking-related disease (LANG et al. 1994) and this may be obvious on HRCT as a reticular element within emphysematous lung (TONELLI et al. 1997).

In addition to the confounding effect of coexisting emphysema, variable ventilation of fibrotic lung composed of cystic air spaces is also likely to weaken the relationship between disease extent and volumetric PFT indices. Although it seems likely that many of such cystic "dead spaces" are ventilated (STRICKLAND et al. 1993), others may air-trap (MINO et al. 1995). For a given extent of fibrotic lung on CT, total gas diffusing capacity (Dlco) is depressed, reflecting a lack of perfusion (irrespective of whether or not the affected lung is ventilated). It is the differing capacity of fibrotic lung to air-trap within and between patients that reduces the usefulness of measures of lung volumes in estimating lung involvement in fibrosing alveolitis (WELLS et al. 1997a).

1.6.2
Predominantly Airway-Centred Disease: Constrictive Obliterative Bronchiolitis

Diseases affecting the bronchioles are reflected as either direct or indirect signs on HRCT and this has led to renewed interest in the imaging of "small airways disease". In its most basic form, bronchospasm of small airways results in decreased ventilation which, in turn, produces a reflex reduction in perfusion, shown as areas of decreased attenuation (or "black lung") on HRCT (GUCKEL et al. 1999). In small airways diseases, there is a gap between the information provided by computed tomography and its correlation with the traditional global information derived from conventional physiological tests: relatively simple correlative studies confirm broadly that the "black lung" component of the mosaic pattern predicts the degree of airflow limitation and probably reflects small airways disease. The lack of histologic studies and limitations in global pulmonary function tests raise the perennial problem of a satisfactory gold standard for elusive diseases of the small airways, typified by constrictive obliterative bronchiolitis.

The identification of individual CT features which most strongly predict airflow obstruction in constrictive bronchiolitis is of relevance to the study of other diseases characterised by more complex pathophysiology in which there are functional elements of restriction and obstruction. In an early study that sought to correlate pulmonary function abnormalities with the extent of CT features of constrictive bronchiolitis, no significant relationships were found except between the FEV_1 and the number of bronchopulmonary segments containing dilated subsegmental bronchi (PADLEY et al. 1993). The lack of any linkage between structure and function in this study probably reflects the inclusion of a substantial subgroup with the CT features of diffuse panbronchiolitis, which is characterised by a mixed restrictive and obstructive pattern on pulmonary function tests (KING 1993); thus any correlation between indices of airflow obstruction and the CT features representing pure constrictive bronchiolitis were lost. Furthermore, expiratory CT scans were not evaluated, and more general functional indices of air-trapping were used rather than specific tests of small airways function such as the $MEF_{25\%}$ (MCFADDEN and LINDEN 1972). A more recent study of patients with constrictive obliterative bronchiolitis has confirmed a strong correlation between the extent of decreased attenuation on expiratory CT and physiologic tests of small airways function (HANSELL et al. 1997).

It has traditionally been argued that patients with intrinsic small airways disease have preserved total diffusing capacity (DLco) in contrast to patients with emphysema (GELB and ZAMEL 1973; GELB et al. 1973). However, Gelb et al. have shown that the total diffusing capacity does not reliably distinguish between emphysema and small airways disease in patients with severe airflow obstruction (GELB et al. 1996, 1998) particularly when the FEV_1 is less than 1 litre (GELB et al. 1993). By contrast, adjusted gas transfer (Kco) is preserved in most patients with severe constrictive bronchiolitis (HANSELL et al. 1997; GELB et al. 1998) (in contradistinction to the depression of Kco that characterises emphysema). This is an important observation because the distinction between constrictive bronchiolitis and emphysema (notably the panacinar of a-1-antitrypsin deficiency) may be difficult on the basis of CT appearances alone (LYNCH 1993). Even when the CT abnormalities are typical of constrictive obliterative bronchiolitis, they maybe erroneously interpreted as the findings of emphysema (see subsequent section, Specific Problems in studies of structure-function). For these reasons, assimilation of the information supplied by both HRCT and adjusted gas transfer is necessary to make the sometimes difficult distinction between emphysema and constrictive obliterative bronchiolitis.

1.6.3
Mixed Interstitial and Airway-Centred Diseases

The potential for varying degrees of involvement of the airways and interstitium in many chronic diffuse lung diseases explains the complex abnormalities found on pulmonary function testing in some conditions; the elucidation of these mixed diseases has been greatly aided by histopathologic-HRCT and, more recently functional-HRCT studies.

1.6.3.1
Hypersensitivity Pneumonitis

Increased density of the lungs (ground-glass opacification) and nodules are the most frequently reported CT abnormalities and areas of decreased attenuation are also a frequent finding (REMY-JARDIN et al. 1993). There is a strong correlation between the CT feature of decreased attenuation (mosaic pattern) and lung function abnormalities indicative of air-trapping (HANSELL et al. 1996). Airflow obstruction in patients with hypersensitivity pneumonitis has been variously ascribed to emphysema (SEAL et al. 1968), bronchiolitis (SUTINEN et al. 1983) and, in a few individuals, chronic bronchitis (BOURKE et al. 1989), or the development of asthma (KOKKARINEN et al. 1993). The distinction between areas of decreased parenchymal attenuation due to early emphysema, before there is the characteristic permeative destruction and distortion of the pulmonary vasculature, and obliterative bronchiolitis may be impossible on HRCT. Because of the coexisting alveolitis in hypersensitivity pneumonitis, the concept that a reduction in gas transfer can discriminate emphysema from small airways disease is less valid (TURTON et al. 1981). Although the identification of emphysema in the presence of interstitial fibrosis is controversial, it is possible that areas of emphysema occur interspersed among pulmonary fibrosis in chronic cases of hypersensitivity pneumonitis. However, in subacute hypersensitivity pneumonitis, bronchiolitis seems to be the predominant cause of airflow obstruction (HANSELL et al. 1996).

1.6.3.2
Sarcoidosis

The spectrum of functional abnormalities in patients with pulmonary sarcoidosis ranges from mild obstruction in early disease through to severe obstruction or restriction in end-stage fibrotic disease (WINTERBAUER and HUTCHINSON 1980). Whether lung restriction or airflow obstruction predominates has traditionally been thought to be dictated by the severity of interstitial fibrosis and airways obstruction respectively. However, the prevalence, anatomical site, and prognosis of airflow limitation in patients with pulmonary sarcoidosis is controversial (MILLER et al. 1974). Postulated mechanisms have included involvement of the large bronchi (UDWADIA et al. 1990) or small airways obstruction either by inflammation or fibrosis (LAMBERTO et al. 1985). The combination of airways and interstitial involvement may result in complex pulmonary function abnormalities which do not, on univariate analysis, correlate consistently with radiographic or histopathologic indices (YOUNG et al. 1966; CARRINGTON et al. 1978). An obstructive ventilatory defect is the most frequent physiologic abnormality at presentation, the frequency of an obstructive defect increases with advancing radiographic stage (HARRISON et al. 1991). Histopathologic evidence of airway involvement is recognized even in the relatively early stages of the disease (DINES et al. 1978; LENIQUE et al. 1995). In support of small airways

involvement in sarcoidosis, there are reports in which expiratory CT demonstrated areas of decreased attenuation, consistent with small airways involvement (GLEESON et al. 1996; DAVIES et al. 2000).

BERGIN et al. (1989), MULLER et al. (1989), and REMY-JARDIN et al. (1994) have reported significant correlations between the global extent of disease on CT and functional impairment. In these studies, patients with evidence of fibrosis on CT tended to have greater functional impairment, and nodular infiltration appeared to have little or no functional significance. A reticular pattern is, in CT/pathologic correlative terms, usually interpreted as representing fibrotic lung (NISHIMURA et al. 1995; AUSTIN et al. 1996) and would thus be expected to have a predominantly restrictive effect. However, it seems that a reticular pattern on CT is the major functional determinant in sarcoidosis and is very strongly associated with airflow obstruction (HANSELL et al. 1998), possibly because lung deranged by fibrosis results in adjacent emphysematous destruction.

1.7
Specific Problems in Studies of Structure-Function

From the simple premise that structural derangements of the lung disturb function, the basic idea that imaging depictions of diseased anatomy will reflect objective measures of function has become established. However, the magnitude of any correlation depends upon the intrinsic strength of the association between two variables and the noise introduced by measurement errors. What follows is a summary of some of the factors that reduce the strength of morphologic-functional correlations.

Nature of pathologic derangement. When considered at a pathologic level, few diffuse lung diseases can be regarded as purely interstitial (restrictive) or airway (obstructive) centred. The images obtained from HRCT are sometimes assumed to be a mirror of microscopic, as opposed to macroscopic, abnormalities; an apparently normal HRCT examination does not absolutely exclude the presence of functionally significant interstitial disease (ORENS et al. 1995). There are many causes for a functional deficit that result from an easily over-looked and extra-pulmonary abnormality. For example, uniform visceral pleural thickening may be responsible for a significant degree of restriction in individuals with asbestos-induced pulmonary fibrosis (JARAD et al. 1991, 1992; COPLEY et al. 2001).

Mismatch of HRCT signs and pathology. The causes of the decreased and increased attenuation components of the mosaic pattern are many and varied: for example, ground glass opacification may reflect many different pathologic states (and indeed may represent normal lung at near residual volume). Conversely, areas of decreased attenuation in patients with severe obstructive airways disease due to constrictive bronchiolitis are sometimes interpreted as "emphysema". In a study of patients with bronchiectasis, it was reported that the widespread areas of decreased attenuation on inspiratory HRCT were caused by emphysema, accounting for the functional gas-trapping (LOUBEYRE et al. 1996). However, the "emphysema" seen in that study was not associated with decreased gas diffusing capacity, the functional hallmark of emphysema. Thus, impaired ventilation in bronchiectasis leading to regional hypoxia may have resulted in regional hypoperfusion by a mechanism called hypoxic vasoconstriction. Hypoperfusion corresponds to decreased regional blood volume which is reflected by decreased attenuation. In non-smoking asthmatic individuals, areas of decreased attenuation, scored visually on HRCT have been ascribed to emphysema (PAGANIN et al. 1992, 1996). However, several strands of evidence suggest that the areas of decreased attenuation identified on expiratory HRCT in asthmatics (GRENIER et al. 1996; LAURENT et al. 2000) reflect air-trapping due to small airway obstruction, rather than emphysematous lung destruction (GEVENOIS et al. 1996b; KING et al. 1999). If decreased attenuation in asthmatics is already appreciated on inspiratory scans, long-standing air trapping has resulted in hypoxic vasoconstriction and hypoperfusion.

Measurement errors. These apply to both imaging and pulmonary function tests. For example, the visual quantitation of disease extent from chest radiography is less accurate and more prone to observer variation than computed tomography (COLLINS et al. 1994). Measurement of gas diffusing capacity is not always standardised and thus is prone to considerable variation between laboratories (KANGALEE and ABBOUD 1992). However robust the method of quantitation, correlations will only be as strong as the intrinsic strength of the measuring tools.

Selection bias. The strength of correlations are heavily dependent on the type of patients selected.

Study groups which are not representative of the whole spectrum of disease under consideration will not yield representative data and may therefore weaken correlations. Furthermore admixed disease, for example, in many patients with fibrosing alveolitis there is often a component of smoking-related emphysema, with consequent airflow obstruction and increased lung volumes, which will influence relationships (WELLS et al. 1997a).

Inappropriate quantification systems. An over-elaborate visual scoring system is likely to be associated with poor observer agreement and may obscure real relationships. Conversely, a coarse scoring system may introduce "noise" and result in an underestimate of the strength of correlations. Furthermore, the differing functional contributions of the upper and lower lung zones may need to be taken into account by weighing zonal scores, particularly in diseases that are not uniformly distributed, such as centrilobular emphysema. Automated methods of quantifying diffuse lung disease on CT may not take account of spurious (technical or physiological) causes of increased or decreased lung density.

Global nature of pulmonary function test results. The single figures of individual pulmonary function test results do not reflect regional differences in disease severity. In addition, pathologic processes that tend to counteract each other will not be reflected. An obvious example is seen in patients with fibrosing alveolitis who have interstitial fibrosis in the lower zones and emphysema in the upper zones, which results in normal lung volumes as measured by plethysmography (DOHERTY et al. 1997).

Differences in physiological status between CT and PFTs. Many measures of pulmonary function are dynamic whereas CT examinations are mostly static. Theoretically, the position of the individual (supine for CT examination, erect for pulmonary function testing) may introduce noise. A further unquantifiable variation may occur because of the labile nature of some physiological phenomena, for example bronchoconstriction, so that unless PFTs and CT are performed near simultaneously, further error will be introduced.

Methods of data analysis. Many different individual morphologic features may affect pulmonary function, and there is a consequent in danger of placing too much reliance on univariate analysis to identify structure-function relationships. The application of multivariate techniques are invaluable in confirming the independence of correlations shown by univariate analysis.

1.8
Conclusion

However the reader defines "functional imaging" the future is exciting: the initial nihilism that resulted from the obviously discordant picture provided by radiographic-functional studies has given way to the thrilling elucidations of HRCT-functional correlative studies. The understanding provided by these studies underpins future work using novel imaging techniques.

Acknowledgement. The material in this chapter is largely based, with the kind permission of the Editor, on an article that appeared in European Radiology (2001; 11:1666–1680).

References

Aberle DR, Hansell DM, Brown K, Tashkin DP (1990) Lymphangiomyomatosis: CT, chest radiographic and functional correlations. Radiology 176:381–387

Adams H, Bernard MS, McConnochie K (1991) An appraisal of CT pulmonary density mapping in normal subjects. Clin Radiol 43:238–242

Arakawa H, Webb WR (1998) Air trapping on expiratory high-resolution CT scans in the absence of inspiratory scan abnormalities: correlation with pulmonary function tests and differential diagnosis. Am J Roentgenol 170:1349–1353

Arakawa H, Webb WR, McCowin M, Katsou G, Lee KN, Seitz RF (1998) Inhomogeneous lung attenuation at thin-section CT: diagnostic value of expiratory scans. Radiology 206:89–94

Austin JHM, Muller NL, Friedman PJ, Hansell DM, Naidich DP, Remy-Jardin M, Webb WR, Zerhouni EA, Austin JH (1996) Glossary of terms for CT of the lungs: recommendations of the nomenclature committee of the Fleischner Society. Radiology 200:327–331

Bankier AA, de Maertelae V, Keyzer C, Gevenois PA (1999) Pulmonary emphysema: subjective visual grading versus objective quantification with macroscopic morphometry and thin-section CT densitometry. Radiology 211:851–858

Beinert T, Behr J, Mehnert F, Kohz P, Seemann M, Rienmuller R Reiser M (1995) Spirometrically controlled quantitative CT for assessing diffuse parenchymal lung disease. J Comput Assist Tomogr 19:924–931

Bergin CJ, Bell DY, Coblentz CL, Chiles C, Gamsu G, MacIntyre NR, Coleman RE, Putman CE (1989) Sarcoidosis: correlation of pulmonary parenchymal pattern at CT with results of pulmonary function tests. Radiology 171:619–624

Bhalla M, Naidich DP, McGuinness G, Gruden JF, Leitman BS, McCauley DI (1996) Diffuse lung disease: assessment with helical CT – preliminary observations of the role of maximum and minimum intensity projection images. Radiology 200:341–347

Bourke S, Anderson K, Lynch P, Boyd J, King S, Banham S, Boyd G (1989) Chronic simple bronchitis in pigeon fanciers. Chest 95:598–601

Brown MS, McNitt-Gray MF, Goldin JG, Greaser LE, Hayward UM, Sayre JW, Arid MK, Aberle DR (1999) Automated measurement of single and total lung volume from CT. J Comput Assist Tomogr 23:632–640

Buist AS (1984) Current status of small airways disease. Chest 86:100–105

Burki NK, Krumpleman JL (1980) Correlation of pulmonary function with the chest roentgenogram in chronic airway obstruction, Am Rev Respir Dis 121:216–223

Cailes JB, Du Bois RM, Hansell DM (1996) Density gradient of the lung parenchyma on CT in patients with lone pulmonary hypertension and systemic sclerosis. Acad Radiol 3:724–730

Carrington CB, Gaensler EA, Mikus JP, Schachter AW, Burke GW, Goff AM (1978) Structure and function in sarcoidosis. Ann NY Acad Sci 29:265–283

Chabat F, Yang GZ, Hansell DM (2003) Texture classification for the differentiation of destructive lung diseases on computed tomography. Radiology 10.114/radio.2283020505

Chinet T, Jaubert F, Dusser D, Danel C, Chretien J, Huchon G (1990) Effects of inflammation and fibrosis on pulmonary function in diffuse lung fibrosis. Thorax 45:675–678

Chotas HG, Ravin CE (1994) Chest radiography: estimated lung volume and projected area obscured by the heart, mediastinum, and diaphragm. Radiology 193:403–404

Chotas HG, Dobbins JT, Ravin CE (1999) Principles of digital radiography with large-area electronically readable detectors: a review of the basics. Radiology 210:595–599

Collins CD, Wells AU, Hansell DM, Morgan RA, MacSweeney JE, Du Bois RM, Rubens MB (1994) Observer variation in pattern type and extent of disease in fibrosing alveolitis on thin section computed tomography and chest radiography. Clin Radiol 49:236–240

Copley SJ, Wells AU, Rubens MB, Chabat F, Sheehan RE, Musk AW, Hansell DM (2001) Functional consequences of pleural disease evaluated with chest radiography and CT. Radiology 220:237–243

Davies CWH, Tasker AD, Padley SPG, Davies RJO, Gleeson FV (2000) Air trapping in sarcoidosis on computed tomography: correlation with lung function. Clin Radiol 55:217–221

Delorme S, Keller-Reichenbecher MA, Zuna I, Schlegel W, Van Kaick G (1997) Usual interstitial pneumonia. Quantitative assessment of high-resolution computed tomography findings by computer-assisted texture-based image analysis, Invest Radiol 32:566–574

Dines DE, Stubbs SE, McDougall JC (1978) Obstructive disease of the airways associated with stage I sarcoidosis. Mayo Clinic Proc 53:788–791

Doherty MJ, Pearson MG, O'Grady EA, Pellegrini V, Calverley PM (1997) Cryptogenic fibrosing alveolitis with preserved lung volumes Thorax 52:998–1002

Edelman RR, Hatabu H, Tadamura E, Li W, Prasad PV (1996) Noninvasive assessment of regional ventilation in the human lung using oxygen-enhanced magnetic resonance imaging. Nat Med 2:1236–1239

Engeler CE, Tashjian JH, Engeler CM, Geise RA, Holm JC, Russell Ritenour E (1994) Volumetric high-resolution CT in the diagnosis of interstitial lung disease and bronchiectasis: diagnostic accuracy and radiation dose. Am J Roentgenol 163:31–35

Epler GR, McLoud TC, Gaensler EA, Mikus JP, Carrington CB (1978) Normal chest roentgenograms in chronic diffuse infiltrative lung disease. N Engl J Med 298:935–939

Epstein DM, Miller WT, Bresnitz EA, Levine MS, Gefter WB (1984) Application of ILO classification to a population without industrial exposure: findings to be differentiated from pneumoconiosis. Am J Roentgenol 142:53–58

Fotheringham T, Chabat F, Hansell DM, Wells AU, Desai SR, Guckel C, Padley SP, Gibson M, Yang GZ (1999) A comparison of methods for enhancing the detection of areas of decreased attenuation on CT caused by airways disease. J Comput Assist Tomogr 23:385–389

Fulmer JD, Roberts WC (1980) Small airways disease and intersitital pulmonary disease. Chest 77:470–472

Fulmer DG, Roberts WC, von Gal ER, Crystal RG (1979) Morphologic-physiologic correlates of the severity of fibrosis and degree of cellularity in idiopathic pulmonary fibrosis. J Clin Invest 63:665–676

Gamsu G, Salmon CJ, Warnock ML Blanc PD (1995) CT quantification of interstitial fibrosis in patients with asbestosis: a comparison of two methods. Am J Roentgenol 164:63–68

Gelb AF, Zamel N (1973) Simplified diagnosis of small-airway obstruction. N Engl J Med 288:395–398

Gelb AF, Gold WM, Wright RR, Bruch HR, Nadel JA (1973) Physiologic diagnosis of subclinical emphysema. Am Rev Respir Dis 107:50–63

Gelb AF, Schein M, Kuei J, Tashkin DP, Muller NL, Hogg JC, Epstein JD, Zamel N (1993) Limited contribution of emphysema in advanced chronic obstructive pulmonary disease. Am Rev Respir Dis 147:1157–1161

Gelb AF, Hogg JC, Muller NL, Schein MJ, Kuei J, Tashkin DP, Epstein JD, Kollin J, Green RH, Zamel N, Elliott WM, Hadjiaghai L (1996) Contribution of emphysema and small airways in COPD. Chest 109:353–359

Gelb AF, Zamel N, Hogg JC, Muller NL, Schein MJ (1998) Pseudophysiologic emphysema resulting from severe small-airways disease. Am J Respir Crit Care Med 158:815–819

Gevenois PA, Estenne M (2001) Can computed tomography predict functional benefit from lung volume reduction surgery for emphysema? Am J Respir Crit Care Med 164:2137–2138

Gevenois PA, de Vuyst P, Sy M, Scillia P, Chaminade L, Maertelae V de, Zanen J, Yernault JC (1996a) Pulmonary emphysema: quantitative CT during expiration. Radiology 199:825–829

Gevenois PA, Scillia P, de Maertelae V, Michils A, de Vuyst P, Yernault JC (1996b) The effects of age, sex, lung size, and hyperinflation on CT lung densitometry. Am J Roentgenol 167:1169–1173

Gleeson FV, Traill ZC, Hansell DM (1996) Expiratory CT evidence of small airways obstruction in sarcoidosis. Am J Roentgenol 166:1052–1054

Gotway MB, Lee ES, Reddy GP, Golden JA, Webb WR (2000) Low-dose dynamic expiratory thin-section CT of the lungs using a spiral CT scanner. J Thorac Imaging 15:168–172

Grenier P, Mourey-Gerosa I, Benali K, Brauner MW, Leung AN, Lenoir S, Cordeau MP, Mazoyer B (1996) Abnormalities of the airways and lung parenchyma in asthmatics: CT observations in 50 patients and inter- and intra-observer variability. Eur Radiol 6:199–206

Guckel C, Wells AU, Taylor DA, Chabat F, Hansell DM (1999) Mechanism of mosaic attenuation of the lungs on computed tomography in induced bronchospasm. J Appl Physiol 86:701–708

Gurney JW, Schroeder BA (1988) Upper lobe lung disease: physiologic correlates. Radiology 167:359–366

Gurney JW, Jones KK, Robbins RA, Gossman GL, Nelson KJ, Daughton D, Spurzem JR, Rennard SI (1992) Regional distribution of emphysema: correlation of high-resolution CT with pulmonary function tests in unselected smokers. Radiology 183:457–463

Gurney JW, Habbe TG, Hicklin J (1997) Distribution of disease in cystic fibrosis: correlation with pulmonary function. Chest 112:357–362

Hansell DM, Wells AU, Padley SPG, Muller NL (1996) Hypersensitivity pneumonitis: correlation of individual CT patterns with functional abnormalities. Radiology 199:123–128

Hansell DM, Rubens MB, Padley SPG, Wells AU (1997) Obliterative bronchiolitis: individual CT signs of small airways disease and functional correlation. Radiology 203:721–726

Hansell DM, Milne DG, Wilsher ML, Wells AU (1998) Pulmonary sarcoidosis: morphologic associations of airflow obstruction at thin-section CT. Radiology 209:697–704

Harrison BDW, Shaylor JM, Stokes TC, Wilkes AR (1991) Airflow limitation in sarcoidosis – a study of pulmonary function in 107 patients with newly diagnosed disease. Respir Med 85:59–64

Hartley PG, Galvin JR, Hunninghake GW, Merchant JA, Yagla SJ, Speakman SB, Schwartz DA (1994) High-resolution CT-derived measures of lung density are valid indexes of intersitital lung disease. J Appl Physiol 76:271–277

Herbert DL, Gur D, Shabason L, Good WF, Rinaldo JE, Snyder JV, Borovetz HS, Mancici MC (1982) Mapping of human local pulmonary ventilation by xenon enhanced computed tomography. J Comput Assist Tomogr 6:1088–1093

Hoffman EA, Tajik JK, Kugelmass SD (1995) Matching pulmonary structure and perfusion via combined dynamic multislice CT and thin-slice high-resolution CT. Comput Med Imaging Graph 19:101–112

Jarad NA, Poulakis N, Pearson MC, Rubens MB, Rudd RM (1991) Assessment of asbestos induced pleural disease by computed tomography – correlation with chest radiograph and lung function. Respir Med 85:203–208

Jarad NA, Wilkinson P, Pearson MC, Rudd RM (1992) A new high resolution computed tomography scoring system for pulmonary fibrosis, pleural disease, and emphysema in patients with asbestos related disease. Br J Industr Med 49:73–84

Johnson MM, Ely EW, Chiles C, Bowton DL, Friemanas RI, Choplin RH, Haponik EF (1998) Radiographic assessment of hyperinflation: correlation with objective chest radiographic measurements and mechanical ventilator parameters. Chest 113:1698–1704

Jones AT, Hansell DM, Evans TW (2001) Pulmonary perfusion in supine and prone positions: an electron-beam computed tomography study. J Appl Physiol 90:1342–1348

Kalender WA, Fichte H, Bautz W, Skalej M (1991) Semi-automatic evaluation procedures for quantitative CT of the lung. J Comput Assist Tomogr 15:248–255

Kangalee KM, Abboud RT (1992) Interlaboratory and intralaboratory variability in pulmonary function testing: a 13-year study using a biologic control. Chest 101:88–92

Kauczor HU, Hofmann D, Kreitner KF, Nilgens H, Surkau R, Heil W, Potthast A, Knopp MV, Otten EW, Thelen M (1996) Normal and abnormal pulmonary ventilation: visualization at hyperpolarized He-3 MR imaging. Radiology 201:564–568

Kauczor HU, Heussel CP, Fischer B, Klamm R, Mildenberger P, Thelen M (1998) Assessment of lung volumes using helical CT at inspiration and expiration: comparison with pulmonary function tests. Am J Roentgenol 171:1091–1095

Kauczor HU, Heitmann K, Heussel CP, Marwede D, Uthmann T, Thelen M (2000) Automatic detection and quantification of ground-glass opacities on high-resolution CT using multiple neural networks: comparison with a density mask. Am J Roentgenol 175:1329–1334

Kauczor HU, Chen XJ, van Beek EJ, Schreiber WG (2001) Pulmonary ventilation imaged by magnetic resonance: at the doorstep of clinical application. Eur Respir J 17:1008–1023

Kazerooni EA, Martinez FJ, Flint A, Jamadar DA, Gross BH, Spizarny DL, Cascade PN, Whyte RI, Lynch JP, Toews G (1997) Thin-section CT obtained at 10-mm increments versus limited three-level thin-section CT for idiopathic pulmonary fibrosis: correlation with pathologic scoring. Am J Roentgenol 169:977–983

Keyzer C, Gevenois PA (1999) Quantitative computed tomography of pulmonary emphysema (in French). Rev Malad Respir 16:455–460

Kido S, Ikezoe J, Tamura S, Nakamura H, Kuroda C (1997) A computerized analysis system in chest radiography: evaluation of interstitial lung abnormalities. J Digit Imaging 10:57–64

Kido S, Arisawa J, Kuriyama K, Kuroda C, Nakamura H (2000) Comparison between computer-aided diagnosis and radiologists: assessment of pulmonary blood flow on chest radiographs. J Thorac Imaging 15:48–55

Kim TK, Doi K, MacMahon H (1988) Image feature analysis and computer-aided diagnosis in digital radiography: detection and characterization of interstitial lung disease in digital chest radiographs. Med Phys 15:311–319

King GG, Muller NL, Pare PD (1999) Evaluation of airways in obstructive pulmonary disease using high-resolution computed tomography. Am J Respir Crit Care Med 159:992–1004

King TE Jr (1993) Overview of bronchiolitis. Clin Chest Med 14:607–610

Kinsella M, Muller NL, Abboud RT, Morrison NJ, DyBuncio A (1990) Quantitation of emphysema by computed tomography using a density mask program and correlation with pulmonary function tests. Chest 97:315–321

Kokkarinen JI, Tukiainen HO, Terho EO (1993) Recovery of pulmonary function in farmer's lung. A five-year follow-up study. Am Rev Respir Dis 147:793–796

Lamberto C, Saumon G, Loiseau P, Battesti JP, Georges R (1985) Respiratory function in recent pulmonary sarcoidosis with special reference to small airways. Bull Eur Physiopathol Respir 21:309–315

Lang MR, Fiaux GW, Gillooly M, Stewart JA, Hulmes DJ, Lamb D (1994) Collagen content of alveolar wall tissue in emphysematous and non-emphysematous lungs. Thorax 49:319–326

Laurent F, Latrabe V, Raherison C, Marthan R, Tunon-de-Lara JM (2000) Functional significance of air trapping detected in moderate asthma. Eur Radiol 10:1404–1410

Lenique F, Brauner MW, Grenier P, Battesti JP, Loiseau A, Valeyre D (1995) CT assessment of bronchi in sarcoidosis: endoscopic and pathologic correlations. Radiology 194:419–423

Lodge T (1946) The anatomy of the blood vessels of the human lung as applied to chest radiology. Br J Radiol 19:1–13

Loubeyre P, Paret M, Revel D, Wiesendanger T, Brune J (1996) Thin-section CT detection of emphysema associated with bronchiectasis and correlation with pulmonary function tests. Chest 109:360–365

Lucidarme O, Coche E, Cluzel P, Mourey-Gerosa I, Howarth N, Grenier P (1998) Expiratory CT scans for chronic airway disease: correlation with pulmonary function test results. Am J Roentgenol 170:301–307

Lynch DA (1993) Imaging of small airways diseases. Clin Chest Med 14:623–634

McAdams HP, Hatabu H, Donnelly LF, Chen Q, Tadamura E, MacFall JR (2000) Novel techniques for MR imaging of pulmonary airspaces. Magn Reson Imaging Clin North Am 8:205–219

McFadden ER, Linden RA (1972) A reduction in maximal end-expiratory flow rate. A spirographic manifestation of small airway disease. Am J Med 52:725–737

McLoud TC, Epler GR, Gaensler EA, Burke GW, Carrington CB (1982) A radiographic classification for sarcoidosis: physiologic correlation. Invest Radiol 17:129–138

Mergo PJ, Williams WF, Gonzalez-Rothi R, Gibson R, Ros PR, Staab EV, Helmberger T (1998) Three-dimensional volumetric assessment of abnormally low attenuation of the lung from routine helical CT: inspiratory and expiratory quantification. Am J Roentgenol 170:1355–1360

Miles KA (1991) Measurement of tissue perfusion by dynamic computed tomography. Br J Radiol 64:409–412

Miller A, Teirstein AS, Jackler I, Chuang M, Siltzbach LE (1974) Airway function in chronic pulmonary sarcoidosis with fibrosis. Am Rev Respir Dis 109:179–189

Miniati M, Pistolesi M, Paoletti P, Giuntini C, Lebowitz MD, Taylor AE, Milne EN (1988) Objective radiographic criteria to differentiate cardiac, renal, and injury lung edema. Invest Radiol 23:433–440

Mink SN, Coalson JJ, Whitley L, Greville H, Jadne C (1984) Pulmonary function tests in the detection of small airways obstruction in a canine model of bronchiolitis obliterans. Am Rev Respir Dis 130:1125–1133

Mino M, Noma S, Kobashi Y, Iwata T (1995) Serial changes of cystic air spaces in fibrosing alveolitis: a CT-pathological study. Clin Radiol 50:357–363

Morishita J, Doi K, Katsuragawa S, Monnier Cholley L, Mac Mahon H (1995) Computer-aided diagnosis for interstitial infiltrates in chest radiographs: optical-density dependence of texture measures. Med Phys 22:1515–1522

Muller CJ, Schwaiblmair M, Scheidler J, Deimling M, Weber J, Loffler RB, Reiser MF (2002) Pulmonary diffusing capacity: assessment with oxygen-enhanced lung MR imaging – preliminary findings. Radiology 222:499–506

Muller NL, Mawson JB, Mathieson JR, Abboud R, Ostrow DN, Champion P (1989) Sarcoidosis: correlation of extent of disease at CT with clinical, functional, and radiographic findings. Radiology 171:613–618

Ng CS, Desai SR, Rubens MB, Padley SPG, Wells AU, Hansell DM (1999) Visual quantitation and observer variation of signs of small airways disease at inspiratory and expiratory CT. J Thorac Imaging 14:279–285

Nishimura K, Itoh H, Kitaichi M, Nagai S, Izumi T (1995) CT and pathological correlation of pulmonary sarcoidosis. Semin Ultrasound CT MRI 16:361–370

Nugent KM, Peterson MW, Jolles H, Monick MM, Hunninghake GW (1989) Correlation of chest roentgenograms with pulmonary function and bronchoalveolar lavage in interstitial lung disease. Chest 96:1224–1228

Ohno Y, Hatabu H, Takenaka D, Adachi S, Cauteren M van, Sugimura K (2001) Oxygen-enhanced MR ventilation imaging of the lung: preliminary clinical experience in 25 subjects. Am J Roentgenol 177:185–194

Orens JB, Kazerooni EA, Fernando JM, Curtis JL, Gross BH, Flint A, Lynch JP (1995) The sensitivity of high-resolution CT in detecting idiopathic pulmonary fibrosis proved by open lung biopsy: a prospective study. Chest 108:109–115

Padley SP, Adler BD, Hansell DM, Muller NL (1993) Bronchiolitis obliterans: high resolution CT findings and correlation with pulmonary function tests. Clin Radiol 47: 236–240

Paganin F, Trussard V, Seneterre E, Chanez P, Giron J, Godard P, Senac JP, Michel FB, Bousquet J (1992) Chest radiography and high resolution computed tomography of the lungs in asthma. Am Rev Respir Dis 146:1084–1087

Paganin F, Seneterre E, Chanez P, Daures JP, Bruel JM, Michel FB, Bousquet J (1996) Computed tomography of the lungs in asthma: influence of disease severity and etiology. Am J Respir Crit Care Med 153:110–114

Pistolesi M, Milne ENC, Miniati M et al (1986) Detection and measurement of pulmonary oedema: the chest radiographic approach. Intensive Crit Care Dig 5:34–36

Pride NB (1971) The assessment of airflow obstruction: role of measurements of airways resistance and of tests of forced expiration. Br J Dis Chest 65:135–169

Remy-Jardin M, Remy J, Wallaert B, Muller NL (1993) Subacute and chronic bird breeder hypersensitivity pneumonitis: sequential evaluation with CT and correlation with lung function tests and bronchoalveolar lavage. Radiology 189: 111–118

Remy-Jardin M, Giraud F, Remy J, Wattinne L, Wallaert B, Duhamel A (1994) Pulmonary sarcoidosis: role of CT in the evaluation of disease activity and functional impairment and in prognosis assessment. Radiology 191:675–680

Rienmuller RK, Behr J, Kalender WA, Schatzl M, Altmann I, Merin M, Beinert T (1991) Standardized quantitative high resolution CT in lung diseases. J Comput Assist Tomogr 15:742–749

Robinson TE, Leung AN, Moss RB, Blankenberg FG, al Dabbagh H, Northway WH (1999) Standardized high-resolution CT of the lung using a spirometer-triggered electron beam CT scanner. Am J Roentgenol 172:1636–1638

Rodriguez LH, Vargas PF, Raff U, Lynch DA, Rojas GM, Moxley DM, Newell JD (1995) Automated discrimination and quantification of idiopathic pulmonary fibrosis from normal lung parenchyma using generalized fractal dimensions in high-resolution computed tomography images. Acad Radiol 2:10–18

Saam BT, Yablonskiy DA, Kodibagkar VD, Leawoods JC, Gierada DS, Cooper JD, Lefrak SS, Conradi MS (2000) MR imaging of diffusion of (3)He gas in healthy and diseased lungs. Magn Reson Med 44:174–179

Seal RME, Hapke EJ, Thomas GO, Meek JC, Hayes M (1968) The pathology of the acute and chronic stages of farmer's lung. Thorax 23:469–489

Sider L, Dennis L, Smith LJ, Dunn MM (1987) CT of the lung parenchyma and the pulmonary function test. Chest 92: 406–410

Simon G, Pride NB, Jones NL, Raimondi AC (1973) Relation between abnormalities in the chest radiograph and changes in pulmonary function in chronic bronchitis and emphysema Thorax 28:15–23

Snider GL, Kleinerman J, Thurlbeck WM, Bengali ZH (1985) The definition of emphysema. Report of a National Heart, Lung, and Blood Institute, Division of Lung Diseases workshop. Am Rev Respir Dis 132:182–185

Snyder JV, Pennock B, Herbert D, Rinaldo JE, Culpepper J, Good WF, Gur D (1984), Local lung ventilation in critically ill patients using nonradioactive xenon-enhanced transmission computed tomography, Crit Care Med 12:46–51

Stender HS, Oestmann J (1997) Thorax. In: Rosenbusch G, Oudkerk M, Ammann E (eds) Radiology in medical diagnostics: evolution of X-ray applications 1895–1995. Blackwell Science, Oxford, pp 67–97

Strickland NH, Hughes JM, Hart DA, Myers MJ, Lavender JP (1993) Cause of regional ventilation-perfusion mismatching in patients with idiopathic pulmonary fibrosis: a combined CT and scintigraphic study. Am J Roentgenol 161:719–725

Sutinen S, Reijula K, Huhti E, Karkola P (1983) Extrinsic allergic bronchiolo-alveolitis: serology and biopsy findings. Eur J Respir Dis 64:271–282

Tonelli M, Stern EJ, Glenny RW (1997) HRCT evident fibrosis in isolated pulmonary emphysema. J Comput Assist Tomogr 21:322–323

Turton CW, Williams G, Green M (1981) Cryptogenic obliterative bronchiolitis in adults. Thorax 36:805–810

Udwadia ZF, Pilling JR, Jenkins PF, Harrison BDW (1990) Bronchoscopic and bronchographic findings in 12 patients with sarcoidosis and severe or progressive airways obstruction. Thorax 45:272–275

Uppaluri R, Mitsa T, Sonka M, Hoffman EA, McLennan G (1997) Quantification of pulmonary emphysema from lung computed tomography images. Am J Respir Crit Care Med 156:248–254

Walsham H, Harrison Orton G (1906) The Röntgen rays in the diagnosis of diseases of the chest. Lewis, London

Watters LC, King TE, Schwarz MI, Waldron JA, Stanford RE, Cherniack RM (1986) A clinical radiographic and physiologic scoring system for the longitudinal assessment of patients with idiopathic pulmonary fibrosis. Am Rev Respir Dis 133:97–103

Wells AU, King AD, Rubens MB, Cramer D, Du Bois RM, Hansell DM (1997a) Lone cryptogenic fibrosing alveolitis: a functional-morphologic correlation based on extent of disease on thin-section computed tomography. Am J Respir Crit Care Med 155:1367–1375

Wells AU, Rubens MB, Du Bois RM, Hansell DM (1997b) Functional impairment in fibrosing alveolitis: relationship to reversible disease on thin section computed tomography. Eur Respir J 10:280–285

Winterbauer RH, Hutchinson JF (1980) Use of pulmonary function tests in the management of sarcoidosis. Chest 78:640–647

Wollmer P, Jakobsson K, Albin M et al (1987) Measurement of lung density by X-ray computed tomography: relation to lung mechanics in workers exposed to asbestos cement. Chest 91:865–869

Yang GZ, Hansell DM (1997) CT image enhancement with wavelet analysis for the detection of small airways disease. IEEE Trans Med Imaging 16:953–961

Young RL, Krumholtz RA, Harkleroad LE (1966) A physiologic roentgenographic disparity in sarcoidosis. Dis Chest 50:81–86

Zagers H, Vrooman HA, Aarts NJ, Stolk J, Schultze KL, Dijkman JH, Van Voorthuisen AE, Reiber JH (1996) Assessment of the progression of emphysema by quantitative analysis of spirometrically gated computed tomography images. Invest Radiol 31:761–767

Zagers R, Vrooman HA, Aarts NJ, Stolk J, Schultze KL, Voorthuisen E van, Reiber JH (1995) Quantitative analysis of computed tomography scans of the lungs for the diagnosis of pulmonary emphysema. A validation study of a semiautomated contour detection technique. Invest Radiol 30:552–562

2 Assessment of Lung Physiology Using Pulmonary Function Tests

Marc Zelter, Christian Straus, André Capderou

CONTENTS

M. Zelter, MD
Professor, Service Central d'Explorations Fonctionnelles Respiratoires, Groupe Hospitalier Pitié Salpêtrière, AP-HP, and Université Paris VI, 47-83 boulevard de l'Hôpital, 75651 Paris Cedex 13, France
C. Straus, MD
Service Central d'Explorations Fonctionnelles Respiratoires, Groupe Hospitalier Pitié Salpêtrière, AP-HP, 47-83 boulevard de l'Hôpital, 75651 Paris Cedex 13, France
A. Capderou, MD
Hôpital Marie Lannelongue, 133 Avenue de la Résistance, 92350 Plessis Robinson, France

2.1 Introduction

The lung combines two basic functions linked to respiration; first gas exchange, consisting of the oxygenation of the incoming desaturated venous blood and the removal of carbon dioxide, thus producing arterialized blood, and second, and consequently, the removal of protons (H+) from this incoming blood, permitting fast regulation of the blood concentration of H+ (pH) (Dejours 1975). To perform this task adequately, ventilation and circulation within the lung must be matched so that the ratio of the distribution of ventilation to that of perfusion is optimum. The ventilation-perfusion ratio ultimately determines the functional performance of the lung in terms of gas exchange at a given level of ventilation (Riley and Cournand 1949). Measurement of blood gases is the most relevant test available to give a global assessment of gas exchange in a given patient. Unfortunately blood gas values, when abnormal, give no clue as to what aspect of lung function is impaired. Furthermore, these values tend to become abnormal only in the late or acute phase of lung diseases, because of the remarkable flexibility of ventilation-perfusion control. The role devoted to pulmonary function testing aims at the identification of the functional site of the respiratory system that is most likely to be related to clinical symptoms, to define the functional syndrome that best characterizes the disease (obstructive, restrictive or both) and to quantify impairment, to allow the follow-up of the disease. However pulmonary function testing cannot per se identify a specific type of pathology. Pulmonary function tests only tell us if a suspected pathology is compatible or not with the functional impairment. Because most measurements are performed at the mouth, they only provide a global evaluation of lung function and do not allow the anatomical location of the disease. Correct interpretation of lung testing may become difficult in cases of very heterogeneous lung disease unless the limitations inherent to each type of test are well integrated. For instance, the helium diffusion technique is highly reliable in normal and restrictive patients to measure residual lung volume but totally irrelevant, because of its physical limitation, in very heterogeneous obstructive lung disease where plethysmography is the tool of reference.

Adequate ventilation implies normal elastic properties of the pulmonary tissue producing normal passive recoil forces, normal muscle function to produce active forces that combines with recoil forces to produce changes in lung volume during ventilation. It is the change in pulmonary volume that produces the change in intra-pulmonary pressure that creates flow between the mouth and the alveoli. Alveolar ventilation, which is the only ventilation factor relevant in terms of gas exchange, depends on the magnitude of flow to and from the alveoli. But flow does not only depend on changes in volumes and therefore in the pressure difference between the mouth and the alveoli but also on the resistance opposed by airways to flow. Airway properties must therefore be tested apart from volume measurements to understand correctly the pressure-flow-volume relationship. Adequate ventilation not only requires adequate integration of active and passive ventilatory functions but also of course a normal drive that needs specific assessment.

Ideally, each of these functions should be tested independently of one another. This implies independent assessment of: 1) the active properties of the respiratory muscles, 2) the passive properties of the thoracic wall, 3) the passive elastic properties of the lung (recoil forces), 4) airway conduction and 5) respiratory control.

The physiological lung volumes that have a functional relevance are determined either by elastic forces alone (functional residual capacity, FRC) or by the interaction of the passive elastic forces and the active forces produced by muscles (total lung capacity, TLC). The conducting airways and their response to various challenges have also to be assessed. In order to proceed, the subject needs to perform so-called forced maneuvers to produce maximum flow. Such maneuvers require full participation of the patient. Correct interpretation also relies on a normal configuration of the thorax. This raises two very important and critical points regarding pulmonary function testing; 1)it is often difficult to discriminate passive from active properties, 2)patient cooperation is critical to obtain correct measurements.

From an engineering point of view, most of the pulmonary function tests define the extreme physical properties of the system and bear little relevance in normal subjects to the requirements of normal function, even during extreme exercise. It is only in pathological situations that there happens to be some consistency between the limitations demonstrated by these tests and the actual possibilities of the patient, as illustrated for instance by the modifications of the maximum effort flow-volume curves (see below).

Because of these complex interactions, performance and interpretation of pulmonary function tests need to follow logical patterns and to obey well-defined standardization rules for both maneuvers and interpretation of the results. A point too often neglected is that the order in which maneuvers are performed may interfere with the results. Finally, attesting normality or abnormality requires comparison with a carefully designed control population. For this it is imperative to follow the statements and guidelines issued by leading scientific societies, basically ERS in Europe and ATS in the USA (ATS 1991, 1995a; EUROPEAN RESPIRATORY SOCIETY 1993). There is no such thing as local laboratory reference values.

"Standard pulmonary function tests", must include spirometry for volume measurements, flow volume loop (maximum effort flow volume curve) and the volume of gas expired during the first second of expiration (FEV1). Study of the bronchial reactivity is warranted when potential airway inflammation is suspected, as in asthma, allergy or COPD where reversibility of obstruction warrants specific therapy. Measurement of diffusion of carbon monoxide is useful if the diffusion barrier may be involved or when it is needed to assess the changes in alveolar exchange surface area as in emphysema. Blood gas measurements should be widely used in chronic diseases, especially now that microneedle sampling procedures make them safe and almost painless in qualified hands. Besides these widely available tests, more specialized tests evaluate specific pathologies. Assessment of the mechanical properties of the lungs and of the chest wall (the results being influenced by respiratory muscle function and drive) are limited to cases where clinical data support either a pathological or functional impairment of the rib cage, the elastic properties of the lung or muscle contraction or control. Assessing respiratory function during exercise is often a mandatory step to complement pulmonary function testing particularly to identify latent dysfunction of ventilation-perfusion matching as seen for instance in some COPD patients where oxygen desaturation of arterial blood at exercise is an index of the extension of the disease.

2.2
Ventilatory Flows and Volumes

The measurement of ventilatory flows and volumes is the first step of pulmonary function testing in clinical practice. These tests are integrative, depending upon the properties of the lungs and of the airways, but also of the chest wall and of the respiratory muscles. Muscular activation is necessary for maximum lung inflation and emptying as well as for the performance of maximal respiratory flow. Complete understanding of the instructions and adequate execution by the patient or subject are required. This again outlines the complex and intricate relations between an apparently simple and basic measurement, such as a volume, and the physiological, technical and human implications involved.

The various volumes of gas contained in the lungs are defined either as volumes (V) or capacities (C), (Fig. 2.1). A lung volume is a volume of gas within the lung related to a specific noteworthy physiological situation. A capacity is the sum of two, or more, lung volumes (QUANJER et al. 1993). The tidal volume (V_T) is the volume of gas inspired or expired during a respiratory cycle. Tidal volume varies with the level of ventilation. The volume of the lungs at the end of a normal expiration, a situation where respiratory muscles are totally inactive is called the relaxation volume (Vr). Vr is determined by the point of equilibrium between the passive elastic recoil forces exerted in opposite direc-

tions at that volume level by the lung and by the chest wall. The functional residual capacity (FRC) is defined as the end-expiratory volume at tidal ventilation. FRC is the sum of the expiratory reserve volume (ERV), and of the residual volume (RV). ERV is the maximal volume that can be expired, starting forced expiration from FRC. ERV decreases in the supine position (AGOSTONI and HYATT 1986; BECKLAKE and PERMUTT 1979). RV is the volume of gas remaining in the lung at the end of a full forced expiration. Therefore, RV cannot be directly measured at the mouth because it cannot be expired. Since techniques are available to measure FRC, RV is usually calculated by subtracting ERV from FRC (QUANJER et al. 1993). In a normal subject, the relaxation volume Vr is equal to FRC. That is not the case in some pathological situations where the end-expiratory volume may depend on other factors than passive forces, such as the intrinsic pulmonary end expiratory pressure (PEEP) due to the inability for passive expiration to terminate before active inflation resumes, as occurs in late COPD. However FRC remains an effective measurement in these instances, to appreciate changes in residual volume and lung distention. The maximal volume that can be inspired from the tidal end-inspiratory level by a maximum forced maneuvers defines the inspiratory reserve volume (IRV). The sum of V_T and IRV defines inspiratory capacity (IC). IC does not change between the sitting and the supine position (AGOSTONI and HYATT 1986; BRULOT et al. 1992). Interestingly, recent findings link changes in IC to dyspnea levels. The vital capacity (VC) is the sum of IC and ERV and corresponds to the

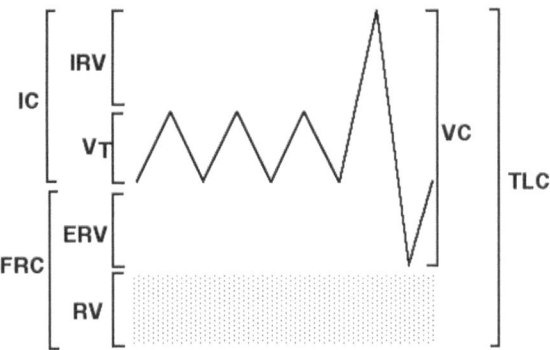

Fig. 2.1. Schematic representation of spirometric volumes (*FRC* functional residual capacity, *IC* inspiratory capacity, V_T: tidal volume, *IRV* inspiratory reserve volume, *ERV,* expiratory reserve volume, *VC* vital capacity, *TLC* total lung capacity, *RV* residual volume). FRC cannot be directly measured with a spirometer. This requires the use of other techniques like either the multiple-breath helium dilution method, the multiple-breath nitrogen washout method or the plethysmographic technique. RV can be computed as FRC-ERV and TLC as FRC+IC

volume change measured at the mouth between full inspiration and complete expiration. The total lung capacity (TLC) is the total volume of gas in the lungs at the end of a full inspiration. It is usually computed as FRC+IC or FRC+V_T+IRV (QUANJER et al. 1993).

Several methods can potentially be used to determine pulmonary volumes and flows depending on the type of equipment available but also and more importantly on the type of suspected pathology as some methods are impaired by lung pathology. For each individual, results must be compared to predicted values (RODENSTEIN et al. 1982; WASSERMAN 1984; QUANJER et al. 1993). When comparing volumes obtained from imaging data it is essential to bear in mind that significant changes in some volumes but not in others, are induced by changes in posture. Imaging data are generally obtained supine, pulmonary function test data in sitting or standing position.

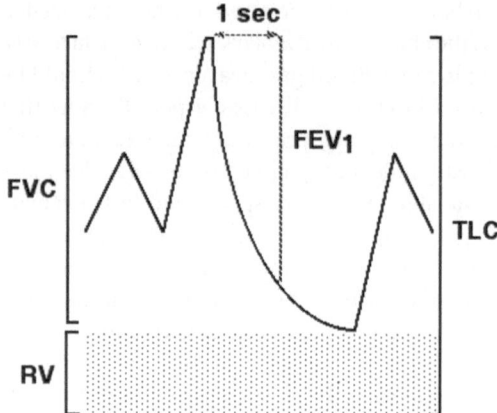

Fig. 2.2. Schematic spirometric tracing during a forced vital capacity (FVC) maneuver. FEV_1 is the maximal volume expired during the first second of a forced expiration

2.2.1
Basic Measurement Methods

2.2.1.1
Spirometry

Spirometry is the most classical method to measure static lung volumes (HUTCHINSON 1846). The device consists of an air-filled bell under which the volume of gas changes with patient ventilation. The water seal around the bell provides good dynamic properties but, recently, dry type spirometers have been developed because they are easier to decontaminate. Ventilation induces displacements of the bell proportional to the mobilized volumes. These displacements can be recorded on paper or stored in a computer. To avoid carbon dioxide accumulation due to the closed circuit, a canister filled with soda lime is placed on the expiratory line. A low flow of oxygen is also supplied into the bell to compensate for oxygen consumption. Measurements are performed at ambient temperature, pressure and water saturation (ATPS) and thus must be corrected to account for the temperature, pressure and water saturation prevailing in the body (BTPS) (QUANJER et al. 1993). Spirometry allows measurement of slow and forced VC, V_T, ERV and IRV and, during a forced expiration, the maximal volume expired during the first second, forced expiratory volume in one second (FEV_1) (Fig. 2.2). Slow VC is obtained through relaxed total expiration while forced VC (FVC) is obtained by maximal effort forced expiration. They do not differ in normal subjects. Slow and forced VC should both be measured

as they may differ in patients with COPD. Slow VC in these subjects tends to be larger and more reproducible. It should be stated if VC or FVC is used in the calculation of the FEV1/VC ratio.

2.2.1.2
Pneumotachography

The pneumotachometer is a flow sensing device positioned at the mouth. The flow signal is integrated over time to compute volumes (QUANJER et al. 1993). Various flow-sensing devices are used. The pressure differential pneumotachometer consists of a tube containing a resistive element causing a pressure drop when gas flows through it. The pressure drop is proportional to the gas flow as long as the latter is laminar. Two types of resistive devices are commonly used, either a bundle of capillary tubes (e.g. the Fleisch pneumotachometer) or a grid (e.g. the Lilly pneumotachometer). The hot wire pneumotachometer uses a heated platinum wire, where the temperature is maintained constant by an electrical current. As the wire is cooled by the gas flow, additional current is needed to maintain the pre-set temperature of the wire. This current is proportional to the flow. In the turbine type pneumotachometer, gas flowing through causes a vane to rotate. It is easy to use at the bedside, but has generally a poor accuracy. Most of these devices require linearization of the calibration curves to be useful over a reasonable range of flows and must be regularly calibrated to give accurate results. The Fleisch pneumotachometer remains the reference instrument but does not always cover the entire range of flows to be measured and causes decontamination problems. Hot wire instruments may prove unreliable at low flow when poorly designed. This is of critical importance when residual volumes are measured by the nitrogen washout technique because reliable assessment of flow at end expiration is needed for good accuracy (see below). These instruments cannot ensure reasonable measurements if flow is not laminar. Pneumotachography alone, as spirometry, cannot measure residual volume or consequently total lung capacity.

Major advantages of the pneumotachometers are their compactness and the fact that they can be used with an open breathing circuit. They are easier to use and to decontaminate than spirometers and are gradually replacing them. They tend, however, to offer higher internal resistance than spirometers, creating potential measurement problems in highly obstructive patients and in patients with muscular disorders. They are the device of choice when spirometry needs to be performed jointly with CT. Some of them have no metallic parts susceptible to interfere with imaging. Because volumes are integrated from flow on a breath-by-breath basis the integrated signal may be used to trigger image acquisition at any preset volume level offering total control on the time of imaging during the breathing cycle.

2.2.1.3
Maximum Effort Flow-Volume Curve

The simultaneous recording of flow and volume during a forced inspiration followed by a forced expiration allows a maximum effort flow-volume curve (MEFV) to be drawn, by plotting instantaneous flow against volume on an XY chart (HYATT and BLACK 1973). Forced expiration is performed from TLC to RV. This maneuver also gives the forced vital capacity (FVC). A complete flow-volume loop is obtained by performing a maximal inspiration immediately after the forced expiration. The tidal loop must systematically be plotted on the same tracing. The volume is plotted on the X-axis, usually from left to right during expiration, i.e. from TLC to RV, from the maximum to the minimum volume contrary to what is intuitive. The flow is plotted on the Y-axis, expiration being upwards from the X axis and inspiration downwards (KNUDSON et al. 1976) (Fig. 2.3).

In addition to FVC, several indices can be computed from flow-volume loops, namely peak inspiratory (PIF) and expiratory (PEF) flows, maximal instantaneous expiratory flow at 25%, 50%, 75% of FVC, commonly reported as $FEF_{25\%}$, $FEF_{50\%}$, $FEF_{75\%}$ and the mean flow between 75 and 25% of FVC, usually reported as FEF_{25-75}. Above 50% of VC the maximal flows are dependent on the expiratory efforts of the patients,

thus the importance of a normal muscular function. Below 50% of VC the maximal expiratory flows are no longer dependent upon the expiratory effort of the patient. They only reflect the elastic and resistive properties of the lungs and in normal subjects decrease linearly with the pulmonary volume just as the flow curve does (PRIDE et al. 1967; LEFF and SCHUMACKER 1993). The computed indices mentioned above give a quantitative but limited vision of the expiratory flow curve still analyzed best by the eye. Besides figures, visual inspection of the shape of the flow-volume loop provides important information. For example, a concave-shaped curve, inducing a decrease in $FEF_{25\%}$ and in $FEF_{50\%}$, suggests distal (small) airways obstruction. In severe obstructive patients, the flow-volume loop obtained during tidal breathing can overlap the loop obtained during a forced maneuver; this paradoxical situation defines flow limitation (Fig. 2.4) (BASS 1973; CHAN and IRVIN 1995). This paradoxical overlap is due to changes in bronchial transmural pressure in flow-limited patients. The effort produced by this type of patients during maximum expiration induces a pleural pressure that overrides the intrabronchial pressure, decreasing bronchial diameter, thus increasing resistance and consequently limiting flow preferentially at the end portion of the descending limb of

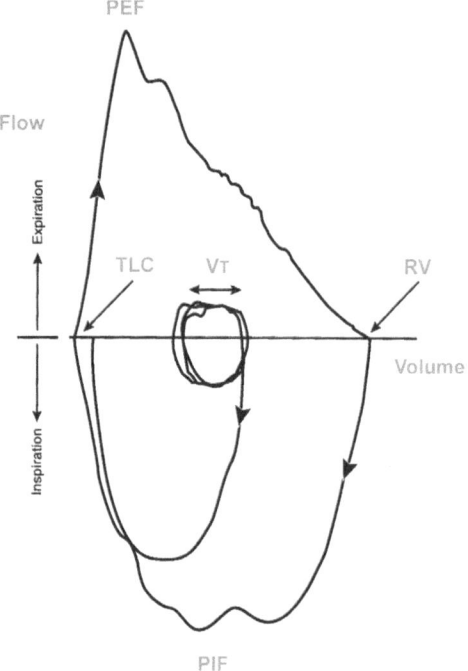

Fig. 2.3. Maximum effort flow-volume curve (MEFV) in a normal subject. X-axis change in lung volume (l) Y-axis change in flow (l/minute), V_T tidal volume, TLC total lung capacity, RV residual volume, PEF peak expiratory flow, PIF peak inspiratory flow

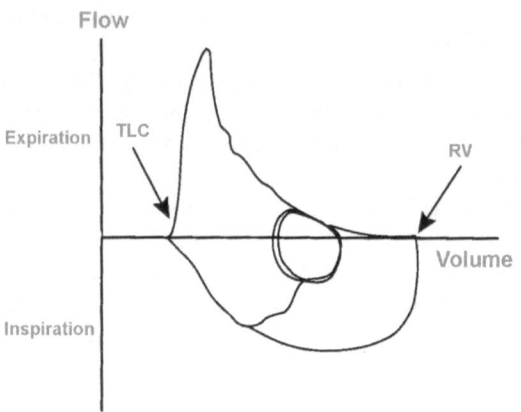

Fig. 2.4. MEFV curve in an obstructive patient. In this example, the forced expiratory flow-volume curve crosses the tidal flow-volume curve, demonstrating flow limitation during a forced expiration maneuver

the curve when intrabronchial pressure is the lowest, far more than occurs during a normal expiration as pleural pressure does not increase as much in this case. Conversely a flattening of the inspiratory limb of the flow-volume loop suggests an extrathoracic obstruction of the airways. A global decrease of the size of the loop, associated with a normal shape, occurs in restrictive diseases.

2.2.1.4
Further Assessment of Flow Limitation by Negative Expiratory Pressure

Although MEFV gives some assessment of expiratory flow limitation when compared to the tidal breathing loop, it is dependent on the patient's ability and commitment to perform a satisfactory maneuver. Also the MEFV curve may be dependent on the volume, time history and airway tone of the lung. In spontaneously breathing subjects it is possible to detect intrathoracic flow limitation by applying a negative pressure (range –3 to –5 cm H_2O) at the mouth during tidal expiration (NEP) (KOULOURIS et al. 1995). NEP induces a widening of the pressure difference between alveoli and airway opening, potentially mimicking the pressure difference generated during a forced expiratory maneuver, but with no active participation required from the patient. In the absence of expiratory flow limitation (EFL) NEP induces an increase in expiratory flow on the expiratory limb of the tidal loop. In the presence of EFL the expiratory flow does not increase throughout the entire part of the tidal expiration during NEP as compared to the flow of the preceding control expiration (TANTUCCI et al. 1998a). NEP

is extremely simple to perform repeatedly, and with no perception from the patient. When applied in early expiration it elicits little or no interference with upper airways reactivity (TTC) (TANTUCCI et al. 1998b).

Quantification and standardization of the method has not yet been carried out. The relationships between flow limitation detected at the mouth during tidal volume, parenchymal and airways structures need to be further studied. However NEP is the most promising, easy and objective test available to detect or monitor EFL in moderate or severe COPD as well as in other pathologies (DUGUET et al. 2000).

2.2.2
Measurements of Functional Residual Capacity and Residual Volume

2.2.2.1
Measurement of FRC by the Multiple-Breath Helium Dilution Method

FRC is the volume of gas remaining in the lungs at the end of a quiet breath. Until now FRC has been measured preferentially by the multiple-breath helium dilution method (HATHIRAT et al. 1970), although the nitrogen washout technique has tended to supplant it gradually, due to breathing circuit decontamination problems. Helium is an inert gas that does not cross the air-blood barrier and therefore equilibrates only in the ventilated compartments of the lungs. To measure FRC, the spirometer is first filled with a known volume of air containing a known amount of helium. Then, the patient is connected to the spirometer and breathes quietly from FRC until complete equilibration of the gas mixture between the spirometer and the lungs. The concentration of helium is measured inside the spirometer and the lung volume in which helium has been diluted is calculated. If the measurement ends at the end of a tidal expiration, the measured helium dilution volume corresponds to FRC. The test is considered completed after a maximum duration of 7-10 min or when the helium fraction monitored by the sensor appears stable, in fact quite a long time to sustain for a breathless patient. The limits of this technique are, in addition to the difficulty to decontaminate the closed breathing circuit, its high sensitivity to leaks in the closed circuit, particularly at the mouthpiece level, leading to overestimation, and the risk of underestimating FRC if some lung compartments are poorly ventilated. Helium will not dilute in these poorly ventilated areas during the duration of the test and therefore will be ignored by the measurement, resulting in underestimation of FRC. This happens essentially during severe COPD.

2.2.2.2
Measurement of FRC by the Multiple-Breath Nitrogen Washout Method

The multiple-breath nitrogen washout method is coupled to the assessment of lung volumes by pneumotachography to provide a measurement of FRC with an open breathing circuit (DARLING et al. 1940).

The patient breathes pure oxygen through a dual valve mouthpiece system. Pure oxygen washes out the non-exchangeable nitrogen originally present in the lungs. Expired nitrogen fraction in the expiratory tubing is measured "breath-by-breath" by using a fast responding analyzer. At the end of each expiration the expiratory valve closes and the other valve opens the inspiratory part of the circuit so that pure oxygen is delivered to the patient during inspiration. Thereby, each breath of pure oxygen washes out some residual nitrogen. The exhaled volume of nitrogen for each breath is calculated from the nitrogen concentration and from the volume of each breath obtained by integrating flow measured at the mouth via the pneumotachometer. Values for each breath are summed up to provide the total volume of washed out nitrogen. The test is carried out either until the nitrogen fraction in the alveolar gas has been reduced to approximately 1% or when the test duration is over 7 min. FRC is calculated by dividing the total volume of nitrogen washed out by the difference in nitrogen concentration from the beginning to the end of the test. Measurements must be corrected for body temperature and water saturation pressure (BTPS conditions) as the actual gas volume within the thorax is water-saturated and at body temperature whereas measurements are made at room temperature and ambient water partial pressure. Measurements must also be corrected for nitrogen excretion from tissue and blood. The resistance of the valve at inspiration may prevent some patients from breathing correctly. Dry mouth due to pure dry oxygen makes the test somewhat uncomfortable for some patients. Technique limitations during heterogeneous lung disease with low ventilated area are similar to that of helium dilution for the same reasons. Nitrogen volumes tend to be slightly lower than helium volumes, by a few percent. There is of course no way to decide which one is best as both techniques do not really measure a volume but the diffusion space for each gas.

2.2.2.3
Measurement of Thoracic Gas Volume by Plethysmography

The plethysmographic technique is the reference method to measure the thoracic gas volume (V_{TG}). Measurement is usually performed at FRC. As previously stated, poorly ventilated lung regions may bias the helium dilution and the nitrogen washout volumes (DUBOIS et al. 1956a). The plethysmographic technique incorporates the entire volume of gas located in the thorax, whether it is located in well-ventilated area or remains trapped (e.g. in bullae). The physical basis of plethysmography is that stated by the law of Boyle–Mariotte on pressure-volume related changes when a constant number, n, of gas molecules are trapped in a constant volume container, V, ($PV=nRT$) or $P_1V_1=P_2V_2$ if temperature, T, remains constant. In order to perform a measurement of the V_{TG}, the patient is seated in the body plethysmograph, which consists in fact of a sealed box. The patient breathes through the mouth in a respiratory circuit connected to ambient air, while wearing a nose-clip. A pneumotachometer and a pressure transducer measure mouth flow and pressure. The airways are occluded by mean of a mouth shutter at the end of a normal expiration. The patient then breathes (pants) against the shutter leading to compression and decompression of the given amount of gas trapped in the chest. Changes of thoracic volume induces mirror changes in the box volume and consequently in box pressure that can be recorded and related to the change in volume, making use again of the Boyle-Mariotte law. Differences in volume between oxygen consumption and carbon dioxide production during the measurement period are considered negligible. The frequency of this "panting" should be around 1–2 cycles per second, the glottis remaining open. The pressure changes in the airway (P) and the volume changes in the body box (V) during the panting maneuver are recorded and used for the calculation of the V_{TG} from the Boyle–Mariotte law (MEAD 1960; QUANJER et al. 1993).

Following this law, $V_{TG}=P_{atmospheric}\times(V/P)$.

In theory the technique measures all the volume of thoracic gas plus all gas in the body that may be affected by pressure changes linked to breathing. This includes abdominal gas, pneumothorax and emphysematous bullae. But in practice the major concern is linked to the assumption that the pressure measured at the mouth is identical to the pressure in the alveoli, assuming that pressure within the thoracic volume is the same everywhere so that gas laws are valid. This may not be the case when airways obstruction becomes severe and elastic properties of the lungs are too inhomogeneous, resulting in different time constants for pressure transmission between various lung compartments and the mouth (RODENSTEIN et al. 1982). In this situation lung volume can be overestimated and the overestimation will increase with obstruction. This error factor can be limited by asking the subject to breathe at lower panting frequency, in practice below 1 Hertz, and at low flow (±300 ml/minute maximum) and tidal volume. The subject must also support his cheeks and mouth floor with his hands to minimize volume changes at the mouth level and reduce cheek compliance. The real limits of the technique when performed properly are the impossibility for some patients to enter the sealed box (e.g. patients on stretcher, claustrophobic or obese patients) and the relative difficulty of the panting maneuver. It remains however, the reference method, in particular for chronic obstructive patients. Keeping this in mind, comparisons between plethysmographic V_{TG} and FRC measured with the helium dilution method in obstructive patients provide a good evaluation of the volume trapped in the lungs (RODENSTEIN and STANESCU 1982).

2.3
Effects of Posture on Volumes

The reduction in VC observed in normal subjects between the standing and the supine positions is small,

usually around 200 ml, and caused by the displacement of gas in the thorax due to the change in repartition of blood in the thorax because of gravity. The fall in TLC is more important and related to the change of the gravitational effects of the content of the abdomen on the diaphragm, reducing the relaxation volume (MORENO and LYONS 1960; AGOSTONI and HYATT 1986; MICHELS et al. 1991). IC is little influenced and residual volume is barely affected. Therefore it is ERV that is most affected by a change in posture (Fig. 2.5). There are no standards available for pulmonary volumes in the supine position. In disease, changes in volumes as a function of posture are important, more specifically to appreciate paralysis of the diaphragm, and may reach 25% of VC. Effect of posture on FRC and ERV are increased further in obese patients.

2.4
Assessment of the Airways

2.4.1
Measurement of Airway Resistance

In normal subjects, the resistance to airflow is mainly due to the large bronchi. In obstructive lung disease, resistance of the small airways become proportionally higher than the resistance of the large airways. Assessing bronchial resistance is particularly relevant in patients having difficulties to perform a reliable forced expiratory maneuver because FEV1 is not reliable, in children for example.

Classically airway resistance is calculated by dividing the difference between alveolar pressure and

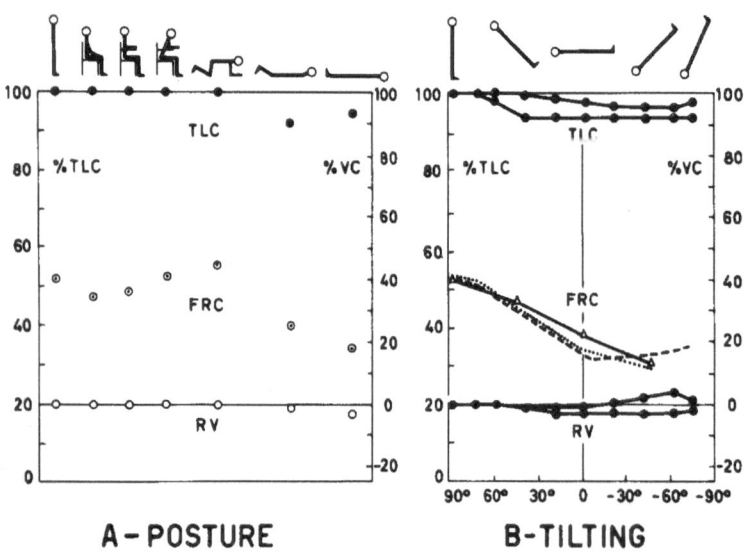

A - POSTURE   **B - TILTING**

Fig. 2.5. Effect of change of posture on pulmonary volumes. *TLC* total lung capacity, *RV* residual volume, *VC* vital capacity, *FRC* functional residual capacity. FRC changes significantly due to change in ERV (expiratory reserve volume). From AGOSTONI and HYATT (1986)

mouth pressure by the gas flow. Measurements are obtained during plethysmographic maneuvers. Mouth flow is measured with a pneumotachometer during panting and alveolar pressure is obtained in the plethysmograph, following derivation from the gas law techniques used to determine the V_{TG} (DuBois et al. 1956b). Airway resistance can also be assessed by the forced oscillation technique. Airways are challenged by external forced oscillations produced at the mouth by a loudspeaker. The analysis of the reflected waveforms allows computation of airway resistance without patient cooperation and permits, at least in theory, to partition proximal and distal resistances (DESAGER et al. 1991). Because the variability is large, a deviation of three times the standard error is required to detect a significant change in resistance, limiting the usefulness of these methods, unless the patient offers little or no cooperation.

2.4.2
Airway Responsiveness to Pharmacological Challenges

2.4.2.1
Bronchial Hyper-responsiveness Challenge

Airway hyper-responsiveness consists of an exaggerated response to a bronchoconstricting stimulus. This response is assessed, most of the time, by monitoring changes in FEV_1 but monitoring airway resistance is also acceptable. The test is particularly relevant in patients reporting transient symptoms consistent with asthma but whose ventilatory flows and volumes are normal. The test is mandatory to assess specific occupational allergens. The high sensitivity and the high negative predictive value of these tests make them particularly suitable to exclude the diagnosis of asthma (BRITTON 1988; COCKCROFT and HARGREAVE 1990; BACKER et al. 1991; COCKCROFT et al. 1992). Absolute contra-indications to a bronchial challenge are: severe baseline airways obstruction ($FEV_1 < 1.2$ l in adults), recent myocardial infarction (<3 months), arterial aneurysms, inability to understand the procedures and the implications of the test. Relative contra-indications are: moderate to severe airways obstruction, recent upper respiratory tract infection (<2 weeks), exacerbation of asthma, arterial hypertension, pregnancy, epilepsy requiring drug treatment. Bronchodilator treatments must have been stopped before the test for a period at least equal to the duration of their action. Particular care needs to be exerted with long-acting substances. The patient inhales increasing concentrations of a bronchoconstrictor agent. For the assessment of non-specific bronchial hyper-responsiveness, the most commonly used pharmacological agent is aerosolized methacholine administered with a inspiration-triggered dosimeter. Doses of methacholine are cumulative and the maximal cumulative dose is usually 1600 µg. Hyper-responsiveness is characterized when a decrease of FEV_1 of more than 20% is obtained by a cumulated dose. It is recommended to standardize the results in term of the cumulative dose of drug theoretically capable of producing a fall of 20% in FEV_1 (PD_{20}). When, as it is likely, the experimental fall in FEV_1 exceeds 20%, PD_{20} needs to be calculated by interpolation (STERK et al. 1993). In the case of suspected exercise-induced asthma, non-specific hyper-responsiveness can be evaluated by inhalation of dry air during exercise performed at 80% of maximum predicted heart rate or less. FEV1 measurement must be performed generally up to 10–15 min after the cessation of the exercise, as bronchial response often occurs late. When occupational asthma is suspected, the airway response to a specific agent like for instance flour, can be assessed using quantified aerosols (STERK et al. 1993). Except for these instances, bronchial challenge is always a non-specific test. It never provides a formal diagnosis for a specific disease, for instance asthma. But conversely it is highly unusual for an asthmatic patient to be a non-responder.

2.4.2.2
Bronchodilator Challenge

Mirroring bronchial challenge, these tests consist of performing spirometry after the administration of a bronchodilator, in order to evaluate the reversibility of airway obstruction. The most frequently used bronchodilators are b_2-agonists, although the effect of either anticholinergic drugs or steroids can also be assessed. Reversibility is defined by an improvement of FEV_1 of at least 12% compared to the predicted value and of at least 200 ml (ATS 1991). Recent data have shown that bronchodilators can increase inspiratory capacity in obstructive patients showing no improvement in terms of FEV_1 (TANTUCCI et al. 1998a). Since inspiratory capacity correlates best to dyspnea in obstructive diseases (O'DONNELL et al. 1998, 1999), its measurement should probably be systematically associated to FEV_1 in order to assess the effects of a bronchodilator. A bronchodilator study is highly recommended in all obstructive patients. It is of some use in treated asthmatic patients to check indirectly the effectiveness of the treatment.

2.5
Functional Semiology

2.5.1
Normal Versus Abnormal Values

Regression equations have been extensively published to allow comparison of volume and diffusion data between a patient and data collected from never-smoking patients (ex-smokers excluded) considered as reference, allowing for age, sex and height. Because this reference equation results from the pooling of data from disparate sources, poorly understood differences exist between some reference data and others. However, reference values need to be used and their reference must be provided each time. Ethnicity must be taken into account, as some subjects from non-European descent tend to have smaller normal lung volumes. Height measurement must be carefully monitored as kyphosis or osteoporosis lead to significant errors. Arm span may be substituted in these cases. The range around the normal mean value provided by the reference equations can be expressed either in terms of "standardized residual", or as "percent predicted". Predicted range changes little with age or sex of the subject of a given gender and may lead to doubtful interpretation at extreme age or size. The predicted value of ±20% generally used is therefore mostly a reasonable rule of thumb to define normality and critical confidence should be given to these figures although they need to be used for the sake of standardizing the presentation of the results. The patient is always his best control for follow-up (EUROPEAN RESPIRATORY SOCIETY 1993).

Ventilatory abnormalities can be grouped in two major syndromes.

2.5.2
Obstructive Syndrome

The obstructive syndrome is characterized by a disproportionate decrease in FEV_1 compared to the decrease in VC (QUANJER et al. 1993) The obstructive syndrome is often defined as a decreased FEV_1/FVC ratio, below 85% of the predicted value. The reduction in FEV_1 is in direct relationship to the severity of the disease and is the best gauge of the severity of obstruction. Current classification specifies that if FEV_1 is greater than or equal to 50% of the predicted value, then obstruction is defined as mild, if FEV_1 is between 35% and 49% of the predicted value it is

defined as moderate. If FEV_1 is less than or equal to 34% of the predicted value, obstruction is defined as severe (ATS 1995b). However this classification is currently challenged as being inadequate. The expiratory flow-volume curve is concave (Fig. 2.4) and expiratory flows are reduced at all volumes (BASS 1973; ULMER et al. 1997). A reduction of flows occurring predominantly at low pulmonary volumes (25–50% of VC) suggests a preferential obstruction of the smallest airways. During the evolution of the disease RV increases, VC first remains normal and then decreases, TLC increases late in the disease (QUANJER et al. 1993). Severe obstructive syndromes are usually associated with hyperinflation, functionally defined as a significant increase of TLC. Hyperinflation seems to be a major cause of dyspnea in obstructive lung diseases (O'DONNELL and WEBB 1992).

2.5.3
Etiology of Obstructive Syndrome

The two most frequent causes of obstructive syndromes are asthma and chronic obstructive pulmonary disease (COPD) (QUANJER et al. 1993). Asthma is characterized by the great variability of airway obstruction. In many patients, pulmonary function is normal or only slightly altered between intermittent episodes. For example, a persistent obstruction of the small airways can be suspected only by a reduction in $FEF_{25\%}$ and $FEF_{50\%}$ or by a concave shape of the flow-volume loop at low volumes. During an asthma attack, however, the FEV_1/FVC ratio decreases sharply but need not be measured. Assessment of non-specific hyper-responsiveness can help eliminate the diagnosis of asthma if the symptoms are transient or atypical. Reversibility of airway obstruction is the rule providing treatment has been correctly discontinued.

In COPD patients, airway obstruction is usually not reversible by bronchodilators. Nonetheless, reversibility should be assessed systematically. Indeed, some degree of bronchial hyper-responsiveness may be associated with COPD. If positive, the test justifies a bronchodilator treatment. Furthermore, bronchodilators can reduce hyperinflation and dyspnea in some COPD patients without necessarily improving FEV_1 (TANTUCCI et al. 1998a; O'DONNELL et al. 1999). A decrease in hyperinflation after administration of bronchodilators, easily assessed by measuring the inspiratory capacity, also justifies a bronchodilator treatment.

2.5.4
Restrictive Syndrome

A restrictive syndrome is defined by a decrease of TLC below 80% of the predicted value (QUANJER et al. 1993). Typically, the flow-volume loop in pulmonary restriction can exhibit a reduced size because flows are decreased proportionally to volumes (RUPPEL 1998) but the shape is normal. In fact it is homothetic to a normal loop. A reduction of VC without a simultaneous reduction of TLC never constitutes an acceptable element in favor of a restrictive syndrome: a decrease in VC can be associated to an increased RV and therefore to a normal or even increased TLC. This occurs in those severe obstructive patients who demonstrate pulmonary gas trapping (RODARTE et al. 1975).

2.5.5
Etiology of Restrictive Syndromes

A typical cause of restrictive syndrome is indeed a quantitative reduction in lung parenchyma, as occurs for example after surgery or due to tuberculosis sequelae. A major parenchymal cause is interstitial lung disease. Restrictive syndromes can also be related to chest wall diseases, such as severe kyphoscoliosis, pleural diseases or neuromuscular disorders (RUPPEL 1998).

2.5.6
Mixed Syndrome

Before retaining the coincidence of obstruction and restriction, which characterizes few respiratory diseases (e.g. bronchiectasis, pneumoconiosis, tuberculosis sequelae or sarcoidosis), it is important to eliminate methodological errors. Careful attention should be paid to the maximum effort flow-volume loop. FRC can be underestimated by the helium dilution method in severe obstruction with major air trapping. The consequence can be an underestimation of TLC, which will wrongly lead to the conclusion that an obstructive and a restrictive syndrome are present simultaneously. To avoid this pitfall, a plethysmographic measurement of the lung volumes should be performed and compared to helium data. But among other major considerations it should always be remembered that smokers may also suffer from interstitial lung disease or something else.

2.6
Lung Compliance

2.6.1
Definition and Measurement

Lung compliance assesses the elastic properties of the lungs, independently of the properties of the chest wall and of muscular effort (RAHN et al. 1946). Lung compliance is defined as the passive volume change per unit of transpulmonary pressure (Ptp) change, Ptp being the difference between airway pressure (Paw) and pleural pressure (Ppl). Ppl is evaluated by measuring the esophageal pressure (Poes) with an esophageal balloon catheter inserted through the nose. Paw is recorded at the mouth. Compliance can be measured under quasi-static (YERNAULT and ENGLERT 1974) or static conditions (RAHN et al. 1946; TURNER et al. 1968). The latter is the reference technique. The subject expires passively from TLC, through the mouth, in a respiratory circuit equipped with a shutter capable of periodically interrupting the expiratory flow. The pressures and the corresponding volume are measured simultaneously during each interruption of flow. The lung compliance is computed from the relationship between transpulmonary pressure and lung volume.

2.6.2
Significance of Lung Compliance

A reduced lung compliance is typical of pulmonary fibrosis, pulmonary edema or the acute respiratory distress syndrome. Compliance monitoring has been suggested with mixed success as a relevant tool to avoid volumetric injury during artificial ventilation. Conversely, an increased compliance is the hallmark of emphysema, illustrating the decrease of the elastic recoil of the lungs. Measurement of lung compliance is semi-invasive and has lost much of its clinical interest since emphysema can be identified more easily by CT. It is now mostly a research tool.

2.7
Assessment of Respiratory Muscles

Respiratory muscles are the effectors of the respiratory system. Ventilation depends upon their phasic and continuous contraction in response to neural commands. Failure of this process leads to alveolar hypoventilation and respiratory distress. Virtually all neuromuscular diseases can affect the respiratory muscles like, for example, myasthenia gravis, Guillain–Barré syndrome, amyotrophic lateral sclerosis, polymyositis and others. The involvement of respiratory muscles in steroid-induced myopathy deserves a special mention as its occurrence has been increasingly recognized in recent years (DEKHUIJZEN and DECRAMER 1992). It can explain an exertional dyspnea or a restrictive ventilatory defect in patients receiving steroids for non-respiratory diseases. In patient with primarily chronic respiratory disease such as asthma or COPD, systemic administration of steroids can alter the load compensation capabilities of respiratory muscles and hence contribute to a worsening of the respiratory status.

2.7.1
Static Pressures

The most common way to assess the respiratory muscles is to measure the maximal inspiratory (Pi max) or expiratory (Pe max) pressure that they can develop (AGOSTONI and RAHN 1960). The airway pressure is recorded at the mouth during a forceful, briefly sustained respiratory effort, either inspiratory or expiratory, against an occluded valve. A small leak in the valve prevents glottal closure and eliminates pressures generated by the cheek muscles by allowing a small amount of gas to enter and to exit the oral cavity. The pressure produced at the mouth by the respiratory muscles indirectly reflects their force.

To account for the force-length relationship of striated muscles, it is mandatory to clearly state at which volume static pressures are measured. Inspiratory muscles are at a mechanical advantage at low lung volume, hence their maximal pressure generation capability is at RV. Conversely, expiratory muscles are at a mechanical advantage at TLC. It is common practice to measure Pi max and Pe max from FRC. Measuring them at different lung volumes increases the amount of information. These tests are very easy to perform but of limited interest due to their strong dependence on the cooperation of the subject and the fact that they assess synergistic muscles without discrimination.

2.7.2
Stimulation

Stimulation techniques alleviate the problem of subject cooperation. They can also separate one muscle from

another. Phrenic nerve stimulation allows specific investigation of the diaphragm. It can be performed using an electric (SIMILOWSKI et al. 1991) or a magnetic stimulator (SIMILOWSKI et al. 1989). Recording the electromyographic response of the diaphragm in relation to phrenic nerve stimulation gives access to phrenic conduction. Recording a pressure in response to phrenic nerve stimulation permits an evaluation of the contractile properties of the diaphragm. The recorded pressure can be either airway pressure at the mouth (YAN et al. 1992) or transdiaphragmatic pressure (SIMILOWSKI et al. 1991). As for the maximal respiratory pressures, the results should be interpreted in relation to lung volume.

2.8
Assessment of the Ventilatory Control

The automatic, phasic and continuous contraction of the respiratory muscles results from the output of neuronal networks located in the brainstem. These networks constitute the ventilatory central pattern generator (CPG) (FELDMAN and SMITH 1995). The ventilatory CPG adapts ventilation to the needs of the body in response to numerous afferent inputs. One of the main goals of ventilation is to provide adequate carbon dioxide elimination and oxygen supply. Chemoreceptors continuously monitor the level of oxygen and carbon dioxide in the blood. As a result, hypercapnia and hypoxemia stimulate the ventilatory CPG and increase the ventilatory drive. In normal subjects, any increase in the respiratory drive leads to an augmentation of ventilation (FITZGERALD and LAHIRI 1986). Therefore, the response to hypercapnia is one of the most common ways to assess the function of the ventilatory CPG.

2.8.1
Measurement of the Respiratory Drive

The output of the ventilatory CPG is conveyed to the respiratory muscles which transform the CPG activity into pressure changes. These pressure variations induce lung volume changes. Therefore the output of the ventilatory CPG should ideally be assessed by recording the electroneurogram of the phrenic nerve because it conveys the inspiratory signal to the diaphragm. Such recording is however impossible to perform in a clinical setting. The quantitative analysis of ventilation is the only practical way to assess the CPG output. The analysis of the spirogram can provide information about the respiratory drive. However, this implies a fully normal chain of transmission and any abnormality of the bronchi, lung, chest wall, respiratory muscles or nerve conduction may interfere with the evaluation of the respiratory central output from a spirogram. To circumvent this problem it has been suggested to take advantage of the fact that no pressure loss due to changes in chest wall shape or to the speed of contraction of the inspiratory muscles occurs during the first 100 ms after the onset of inspiration when a subject breathes against an infinite load, in other words when the mouth is occluded. If this is true, then the change of pressure observed at the mouth during the first 100 ms of inspiration after occlusion (P0.1) is a direct assessment of the CPG activity (DERENNE et al. 1976; WHITELAW et al. 1976; WHITELAW and DERENNE 1993).

The subject breathes through a mouthpiece in a two-way valve. During expiration, the inspiratory limb is occluded with a shutter, the subject being unaware of this occlusion. The subsequent inspiration is performed against an infinite resistance due to occlusion and produces a negative pressure in the respiratory circuit. Because a conscious subject does not perceive the occlusion before about 200 ms following the beginning of the effort, the pressure measured in the airway at 100 ms is believed to reflect only the pressure output of the automatic ventilatory CPG. P0.1 can be measured independently of flow or volume changes. Since P0.1 is usually measured at FRC, muscle length and therefore muscle force is standardized. The measurement of P0.1 is easy and non-invasive.

There are pitfalls to the P0.1 technique that should be taken into account when evaluating the ventilatory CPG with this method (WHITELAW and DERENNE 1993). An increase in FRC, as it can occur in emphysema, decreases the force generation capability of the respiratory muscles, hence the risk to underestimate P0.1. An increased airway time constant can delay the transmission of pleural pressure to the mouth, hence again a risk of underestimation. Therefore, in patients with chronic obstructive pulmonary disease, occlusion pressure should not be measured in the airway but with an esophageal balloon-tipped catheter. Paradoxical motion of the rib cage and abdomen may occur during an occluded effort, resulting in marked change in muscle length despite no change in lung volume. As a consequence, inspiratory muscle contraction is no longer isometric and depends upon force-length and force-velocity factors. Finally a modification of the shape of the pressure wave can also alter the measurement of P0.1. For example, positive pressure ventilation can make this wave more convex, a factor leading to an overestimation of P0.1.

2.8.2
Stimulation of the Ventilatory Central Pattern Generator

Routine assessment of the ventilatory CPG involves the evaluation of its response to a stimulation. The induction of hypercapnia is the most commonly used method. In the open-circuit or steady-state technique (FENN and CRAIG 1963), the subject breathes various concentrations of carbon dioxide from a reservoir until a steady state is reached. In the rebreathing technique (READ 1967), which is the most commonly used, the subject rebreathes from a one-way circuit containing a reservoir of 7% carbon dioxide in oxygen. Because of the closed nature of the circuit, ventilation fails to eliminate carbon dioxide and arterial partial pressure of CO_2 ($PaCO_2$), estimated from end tidal PCO_2, increases. The subject rebreathes until the concentration of end tidal PCO_2 exceeds 9% or until 4 min have elapsed. The response of the ventilatory CPG to hypercapnia is evaluated by measuring the total ventilation and P0.1, which normally increase. The normal ventilatory response is linear with a slope between 1.5 and 3 l/min/mmHg in 80% of the subjects. This slope characterizes the sensitivity of the response and is similar for the two techniques. The response to hypoxia can also be assessed (DEJOURS et al. 1957; REBUCK and CAMPBELL 1973). However, this test is more rarely performed because exposure to hypoxia may be hazardous.

2.8.3
Significance of Ventilatory Control Assessment

A decrease in the response to hypercapnia characterizes the hypoventilation syndromes, either congenital or acquired (KRACHMAN and CRINER 1998).

Investigation of the respiratory drive remains very dependent upon the respiratory muscles. Interpretation of the results should therefore take into account their function. For example, in the case of myasthenia gravis, the respiratory drive is likely to be normal but the muscular function is impaired, possibly leading to an apparent decrease in the response to hypercapnia (SPINELLI et al. 1992). In Duchenne and Steinert myopathies, the ventilatory response to hypercapnia is reduced while the rise of P0.1 with increasing hypercapnia remains normal (SCANO et al. 1993).

2.9
Assessment of Gas Exchanges

The purpose of ventilation is to provide adequate gas exchanges to the body. The assessment of these exchanges is therefore an important part of pulmonary function testing. The three main methods used to evaluate the gas exchanges are: measurement of blood gases at rest, assessment of the diffusing capacity and the exercise test. Direct evaluation of the ventilation-perfusion ratio is indeed the best integrative approach to gas exchange but unfortunately remains a difficult semi-invasive and cumbersome procedure limited to research protocols.

2.9.1
Blood Gases

2.9.1.1
Specimen Collection and Measurement

Measurement of blood gases is performed on an arterial sample preferably obtained from the radial artery but may also be drawn from brachial and femoral arteries. The adequacy of collateral circulation to the hand via the ulnar artery should be assessed before a radial artery puncture is performed using the modified Allen's test: first both radial and ulnar arteries are occluded by pressing down over the wrist. The subject is then instructed to make a fist and then to open and to relax the fingers. At that time, the palm of the hand is pale and bloodless. The ulnar artery is released while the radial artery remains occluded. If the hand is reperfused in less than 10 s, it is concluded that the ulnar supply is adequate and that the radial artery can be punctured. Blood is collected in a heparinized syringe which should not contain any air bubbles and which should be rapidly sealed from the atmosphere. Analysis of the blood sample must be performed as quickly as possible. Capillary samples are useful when arterial puncture is impractical, especially in children. The ear lobe is often chosen. The region should be heated by warm compresses and lanced.

Commercially available instruments measure the partial oxygen pressure (PO_2), the partial carbon dioxide pressure (PCO_2) and the pH in arterial blood samples. Bicarbonate concentration and oxygen saturation of hemoglobin (SaO_2) are calculated from these data. Calculated SaO_2 is acceptable only if the hemoglobin concentration is normal. Therefore devices that also measure hemoglobin concentration and SaO_2 must be preferred (AARC 1993). These instruments also measure carboxyhemoglobin. This information is required for correct interpretation of the carbon monoxide diffusing capacity (see below).

Pulse oximetry provides a non-invasive assessment of oxygen saturation (SpO_2). SpO_2 estimates SaO_2 by analyzing absorption of light passing through a capillary bed. The agreement of SpO_2 with SaO_2 is within 2% from 85% to 100% but deteriorates quickly below this value. Most pulse oximeters use a sensor that attaches to a finger or an ear lobe (SEVERINGHAUS and KELLEHER 1992).

2.9.1.2
Acid-Base Equilibrium

The measurement of the arterial pH assesses the acid-base equilibrium (Table 2.1). In healthy adults, the pH averages 7.40, ranging from 7.38 to 7.42. Arterial pH below 7.38 defines acidemia, a pH above 7.42 defines alkalemia. Although pH is generally used, it is an artificial representation of the ionic blood status and H+ concentration should ideally be preferred. Respiratory acid-base disorders are caused by changes in $PaCO_2$: an increase in $PaCO_2$ induces a respiratory acidosis, a decrease in $PaCO_2$ provokes a respiratory alkalosis. Normal $PaCO_2$ ranges from 35 mmHg to 45 mmHg (4.7–6.0 kPa). Metabolic acid-base disorders are due to changes in bicarbonate concentration. A respiratory acid-base disorder can be compensated by bicarbonate adaptation and a metabolic acid-base disorder can be compensated by ventilatory changes.

Hypercapnia and the related respiratory acidosis are the result of alveolar hypoventilation. The latter can be due to various neurological or neuromuscular diseases affecting the respiratory pump. Alveolar hypoventilation can also be caused by an increase in the dead space-tidal volume ratio (VD/VT) (KRACHMAN and CRINER 1998). On the contrary, respiratory alkalosis is due to hyperventilation like, for instance, in the hyperventilation syndrome (GARDNER 1996). Metabolic acid-base disorders are beyond the scope of this chapter.

2.9.1.3
Hypoxemia

The PaO_2 of healthy adults at sea level varies from 85 to 100 mmHg (11.3–13.3 kPa). Hypoxemia is defined as an arterial PaO_2 less than 85 mmHg (11.3 kPa) (Table 2.1). Although a value above 85 mmHg (11.3 kPa) may still be abnormal in some instances, a value below that threshold is always abnormal.

Hypoxemia etiologies are numerous. Contrary to a strongly anchored credo PaO_2 does not decrease with age in subjects with normal heart and lung functions (DELCLAUX et al. 1994; GUENARD and MARTHAN 1996) but decreases after a few days in the supine position. The most frequent mechanism is probably an abnormality of the ventilation-perfusion matching, resulting in venous admixture. Other possible causes include alterations of the alveolocapillary diffusing capacity, a right-to-left shunt or alveolar hypoventilation. In the case of pure alveolar hypoventilation, the sum of $PaO_2 + PaCO_2$ exceeds 130 mmHg (17.3 kPa). A right-to-left shunt can be differentiated from an increased venous admixture by inhalation of pure oxygen for 20 min or longer. If there is no true right-to-left shunt, the PaO_2 under these conditions exceeds 550 mmHg (73.3 kPa). In other words, pure oxygen inhalation corrects hypoxemia caused by venous admixture. The blood gases collected in a subject breathing pure oxygen also permit an estimation of the right-to-left shunt (WEST 1990). Cardiac or hematological etiologies are not considered here.

2.9.2
Assessment of Diffusing Capacity

2.9.2.1
Method

An alteration of the capacity of oxygen to diffuse across the alveolocapillary membrane can lead to hypoxemia. The diffusing capacity of the membrane is evaluated by using a small amount of carbon monoxide (CO) mixed to inspired air because membrane behavior of CO matches that of O_2 so that it can be used as an external tracer. CO diffuses from the alveoli to the bloodstream and then remains fixed on hemoglobin, its affinity for hemoglobin being approximately 210 times higher than the affinity for oxygen (FORSTER 1987). This second property considerably simplifies mathematical analysis of tracer kinetics.

The measurement of CO diffusion can be performed during steady-state ventilation. However this method is poorly standardized and the results depend upon the ventilation-perfusion ratio. The most commonly used method is the single breath, or breathhold technique (FORSTER 1987). At the end of a maximal expiration, the subject inspires up to TLC, a gas mixture containing CO in low concentration (0.3%), an inert gas not normally present in the body (for example 10% of helium), 21% of oxygen and nitrogen. Then the breath

Table 2.1. Blood gas abnormalities

Hypoxemia:	$PaO_2 < 85$ mm Hg (11.3 kPa) (Ambient air at sea level)
Shunt or venous admixture:	Decreased $PaO_2 + PaCO_2$ (<130 mmHg, 17.3 kPa)
Alveolar hypoventilation:	Preserved $PaO_2 + PaCO_2$ (~130–140 mmHg), (17.3–18.7 kPa)
Acidosis:	pH<7.38
Respiratory:	$PaCO_2 > 45$ mmHg (6.0 kPa) (hypercapnia)
Metabolic:	$HCO_3^- < 25$ mmol/l
Alkalosis:	pH>7.42
Respiratory:	$PaCO_2 < 35$ mmHg (4.7 kPa)
Metabolic:	$HCO_3^- > 27$ mmol/l

is held at TLC for approximately 10 s during which CO diffuses across the alveolocapillary membrane and the inert gas dilutes in the ventilated compartment of the lungs. At the end of this apnea, the expiratory gases are analyzed. The diffusing capacity, also called the transfer factor (T_LCO), is calculated from the measurement of the amount of expired CO, knowing the amount of inhaled CO. The investigated lung volume (V_A) is calculated from the dilution of the inert gas.

In fact this measurement tests the transfer mechanisms of CO from the alveolar space to its combination with hemoglobin, thus including diffusion from the plasma to the red cell cytoplasm and then the kinetics of CO fixation to hemoglobin. Many determinants, some not related to membrane properties, can therefore influence the measurement of the transfer factor: the lung volume, the size of the exchange surface area, the alveolocapillary membrane, the capillary volume, the concentration of hemoglobin, the distribution of the ventilation-perfusion ratios and the partial pressure of CO in arterial blood (PaCO) at the beginning of the measurement (FORSTER 1987). Tobacco consumption, for instance, can substantially increase PaCO and should be taken into account when calculating T_LCO. An arterial blood sample must be collected and PaCO measured before the assessment of T_LCO. T_LCO increases from the upright to the sitting position and from the sitting to the reclining positions. Gas diffusion across the alveolocapillary membrane is better described by the transfer coefficient K_{CO}, the ratio of T_LCO to V_A ($K_{CO}=T_LCO/V_A$). K_{CO} is by nature less influenced by the exchange surface than T_LCO. Thus a decrease of the membrane surface area will result in drop of the transfer factor but no change in the transfer coefficient as the membrane is intact. A disease of the membrane with no change in surface area will result in a decrease of both. The value of T_LCO or K_{CO} should also be corrected for hemoglobin concentration because anemia decreases and polyglobulia increases T_LCO. Although diffusion procedures are now well standardized (COTES et al. 1993) it is unfortunate that normal control population values are still incomplete. It is therefore advisable that the patient be its own control during a follow-up procedure and those repetitive studies be performed preferably in the same laboratory. This especially matters when patients receive drugs potentially harmful for the lungs, such as bleomycin, in which case a baseline measurement needs to be performed before the drug is administered.

2.9.2.2
Significance of Diffusing Capacity

Measurement of T_LCO is usually performed following the assessment of lung volumes in patients with suspected or confirmed disease of the lung parenchyma. The measurement of T_LCO is useful for both diagnosis and follow-up purposes. Numerous etiologies and various mechanisms can induce a decrease in T_LCO (COTES et al. 1993). T_LCO is reduced in interstitial fibrosis, pneumoconiosis, widespread granulomas, and pulmonary edema. Impairment of the alveolocapillary membrane probably explains the reduction of T_LCO in these diseases. Lung resection, for cancer or other reasons, typically results in decreased T_LCO but affects K_{CO} much less. T_LCO is also reduced in emphysema because of the destruction of the alveolocapillary membrane but in this example, K_{CO} is indeed also decreased. Before CT became available, the diagnosis of emphysema relied on the association of an obstructive syndrome, a decreased diffusing capacity and an increased pulmonary compliance (YERNAULT and PAIVA 1986). A decrease in T_LCO can also result from a mismatching of the ventilation-perfusion ratio. T_LCO can be diminished in pulmonary vascular disorders including pulmonary hypertension. The capability of a decrease in T_LCO to predict oxygen desaturation during exercise remains controversial (RUPPEL 1998). The test is always a worthy non-invasive window on the pulmonary circulation. A low value accompanied by normal lung volumes is always suggestive of a pulmonary vascular pathology.

T_LCO can increase in some circumstances such as asthma, in some bronchiectasis and in the very early stages of congestive heart failure. T_LCO is also increased during intrapulmonary hemorrhage because carbon monoxide combines with the intra-alveolar hemoglobin (COTES et al. 1993). In this case intermittent rises of K_{CO} are highly specific.

2.9.3
Assessment of the Ventilation-Perfusion Ratio

The mismatch of ventilation and perfusion is the most common cause of arterial hypoxemia due to lung dysfunction. This justifies the interest put into finding an adequate technique to measure it properly. The global ventilation-perfusion ratio in the normal lung is about 0.8, against a theoretical perfect value that should be 1. This discrepancy mostly originates from the difference in vertical distribution of ventilation and perfusion within the lung. Vertical gradient in ventilation results from the elastic structure of the lung producing an increase in ventilation per unit of lung volume from top to bottom. The vertical gradient in perfusion is due to the gradient of the difference in transmural pressure at the level of the microvessels exposed to an almost constant alveolar pressure on the outside and to a hydrostatic pressure rising from top to bottom on the inside. This results in an increase of the perfusion per unit of lung volume from top to bottom. However the vertical increases in ventilation and perfusion do not exactly match each other resulting in a slight mismatch, thus the value of 0.8. The small normal physiological shunt also adds to this. For a given ratio of CO_2 production to O_2 consumption R, the values of arterial O_2 and CO_2 are determined by the ventilation-perfusion ratio. In other words when metabolism reaches a steady state, there is only one possible CO_2 value for a given PaO_2 and vice-versa because of the ventilation-perfusion ratio. Gross patterns of ventilation and perfusion distribution have been originally assessed by gaseous radionuclides (WEST and DOLLERY 1960). However topographical studies show little abnormalities, if any, in the face of significant blood gas abnormalities during disease because the actual level of mismatch at the acinus level, the effective unit of gas exchange (ca. 0.06 ml) of which there are around 50,000, is beyond the resolution of the gamma camera or any other external imaging technique, save possibly ^{3}He MRI. Proper assessment of the VA/Q distribution is feasible by the multiple inert gas elimination technique (MIGET). Six inert gases with a wide range of solubilities characterized by their capacitance b and a linear dissociation curve, are infused intravenously for about 30 min (WAGNER et al. 1974). Arterial, mixed venous and mixed expired gas samples are analyzed by gas chromatography to compute arterial retention and alveolar excretion ratio for each gas. There is a unique relationship between the Pa/Pv retention value and the gas solubility, (b), for each VA/Q ratio. Plots of ventilation and blood flow versus VA/Q ratio are derived as the best fit for the excretion and retention data and fitted from a 50-compartments model, one of them incorporating the shunt (VA/Q=0) and another the alveolar dead space (VA/Q=∞). Although the 50 compartments approach is a somewhat arbitrary model, much information can be gained from it, and constitutes a major step forward from the original three-compartment model of RILEY and COURNAND (1949).

Normal subjects present a unimodal narrow distribution (CARDUS et al. 1997). The degree of VA/Q abnormalities is already high in mild to moderate obstructive patients. The distribution of ventilation and perfusion becomes bimodal in severe COPD. The first population of lung units appears approximately normal but the second population shows either predominantly high VA/Q ratios, demonstrating little perfusion, or predominantly low VA/Q ratios, demonstrating little ventilation. A minority of patients shows both high and low patterns of VA/Q ratios (PAIN et al. 1977; WAGNER et al. 1977; MARTHAN et al. 1985). Excellent correlations have been demonstrated in obstructive patients between PaO_2 and the averaged VA/Q computed from MIGET, in contrast to the extremely poor relationship that exists between PaO_2 and FEV1 in the same subjects (WAGNER et al. 1977). Quite interestingly, studies of the distribution of pulmonary blood flow by first pass angioscintigraphy, performed independently of MIGET studies, demonstrate very significant changes of blood flow distribution, even in the early stages of COPD (CAPDEROU et al. 2000).

2.9.4
Exercise Testing

2.9.4.1
Method

Exercise testing as performed for pulmonary function testing, has various purposes: unmasking hypoxemia, quantifying dyspnea, assessing effort limitation factors, either ventilatory, cardiovascular or peripheral (deconditioning), measuring oxygen consumption and metabolic demand.

During an exercise test, the patient performs a quantified workload. This workload is standardized by using a treadmill or a cycle. Blood samples are collected during the test for blood gas determination and the inspiratory and expiratory fractions of oxygen and carbon dioxide are usually measured breath by breath, with fast analyzers, although values averaged over several respiratory cycles are quite sufficient and often more reliable in practice. These measurements allow the calculation of oxygen consumption and of carbon dioxide production. Electrocardiogram, blood pressure and pulse oximetry must be monitored during the duration of the test.

Major contraindications to exercise testing include severe hypoxemia breathing room air, $PaCO_2$ greater than 70 mmHg (9.3 kPa), FEV_1 less than 30% of predicted, recent (within 4 weeks) myocardial infarction, unstable angina pectoris, second or third-degree heart block, rapid ventricular or atrial arrhythmia, orthopedic impairment, severe aortic stenosis, congestive heart failure, uncontrolled hypertension, limiting neurologic disorders, dissecting or ventricular aneurysm, severe pulmonary hypertension, thrombophlebitis or intracardiac thrombi, recent systemic or pulmonary embolus, acute pericarditis (RUPPEL 1998).

2.9.4.2
Goals of Exercise Testing in the Pulmonary Function Laboratory

Unmasking hypoxemia is one of the main goals of exercise testing (AARC 1992). Effort-related desaturation can be due to a decrease in diffusing capacity. It can be observed in emphysema and in interstitial pneumonia and requires further investigations.

Exercise testing also permits assessment of dyspnea. During the test, breathlessness is quantified by using either a visual analog scale or a "Borg-type" scale (HANSEN 1984). The test permits respiratory, cardiovascular and peripheral causes of dyspnea to be discriminated.

Measurement of exhaled gases during exercise allows a non-invasive estimate of the anaerobic threshold, also referred to as the ventilatory threshold. Anaerobic threshold can be determined by plotting oxygen consumption against carbon dioxide production during an exercise of increasing intensity (WASSERMAN 1984). The anaerobic threshold is defined as the inflection point of the curve, indicating an abrupt increase in carbon dioxide production. The cardiac frequency corresponding to the anaerobic threshold provides the guidance parameter for pulmonary rehabilitation programs. This frequency should be reached but not surpassed during the training to ensure that the patient is trained at his maximum aerobic level (ACCP/AACVPR 1997).

2.10
Conclusion

The information provided by pulmonary function tests should be part of a global strategy of diagnosis, including clinical, radiological and biological approaches. The strategy of pulmonary function testing should follow a logical pattern such as that suggested in Fig. 2.6. Patient cooperation is required for most of these tests. Because most of them are non-invasive they can easily be repeated. However, they may prove to be quite exhausting for very sick patients, limiting the feasibility and the reproducibility required for good laboratory practice. Finally interpretation of pulmonary function tests is, needless to say, best performed when both clinical and thorax imaging data are provided to the investigator. When some of these tests need to be related to or performed during CT, evaluation of the influence of the supine position on measurements must be assessed.

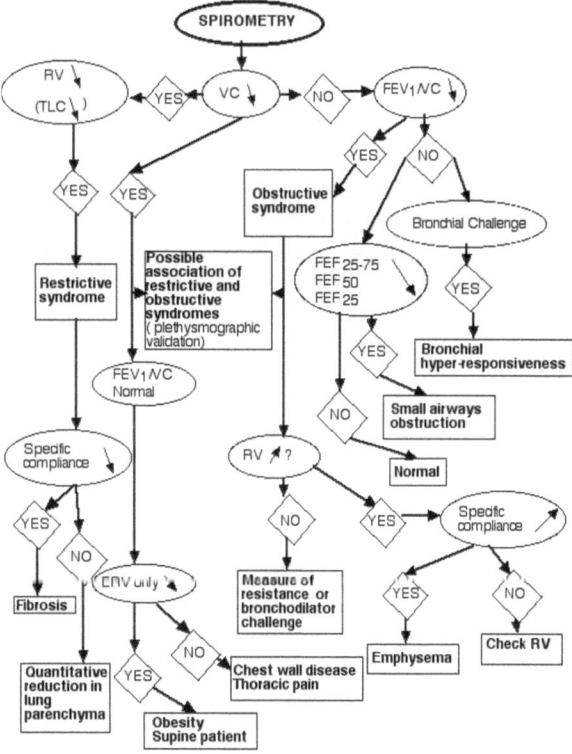

Fig. 2.6. Pulmonary function testing strategy (From C. Straus and M. Zelter (2001), in *Pneumologie pour le praticien*, G. Huchon ed, Masson, Paris). *VC* vital capacity, *FEV₁* forced expiratory volume in one second, *RV* residual volume, *TLC* total lung capacity, *ERV* expiratory reserve volume, *FEF* forced expiratory flow

References

AARC (1992) Clinical practice guideline: exercise testing for evaluation of hypoxemia and/or desaturation. Respir Care 37:907–912

AARC (1993) In-vitro pH and blood gas analysis and hemoximetry. Respir Care 38:505–510

ACCP/AACVPR (1997) Pulmonary rehabilitation. Joint ACCP/AACVPR evidence-based guidelines. Chest 112:1363–1396

Agostoni E, Rahn H (1960) Abdominal and thoracic pressures at different lung volumes. J Appl Physiol 15:1087–1092

Agostoni E, Hyatt RE (1986) Static behaviour of the respiratory system. In: Fishman AP (ed) Handbook of physiology, ssection 3: the respiratory system. American Physiological Society, Bethesda, pp 113–130

ATS (1991) Lung function testing: selection of reference values and interpretative strategies. Am Rev Respir Dis 144:1202–1218

ATS (1995a) Standardization of spirometry, 1994 update. Am J Respir Crit Care Med 152:1107–1136

ATS (1995b) Standards for the diagnosis and care of patients with chronic obstructive pulmonary disease. Am J Respir Crit Care Med 152:S77–S121

Backer V, Groth S, Dirksen A, Bach-Mortensen N, Hansen KK, Laursen EM, Wendelboe D (1991) Sensitivity and specificity of the histamine challenge test for the diagnosis of asthma in an unselected sample of children and adolescents. Eur Respir J 4:1093–1100

Bass H (1973) The flow volume loop: normal standards and abnormalities in chronic obstructive pulmonary disease. Chest 63:171–176

Becklake MR, Permutt S (1979) Evaluation of tests of lung function for "screening" for early detection of chronic obstructive lung disease. In: Macklem PT, Permutt S (eds) The lung in the transition between health and disease. Dekker, New York, pp 345–387

Britton J (1988) Is hyperreactivity the same as asthma? Eur Respir J 1:478–479

Brulot N, Kadas V, Grassino A, Milic-Emili J (1992) Positional variation in lung volumes in COPD. Am Rev Respir Dis 145:A764

Capderou A, Aurengo A, Derenne JP, Similowski T, Zelter M (2000) Pulmonary blood flow distribution in stage 1 chronic obstructive pulmonary disease. Am J Respir Crit Care Med 162:2073–2078

Cardus J, Burgos F, Diaz O, Roca J, Barbera JA, Marrades RM, Rodriguez-Roisin R, Wagner PD (1997) Increase in pulmonary ventilation-perfusion inequality with age in healthy individuals. Am J Respir Crit Care Med 156:648–653

Chan ED, Irvin CG (1995) The detection of collapsible airways contributing to airflow limitation. Chest 107:856–859

Cockcroft DW, Hargreave FE (1990) Airway responsiveness. Relevance of random population data to clinical usefulness. Am Rev Respir Dis 142:497–500

Cockcroft DW, Murdock KY, Berscheid BA, Gore BP (1992) Sensitivity and specificity of histamine PC20 determination in a random selection of young college students. J Allergy Clin Immunol 89:23–30

Cotes JE, Chinn DJ, Quanjer PH, Roca J, Yernault J-C (1993) Standardization of the measurement of transfer factor (diffusing capacity). Eur Respir J 6 [Suppl 16]:41–52

Darling RC, Cournand A, Richards DW Jr (1940) Studies on

intrapulmonary mixture of gases. Open circuit methods for measuring residual air. J Clin Invest 19:609–618

Dejours P (1975) Principles of comparative respiratory physiology. North Holland/American Elsevier, New York

Dejours P, Labrousse Y, Raynaud J, Teillac A (1957) Stimulus oxygène chémo-réflexe de la ventilation à basse altitude (50 m) chez l'homme. I. Au repos. J Physiol (Paris) 49:115–120

Dekhuijzen PNR, Decramer M (1992) Steroid-induced myopathy and its significance to respiratory disease: a known disease rediscovered. Eur Respir J 5:997–1003

Delclaux B, Orcel B, Housset B, Withelaw WA, Derenne JP (1994) Arterial blood gases in elderly persons with chronic obstructive pulmonary disease (COPD). Eur Respir J 7:856–861

Derenne J-P, Couture J, Iscoe S, Whitelaw W, Milic-Emili J (1976) Occlusion pressure in man rebreathing CO_2 under methoxyflurane anesthesia. J Appl Physiol 40:805–814

Desager KN, Buhr W, Willemen M, Van Bever HP, Backer W de, Vermeire PA, Landser FJ (1991) Measurement of total respiratory impedance in infants by the forced oscillation technique. J Appl Physiol 71:770–776

DuBois AB, Botelho SY, Bedell GN, Marshall R, Comroe JH Jr (1956a) A rapid plethysmographic method for measuring thoracic gas volume; a comparison with a nitrogen washout method for measuring functional residual capacity. J Clin Invest 35:322–326

DuBois AB, Botelho SY, Comroe JH Jr (1956b) A new method for measuring airway resistance in a man using a body plethysmograph: values in normal subjects and in patients with respiratory disease. J Clin Invest 35:327–335

Duguet A, Tantucci C, Lozinguez O, Isnard R, Thomas D, Zelter M, Derenne JP (2000) Expiratory flow limitation as a determinant of orthopnea in acute left heart failure. J Am Coll Cardiol 35:690–700

European Respiratory Society (1993) Standardized lung function testing. Eur Respir J 6 [Suppl 16]

Feldman JL, Smith JC (1995) Neural control of respiratory pattern in mammals: an overview. In: Dempsey JA, Pack AI (eds) Regulation of breathing. Dekker, New York, pp 39–70

Fenn WO, Craig AB (1963) Effect of CO_2 on respiration using a new method of administering CO_2. J Appl Physiol 18:1023–1024

Fitzgerald RS, Lahiri S (1986) Reflex response to chemoreceptor stimulation. In: Geiger SR, Widdicombe JG, Cherniack NS, Fishman AP (eds) Handbook of physiology, section 3: the respiratory system. American Physiological Society, Bethesda, pp 313–362

Forster RE (1987) Diffusion of gases across the alveolar membrane. In: Fahri LF, Tenney SM (eds) Handbook of physiology, section 3: the respiratory system. Gas exchange. American Physiological Society, Bethesda, pp 71–88

Gardner NG (1996) The pathophysiology of hyperventilation disorders. Chest 109:516–534

Guenard H, Marthan R (1996) Pulmonary gas exchange in elderly subjects. Eur Respir J 9:2573–2577

Hansen JE (1984) Exercise instruments, schemes, and protocols for evaluating the dyspneic patient. Am Rev Respir Dis 129 [Suppl]:S25–S27

Hathirat S, Renaetti AD Jr, Mitchell M (1970) Measurement of the total lung capacity by helium dilution in a constant volume system. Am Rev Respir Dis 102:760–770

Hutchinson J (1846) On the capacity of the lungs, and on the respiratory movements, with the view of establishing a precise and easy method of detecting disease by the spirometer. Lancet i:630–632

Hyatt RE, Black LF (1973) The flow volume curve. A current perspective. Am Rev Respir Dis 107:191–199

Knudson RJ, Slatin RC, Lebowitz MD, Burrows B (1976) The maximal expiratory flow-volume curve. Normal standards, variability, and effects of age. Am Rev Respir Dis 113:587–600

Koulouris NG, Valta P, Lavoie A, Corbeil C, Chasse M, Braidy J, Milic-Emili J (1995) A simple method to detect expiratory flow limitation during spontaneous breathing (see comments). Eur Respir J 8:306–313

Krachman S, Criner GJ (1998) Hypoventilation syndromes. Clin Chest Med 19:139–155

Leff AR, Schumacker PT (1993) Respiratory physiology. Basics and applications. Saunders, Philadelphia

Marthan R, Castaing Y, Manier G, Guenard H (1985) Gas exchange alterations in patients with chronic obstructive lung disease. Chest 87:470–475

Mead J (1960) Volume displacement body plethysmograph for measurements on human subjects. J Appl Physiol 15:736–740

Michels A, Decoster K, Derd L, Vleurinck C, Woestijne KP van de (1991) Influence of posture on lung volumes and impedance of respiratory system in healthy smokers and non smokers. J Appl Physiol 71:294–299

Moreno F, Lyons HA (1960) Effect of body posture on lung volumes. J Appl Physiol 16:27–29

O'Donnell DE, Webb KA (1992) Breathlessness in patients with severe chronic airflow limitation. Physiologic correlations. Chest 102:824–831

O'Donnell DE, Lam M, Webb KA (1998) Measurement of symptoms, lung hyperinflation, and endurance during exercise in chronic obstructive pulmonary disease. Am J Respir Crit Care Med 158:1557–1565

O'Donnell DE, Lam M, Webb KA (1999) Spirometric correlates of improvement in exercise performance after anticholinergic therapy in chronic obstructive pulmonary disease. Am J Respir Crit Care Med 160:542–549

Pain MC, Glazier JB, Simon H, West JB (1977) Regional and overall inequality of ventilation and blood flow in patients with chronic airflow obstruction. Thorax 22:453–461

Pride NB, Permutt S, Riley RL, Bromberger-Barnea B (1967) Determinants of maximum expiratory flow from the lungs. J Appl Physiol 23:646–662

Quanjer PH, Tammeling GJ, Cotes JE, Pedersen OF, Peslin R, Yernault J-C (1993) Lung volumes and forced ventilatory flows. Eur Respir J 6 [Suppl 16]:15–40

Rahn H, Otis AB, Chadwick L, Fenn O (1946) The pressure volume diagram of the thorax and lung. Am J Physiol 146:161–178

Read DJC (1967) A clinical method for assessing the ventilatory response to carbon dioxide. Australas Ann Med 16:20–32

Rebuck AS, Campbell EJM (1973) A clinical method for assessing the ventilatory response to hypoxia. Am Rev Respir Dis 109:345–350

Riley RL, Cournand A (1949) "Ideal" alveolar air and the analysis of ventilation-perfusion relationships in the lungs. J Appl Physiol 1:825–847

Rodarte JR, Hyatt RE, Cortese DA (1975) Influence of expiratory flow on closing capacity at low expiratory flow rates. J Appl Physiol 39:60–65

Rodenstein DO, Stanescu DC (1982) Reassessment of lung volume measurement by helium dilution and by body plethysmography in chronic air-flow obstruction. Am Rev Respir Dis 126:1040–1044

Rodenstein DO, Stanescu DC, Francis C (1982) Demonstration

of failure of body plethysmography in airway obstruction. J Appl Physiol 52:949–954

Ruppel GL (1998) Manual of pulmonary function testing, 7th edn. Mosby, Saint Louis

Scano G, Gigliotti F, Duranti R, Gorini M, Fanelli A, Marconi G (1993) Control of breathing in patients with neuromuscular diseases. Monaldi Arch Chest Dis 48:87–91

Severinghaus JW, Kelleher JF (1992) Recent developments in pulse oximetry. Anesthesiology 76:1018–1038

Similowski T, Fleury B, Launois S, Cathala HP, Bouche P, Derenne JP (1989) Cervical magnetic stimulation: a new and painless method for bilateral phrenic nerve stimulation in conscious humans. J Appl Physiol 67:1311–1318

Similowski T, Yan S, Gauthier AP, Macklem PT, Bellemare F (1991) Contractile properties of the human diaphragm during chronic hyperinflation. N Engl J Med 325: 917–923

Spinelli A, Marconi G, Gorini M, Pizzi A, Scano G (1992) Control of breathing in patients with myasthenia gravis. Am Rev Respir Dis 145:1359–1365

Sterk PJ, Fabbri LM, Quanjer PH, Cockcroft DW, O'Byrne PM, Anderson SD, Juniper EF, Malo J-L (1993) Airway responsiveness. Standardized challenge testing with pharmacological, physical and sensitizing stimuli in adults. Eur Respir J 6 [Suppl 16]:53–83

Tantucci C, Duguet A, Similowski T, Zelter M, Derenne JP, Milic-Emili J (1998a) Effect of salbutamol on dynamic hyperinflation in chronic obstructive pulmonary disease patients. Eur Respir J 12:799–804

Tantucci C, Mehiri S, Duguet A, Similowski T, Arnulf I, Zelter M, Derenne JP, Milic-Emili J (1998b) Application of negative expiratory pressure during expiration and activity of genioglossus in humans. J Appl Physiol 84:1076–1082

Turner JM, Mead J, Wohl ME (1968) Elasticty of human lung in relation to age. J Appl Physiol 25:664–671

Ulmer W, Kowalski J, Schmidt EW (1997) The flow-volume curve in patients with obstructive airway diseases partial analysis and functional importance. Pneumonol Alergol Pol 65:435–445

Wagner PD, Saltzman HA, West JB (1974) Measurement of continuous distributions of ventilation-perfusion radios: theory. J Appl Physiol 36:585–592

Wagner PD, Dantzker DR, Dueck R, Clausen JL, West JB (1977) Ventilation-perfusion inequality in chronic obstructive pulmonary disease. J Clin Invest 59:203–216

Wasserman K (1984) The anaerobic threshold measurement in exercise testing. Clin Chest Med 5:77–88

West JB (1990) Respiratory physiology. The essentials, 4th edn. Williams and Wilkins, Baltimore

West JB, Dollery CT (1960) Distribution of blood flow and ventilation perfusion ratio in the lung with radioactive CO_2. J Appl Physiol 15:405–410

Whitelaw WA, Derenne JP (1993) Airway occlusion pressure. J Appl Physiol 74:1475–1483

Whitelaw WA, Derenne JP, Milic-Emili J (1976) Occlusion pressure as a measure of respiratory centre output in conscious man. Respir Physiol 23:181–199

Yan S, Gauthier AP, Similowski T, Macklem PT, Bellemare F (1992) Evaluation of human diaphragm contractility using mouth pressure twitches. Am Rev Respir Dis 145: 1064–1069

Yernault JC, Englert M (1974) Static mechanical lung properties in young adults. Bull Eur Physiopathol Respir 10:435–450

Yernault JC, Paiva M (1986) In vivo diagnosis of pulmonary emphysema: an uncompletely resolved issue. Bull Eur Physiopathol Respir 22:95–97

3 Large Airways at CT: Bronchiectasis, Asthma and COPD

Philippe A. Grenier, Catherine Beigelman-Aubry, Catalin Fetita, Françoise Preteux

CONTENTS

P. A. Grenier, MD
Professor, Service de Radiologie, Hôpital de la Pitié-Salpêtrière, 47-83, boulevard de l'Hôpital, 75651 Paris Cedex 13, France
C. Beigelmann-Aubry, MD
Service de Radiologie, Hôpital de la Pitié-Salpêtrière, 47-83, boulevard de l'Hôpital, 75651 Paris Cedex 13, France
C. Fetita, PhD
Department ARTEMIS, Institut National des Télécommunications, 9, rue Charles Fourier, 91011 Evry Cedex, France
F. Preteux, PhD
Department ARTEMIS, Institut National des Télécommunications, 9, rue Charles Fourier, 91011 Evry Cedex, France

3.1 Introduction

Combining helical volumetric CT acquisition and thin-slice thickness during breath hold provides an accurate assessment of both focal and diffuse airway diseases. With multiple detector rows, compared to single slice helical CT, multislice CT can cover a greater volume during a simple breath hold, and with better longitudinal and in-plane spatial resolution and improved temporal resolution. The result in data set allows the generation of superior multiplanar and 3-dimensional images of the airways, including those obtained from techniques developed specifically for airway imaging, such as CT bronchography and virtual bronchoscopy. Improvement in image analysis techniques and the use of spirometric control of lung volume acquisition have made possible accurate and reproducible quantitative assessment of airway wall, lumen areas and lung density. This quantitative assessment of the airways will lead to the increasing use of CT as a research tool for better insights in physiopathology of obstructive lung disease, particularly in COPD and asthma, with an ultimate benefit in clinical practice.

3.2 Acquisition Parameters and Image Processing

3.2.1 Scanning Protocol

The introduction of the multidetector-row technology now allows the performance of fast acquisition of around 20 s for the entire thorax while using 1.25 mm-slice thickness. Because slice thickness has the greatest impact on the quality of 2D and 3D reconstructions, the overall quality of the 3D simulations of the bronchial tree is increased (Schaefer-Prokop and Prokop 1996; Hopper et al. 1998; Summers et al. 1998; Ferretti et al. 2001).

The z-axis resolution is directly related to the effective slice thickness. The smaller the effective slice thickness, the more precise are 2D reformats or 3D reconstructions. The optimal effective slice thickness is therefore of 1.25 mm or less (0.6 or 0.75 mm when using a 16-detector rows CT scanner) rendering isotropic or near isotropic image reconstruction feasible. In all cases, overlapping transaxial image reconstruction by 50% is recommended to provide the highest morphologic detail (FERRETTI et al. 2001).

Because of the great natural contrast between the airways and their environment, relatively low kilovoltage (100–120 kV) and milliamperage (60–100 mA) may be used (FERRETTI et al. 2001). CHOI et al. (2002) showed recently that image quality of surface-rendered 3D images of the central airways is preserved when the tube current decreases from 240 to 50 mA.

An advantage of multidetector row CT (MDCT) is its ability to allow slice thickness to be changed retrospectively, thus enabling one to obtain high quality reconstruction images from routine CT studies.

3.2.2
Expiratory CT

Airway imaging is routinely performed at end-inspiration during a simple breath hold. Additional scanning is also required during continuous or suspended expiration in patients with either suspected tracheobronchomalacia to assess the degree of abnormal expiratory collapse of the proximal airways at expiration, or those with suspicion of small airway disease to assess the presence and extent of expiratory air trapping. In such cases, low dose technique (40 mA) is recommended to decrease radiation exposure (GOTWAY et al. 2000).

The most commonly used technique for the assessment of air trapping at CT is based on post-expiratory CT scans, obtained during suspended respiration following a forced exhalation. However a dynamic expiratory maneuver performed during helical CT acquisition, particularly during the last part of the expiratory maneuver, may provide with good results without too many artifacts. This is called continuous expiratory CT (LUCIDARME et al. 2000). Patients can have greater difficulty maintaining the residual volume after an exhalation than during an active exhalation when they have to continue the expiratory effort until the end of the acquisition. This technique is proposed as an alternative when patients have difficulty performing the suspended end-expiration maneuver adequately (LUCIDARME et al. 2000).

3.2.3
Cine-Viewing

Visualization of the overlapped thin axial images sequentially in a cine-mode allows the bronchial divisions to be followed from the segmental origin to the distal bronchial lumens down to the smallest bronchi which can be identified on thin section images. This viewing technique helps indicate the segmental and subsegmental distribution of any airway lesion and may serve as a roadmap for the endoscopist. Moving up and down through the volume at the monitor has become an alternative to film-based review (GRENIER et al. 2002).

3.2.4
Multiplanar Reformations

They are the easiest reconstructions to generate and can be interactively performed in real time at the console or at the dedicated console or workstation. They allow to obviate the underestimation of the limits of the craniocaudal extent of a vertically oriented disease as tracheobronchial stenosis (QUINT et al. 1995). They are of particular value for better detection and evaluation of mild focal stenosis. Whereas the thickness of the displayed planar image is 0.6–0.8 mm, depending on the dimension of the field of view, multiplanar volume reconstruction (MPVR) consists of a slab with a thickness of several pixels and in a less noisy reformation. Because underestimation of a stenosis may occur if the reformation plane is not adequately chosen, simultaneous reading of the native cross sectional images and a selection of reformation plane from the 3-dimensional reconstructed image of the airways are recommended. Multiplanar volume reformation may be associated with the use of the intensity projection techniques (REMY-JARDIN et al. 1996).

3.2.5
Minimum and Maximum Intensity Projection

The minimum intensity projection (mIP) technique projects the tracheobronchial air column into a viewing plane. It is applied to a selected volume of the thorax containing the airways under evaluation (MPVR-mIP technique) (RUBIN 1996). Pixels encode the minimum voxel value encountered by each ray. However, numerous drawbacks have limited the indications of mIP in assessing airway disease. The technique is very vulnerable to varying width of volume of interest and to partial volume effects.

In the maximum intensity projection (MIP) technique, pixels encode the maximum voxel value encountered by each ray. This benefits in the display of the mucoid impactions seen in dilated bronchi, or the display of the small centrilobular nodular and/or linear branching opacities expressing infectious or inflammatory bronchiolitis. The thickness of the slab is selected interactively at the console or workstation (MPVR-MIP technique) (RÉMY et al. 1998).

3.2.6
3D Surface Rendering Technique with Shaded Display (3D-SSD)

The surface of the volume of air contained in the airways is isolated from the initial volume data by thresholding segmentation. The technique allows for a better understanding of the longitudinal extent of airway stenoses than axial CT images (KAUCZOR et al. 1996; RÉMY-JARDIN et al. 1996; RÉMY et al. 1998; BOISELLE et al. 2002). The 3D images offer an overview of the pathology, particularly appreciated in complex airway anatomy. The major limits of this technique are related to thresholding range. The results of reconstructions are sensitive to partial volume artifacts and motion-related artifacts.

3.2.7
CT Bronchography

Volume rendering techniques are preferable to shaded surface display to image the airways. CT bronchography is a new functionality consisting of a segmentation of the lumen-wall interface of the airways. RÉMY-JARDIN et al. (1998a) used a continuous rim of peripheral voxels and the volume rendering algorithm. This technique has proven to be of particular interest in diagnosing mild changes in airway caliber and understanding complex tracheobronchial abnormalities (RÉMY-JARDIN et al. 1998b). Using the same concept, we developed a fully automatic method for 3D bronchial tree reconstruction based on bronchial lumen detection within the thoracic volume data set obtained from thin multidetector CT acquisition and thin collimation during breath hold (FETITA et al. 1999). The 3-dimensional reconstruction of the bronchial tree is achieved up to the 7th order division and the bronchial tree is visualized using a semi-transparent volume rendering technique (GRENIER et

al. 2002) (Fig. 3.1). In addition, automatic delimitation and indexation of anatomical segments make possible local and reproducible analysis at a given level of the bronchial tree, and an automatic extraction of the central axis of the bronchial tree which simplifies the interactivity during the navigation within CT bronchography or virtual endoscopy modes (Fig. 3.2). These tools can also be used for quantitative assessment of the wall and lumens of the airways on cross section images of the bronchi reconstructed perpendicular to their central axis (GRENIER et al. 2002).

3.2.8
Virtual Bronchoscopy

Virtual bronchoscopy provides an internal rendering of the tracheobronchial walls and lumen. Owing to a perspective rendering algorithm, this simulates an endoscopist's view of the internal surface of the airways. The observer may interactively move through the airways. This technique may be obtained from both 3D surface rendering and volume rendering techniques (RÉMY-JARDIN et al. 1996; RÉMY et al. 1998). The volume rendering technique is less sensitive to partial volume effects than surface rendering. Powerful computers permit real-time rendering (15–25 images/s) making flying within the airways in a virtual manner possible. The technique allows

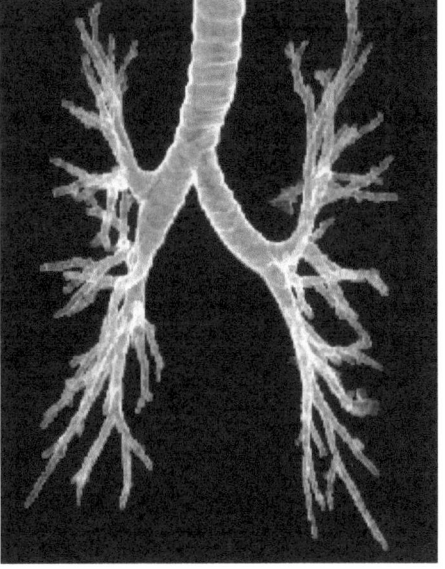

Fig. 3.1. CT bronchography (semi-transparent volume rendering of the segmented airways after multidetector row CT acquisition with 1.25 mm slice thickness and 0.6 mm overlap in reconstruction) in a coronal view. Normal appearance of the bronchial tree in a mild intermittent asthmatic patient

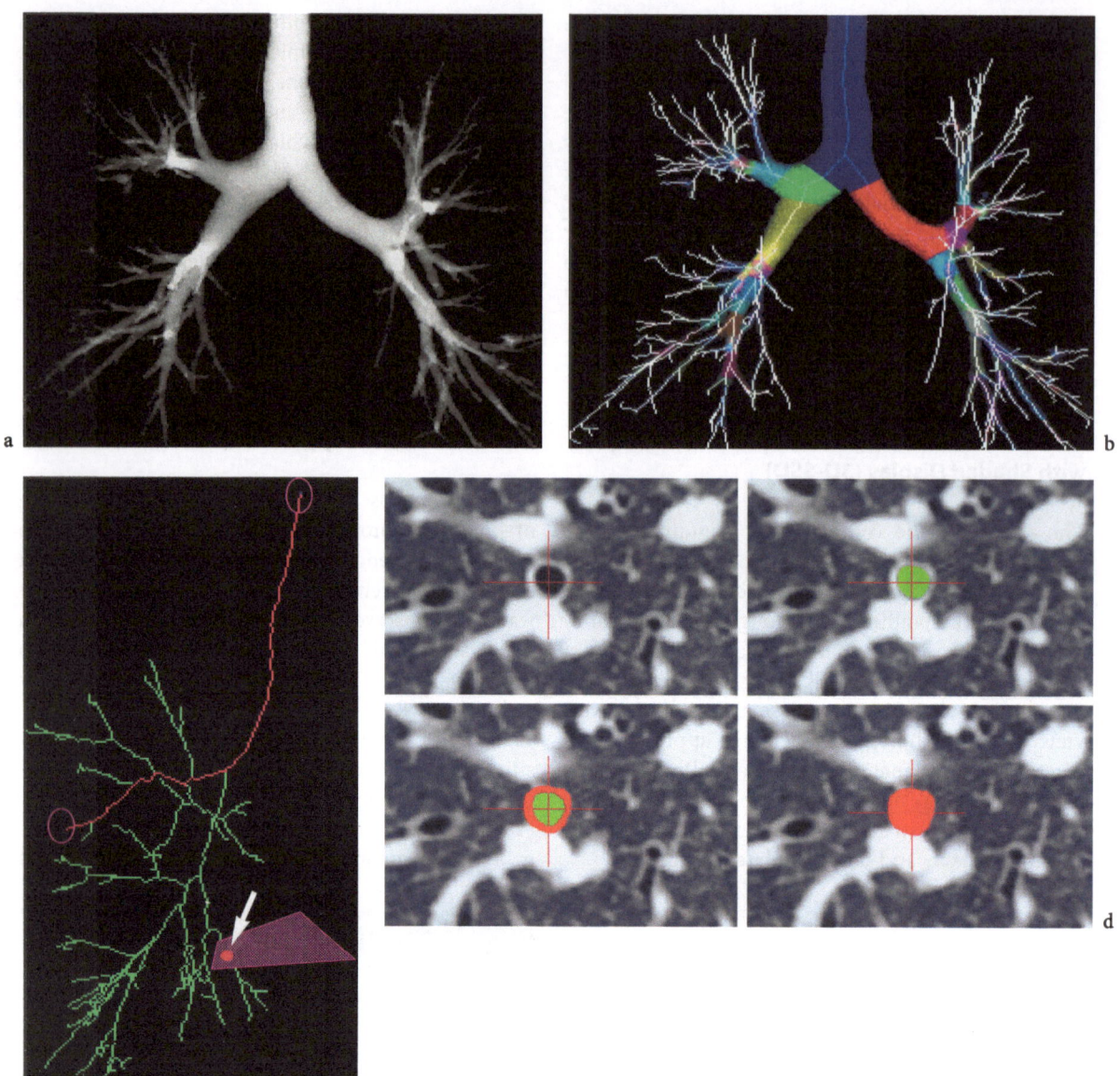

Fig. 3.2. Quantitative assessment of the airways. **a** 3D reconstruction of the tracheobronchial tree after multidetector row CT acquisition (CT bronchography). **b** Automatic indexation of the trachea, main and lobar bronchi (*different colors*) and 3D representation of the central axis of the tracheobronchial tree. **c** Oblique view of the central axis of the bronchial tree in the right middle and lower lobes (*green*). Red color simulates navigation along the bronchial central axis from the trachea to a distal bronchus in the superior segment of the right lower lobe. The pink dot (*arrow*) represents the selected point on the bronchial central axis for quantitative assessment of the airways. The plane of section perpendicular to the central axis is represented in pink. The cross section of the corresponding bronchus is seen in **d** cross section of the selected bronchus (*top left*). Automatic detection of the inner contour of the bronchial wall and segmentation of the bronchial lumen area (*top right*). Bottom right shows the automatic detection of the outer contour of the bronchial wall and segmentation of the bronchial cross section area. Bottom left shows by subtraction the segmentation of the bronchial wall area

accurate reproduction of major endoluminal abnormalities with an excellent correlation with fiberoptic bronchoscopy results regarding the location, severity, and shape of airway narrowing (McAdams et al. 1998). Virtual bronchoscopy is also able to visualize the bronchial tree beyond an obstructive lesion and thus to perform a retroscopy when looking back toward the distal part of the stenosis. Despite these appreciable abilities, virtual bronchoscopy remains very sensitive to the partial volume averaging effect and motion artifacts. In addition, this technique is unable to identify the causes of bronchial obstruction; mild stenosis, submucosal infiltration, and superficial spreading tumors are not identified (Rémy-Jardin et al. 1996).

3.2.9
Quantitative Assessment

Airway lumen and airway wall areas may be quantitatively assessed on CT images by using specific techniques that must be reproducible as well as accurate in order to compare the airways pre-intervention and post-intervention (bronchoprovocation, bronchodilatation, therapeutic response) and to carry out longitudinal studies of airway remodeling. Airway lumen and wall areas measured on axial images depend on the lung volume, and angle between the airway central axis and the plane of section. Volumetric acquisition at controlled lung volume is required in order to precisely match the airways of an individual on repeated studies. The control of lung volume at CT is obtained by spirometric triggering (Fig. 3.3). During exhalation the spirometer and associated microcomputer measure the volume of gas expired and trigger CT after the specific volume is reached. When the trigger signal is generated, airflow is inhibited by closure of mechanical occlusion device attached to the spirometer and scanning starts.

Measurements of airway lumen and wall area have to be restricted to airways that appear to have been cut in cross section based on the apparent roundness of the airway lumen. Measuring airway lumen and airway walls when they are not perpendicular to the scanning plane may lead to significant errors, the magnitude of which will depend on how acutely the airways are angled, the collimation and the field of view. The larger the angle and field of view and the thicker the collimation, the greater the overestimation of airway wall area. Most of the airways examined in axial CT slices are also more likely to be running obliquely to the plane of the section, rather than perpendicularly owing to the anatomy of the lungs. With the new generation of multislice CT scanner, it becomes possible to acquire a volume of lung with 0.6–0.75 mm slice thickness or less and to reconstruct axial images every 0.6 mm or less by interpolation; in such a maneuver, the CT voxels may be converted into cubic dimension (isotropic voxels). Then the segmentation of bronchial lumens and reconstruction of the airways in 3D allow the central axis of the airways to be determined and to reconstruct the airway cross section in a plane perpendicular to this axis (Fig. 3.2). This analysis technique overcomes the major limitation to the use of HRCT

in quantitative analysis, which is that accurate or true airway lumen and airway wall area can only be measured from airways which are oriented approximately perpendicular to the plane of scanning.

Different image analysis techniques have been developed to make measurements of airways dimension on CT scans possible. McNamara et al. (1992) modified a method developed by Webb et al. (1984) based on visual analysis of photographed images. They found that it was crucial to use a window level of –450 HU. Amirav et al. (1993) developed a computerized algorithm for measuring airway lumen area, based on an edge detection method using the full width at half maximum principle, that has advantages of less subjectivity and greater speed than the method of McNamara et al. (1992). With Preteux et al. (1999) we developed an automatic method for segmentation and calculation of airway lumen areas based on mathematical morphology theory, marking techniques derived from the concept of connection cost, and conditional watershed segmentation. Wood et al. (1995) developed an algorithm to measure airway lumens and the airway angle of orientation using a 3-D reconstruction of the lungs. This technique allowed cross sectional images of the airways to be generated irrespective of the orientation of the airway. King et al. (2000) more recently developed an automatic computed tomographic image analysis algorithm to measure not only the airway lumen areas but also the wall areas and angle of orientation of airways.

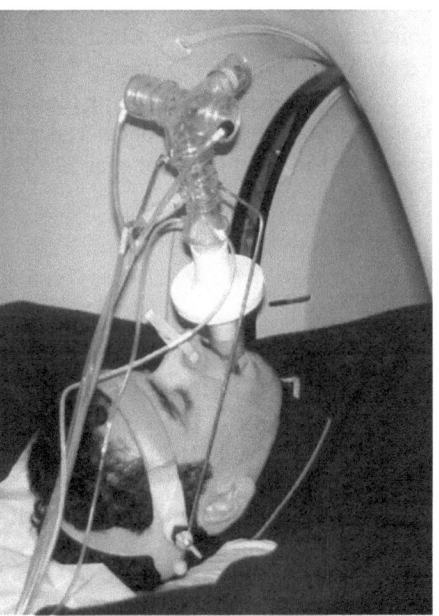

Fig. 3.3. Spirometrically triggered CT. View of a patient breathing through a small spirometer while positioned on the CT scan table

PEROT et al. (2001) had a different approach with similar results as KING et al. (2000). These results proved to be more accurate than those obtained with manual methods. All these analysis algorithms have been validated using data from phantom studies (MCNAMARA et al. 1992; AMIRAV et al. 1993; WOOD et al. 1995; PRETEUX et al. 1999; KING et al. 2000) and excised animal lungs (KING et al. 2000; PEROT et al. 2001), or by developing a realistic modeling of airways and pulmonary arteries included in CT scans of animal lungs obtained in vivo (PRETEUX et al. 1999). Their accuracy in measuring the airway lumens (MCNAMARA et al. 1992; AMIRAV et al. 1993; PRETEUX et al. 1999; KING et al. 2000) and wall (KING et al. 2000; PEROT et al. 2001) areas was very good only for bronchi measuring at least 2 mm in diameter. These techniques have been used to quantify the magnitude and distribution of airway narrowing in excised lung animals and in animals lungs in vivo as well as in normal and asthmatic subjects (MCNAMARA et al. 1992; AMIRAV et al. 1993; BROWN et al. 1997, 1998; BEIGELMAN-AUBRY et al. 2002). Although they have been used almost exclusively for research purposes, they will, with continued refinements eventually be of benefit in the clinical practice of radiology (KING et al. 1999).

3.3
Bronchiectasis

In spite of its decreased prevalence in developed countries, bronchiectasis remains an important cause of hemoptysis and chronic sputum production. It is now generally accepted that HRCT is the imaging modality of choice to establish the presence of bronchiectasis and to determine its precise extent (FRASER et al. 2000).

3.3.1
CT Characteristics

The characteristic CT findings of bronchiectasis include the following (MCGUINNESS et al. 1993; KIM et al. 1997b):

1. An internal bronchial diameter greater than that of the adjacent pulmonary artery
2. Lack of bronchial tapering, defined as a bronchus that has the same diameter as its parent bronchi for a distance of more than 2 cm
3. Visualization of bronchi abutting the mediastinal pleura
4. Visualization of bronchi within 1 cm of the costal pleura.

A bronchial wall thickening is also often present but this abnormality is a non-specific finding that may also be seen in other conditions, particularly asthma, and in asymptomatic smokers.

The HRCT appearance of dilated bronchi varies depending on the type of bronchiectasis and the orientation of the airways relative to the plane of the HRCT scan (Fig. 3.4).

In cylindrical bronchiectasis, bronchi coursing parallel and in the plane of section are visualized as parallel lines whereas bronchi coursing perpendicular to the plane of scanning appear as circular lucencies larger than the diameter of the adjacent pulmonary artery, resulting in a signet ring appearance. Varicose bronchiectasis is characterized by the presence of non-uniform bronchial dilatation whereas cystic bronchiectasis results in a cluster of cystic spaces sometimes containing air-fluid levels.

Secretion accumulation within bronchiectatic airways is generally easily recognizable as lobulated glove-finger, V-shaped or Y-shaped densities. When oriented perpendicular to the scanning plane the filled dilated bronchi are visualized as nodular opacities and recognized by the observation of the homologous pulmonary arteries, whose diameters are smaller than those of the dilated filled bronchi.

The size of the ectatic bronchi decreases on end expiratory as compared with inspiratory HRCT scans. Occasionally, they collapse completely because of their increased compliance.

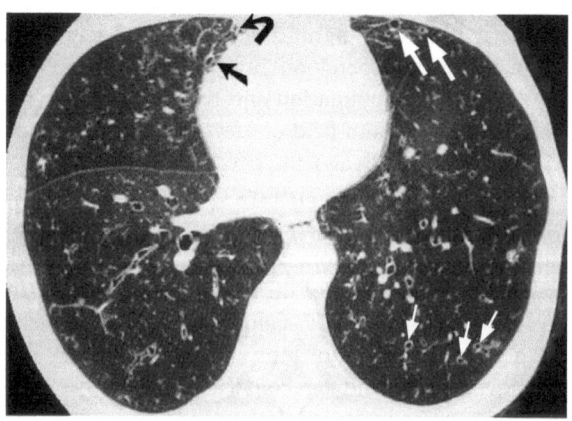

Fig. 3.4. Cylindrical bronchiectasis. Axial thin section CT scan shows cylindrical bronchiectasis on the basis of the visibility of bronchial lumens abutting the mediastinal pleura (*black and curved arrows*) in the right middle lobe, and bronchial lumen within the 1 -cm from the costal pleura (*white arrows*) in the lingula. The signet ring sign is also present in the posterobasal segment of the left lower lobe (*white small arrows*). From GRENIER et al. (2002), with permission

In addition to bronchiectasis itself a number of other abnormalities are seen with increased frequency in patients who have the disease, including areas of decreased lung attenuation and perfusion, expiratory air trapping, tracheomegaly and mediastinal lymph node enlargement.

3.3.2
CT Accuracy

HRCT has proven to be a reliable and non-invasive method for the assessment of bronchiectasis and has largely eliminated the need for bronchography (GRENIER et al. 1986). HRCT is not however 100% sensitive and specific and several limitations of the technique need to be recognized (KANG et al. 1995). These limitations include: 1) artifacts resulting from both respiratory and cardiac motion, 2) overlooking of areas of focal bronchiectasis located exclusively in skipped areas, 3) difficulty of perceiving the slight dilatation of mild cylindrical bronchiectasis.

Motion degradation of bronchial images can be reduced by using a 180° interpolation algorithm to reconstruct axial images after helical scanning. ECG gating may be used to reduce cardiac motion artifacts that may mimic disease (SCHOEPF et al. 1999). In addition, volumetric helical acquisition during breath holding eliminates the potential risk of missing small subtle bronchiectasis in areas skipped by the interspacing between thin-section CT scans (LUCIDARME et al. 1996). Because of many factors that can cause transient or permanent changes in diameter of the relatively compliant pulmonary arteries, invalidating the finding of cylindrical bronchiectasis based on the bronchoarterial diameter ratio >1, the lack of tapering of bronchial lumen has proven to be the most reliable sign of cylindrical bronchiectasis (KANG et al. 1995). However this finding is difficult to perceive on successive spaced thin-section CT scans and its assessment is significantly improved by using helical CT with thin collimation and viewing the contiguous scans in a cine-mode. As mucoid impaction filling bronchiectatic bronchi can simulate pulmonary nodules or masses on a single thin-section CT scan and pulmonary arteries are not recognized because of intense vasoconstriction in hypoventilated areas, helical CT with thin collimation provides a multiplanar display of the radiopaque tubular structures converging toward the hilum in a segmental or subsegmental distribution, corresponding to bronchiectasis.

By comparing helical CT with 3 mm collimation and HRCT scans with 10 mm intervals in a series of 50 patients with suspicion of bronchiectasis, LUCIDARME et al. showed that the number of patients, segments and bronchi affected with bronchiectasis, was higher on helical CT than HRCT scans. In addition, the interobserver agreement was significantly better with helical CT than with HRCT scans for the presence or absence of bronchiectasis on a per-segment basis, and also for assessing the extent of bronchiectasis in a given lobe and the distribution of disease in a given segment (LUCIDARME et al. 1996). Such an accurate assessment of extent of the disease is particularly required before surgery for focal bronchiectasis. Indeed, the surgeon must identify with certainty which segments are diseased, since surgical techniques frequently permit preservation of one or more normal pulmonary segments from a lobe in which bronchiectasis is present. For all these reasons, MDCT with thin-slice thickness using low dose should become the recommended routine protocol for the diagnosis and assessment of bronchiectasis. In addition, CT bronchography reconstructed after MDCT acquisition with thin-slice thickness should permit an increase in the confidence level of diagnosis of mild cylindrical bronchiectasis or to improve the visual assessment of the extent of bronchiectasis (GRENIER et al. 2002) (Fig. 3.5).

3.3.3
Correlation with Lung Function

The obstructive defect found at pulmonary function tests in patients with bronchiectasis, seems not to be related to the degree of collapse of large airways on expiratory CT or the extent of mucous plugging of the airway, but the consequence of an obstructive involvement of the peripheral airways (obliterative bronchiolitis) (ROBERTS et al. 2000). The extent of CT evidence of small airway disease (decreased lung attenuation and expiratory air trapping) commonly present in patients with bronchiectasis has proven to be the major determinant of airflow obstruction (ROBERTS et al. 2000). However, in a recent study based on a quantitative HRCT protocol in 60 patients with steady-state bronchiectasis, the extent of bronchiectasis, bronchial wall thickening and mosaic lung attenuation were all inversely related to indices of airway obstruction, and in addition, multiple regression analysis demonstrated bronchial wall thickening was the most significant determinant of all pulmonary function test indices (OOI et al. 2002).

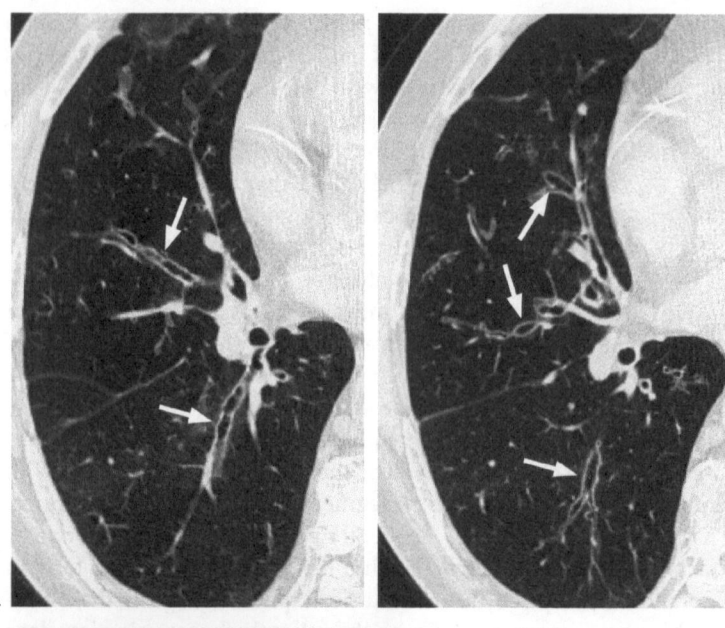

a

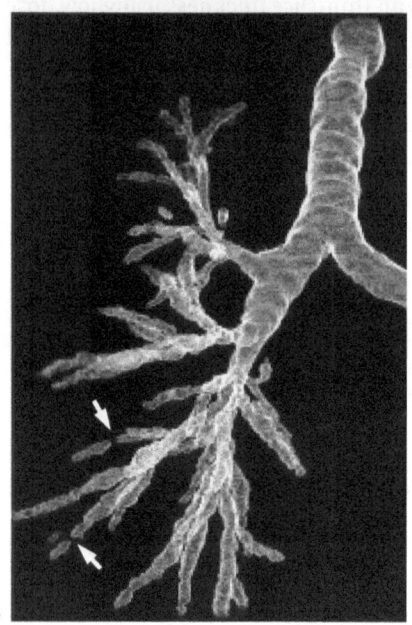

b

Fig. 3.5. Patient with cylindrical bronchiectasis in the right lung. **a** MDCT axial scans targeted on the right lung showing cylindrical bronchiectasis in both segments of the right middle lobe and the inferomedial and lateral subsegments of the superior segment of the right lower lobe (*arrows*). **b** Coronal view of CT bronchography after multislice CT acquisition and thin slice thickness targeted on the right lung. Cylindrical bronchiectasis is seen involving several bronchi in the right middle and lower lobes. Irregularities in bronchial caliber are well displayed. Discontinuities in bronchial lumen (*white arrows*) reflect the filling of bronchial lumen by retained secretions. From GRENIER et al. (2002), with permission

3.4
Asthma

Asthma is a chronic inflammatory condition involving the airways. The precise component of this inflammation remains to be elucidated and the causes are uncertain. This inflammation of the airways causes increases in the existing bronchial hyperresponsiveness responding to a variety of stimuli. This is commonly used in practice to confirm the clinical diagnosis of asthma. In susceptible individuals, this inflammation induces recurrent episodes of wheezing, chest tightness, breathlessness and coughing usually associated with widespread but variable airflow obstruction that is often reversible either spontaneously or with treatment. A chronic inflammation process leads to structural changes such as fibrosis, airway smooth muscle thickening and new vessel formation which may result in irreversible airway narrowing. The current therapy is based in most instances on an inhaled steroid as a controller medication (anti-inflammatory). When necessary a bronchodilator (b_2 agonist) is used.

3.4.1
HRCT Abnormalities

HRCT has made possible the investigation of the size, magnitude, and distribution of airway abnormalities in vivo. The most common abnormalities seen on HRCT are bronchial wall thickening, narrowing of the bronchial lumen, bronchial dilatation, patchy areas of decreased lung attenuation and vascularity, and air trapping (PAGANIN et al. 1992; LYNCH et al. 1993; GRENIER et al. 1996; PARK et al. 1997; LYNCH 1998) (Fig. 3.6). The distribution of bronchial abnormalities is often heterogeneous; some airways have normal thickness and diameters, while others have thick walls and are narrowed or dilated. The prevalence of HRCT abnormalities increases with increased severity of symptoms (GRENIER et al. 1996; PAGANIN et al. 1996). Considerable variations exist however in the reported frequency of abnormalities. This variation is related to differences in diagnostic criteria and patient selection (PAGANIN et al. 1992; LYNCH et al. 1993; GRENIER et al. 1996; PARK et al. 1997; LYNCH 1998).

PARK et al. (1997) demonstrated that only three findings were significantly more frequent in asthmatic patients than in normal individuals: bronchial wall thickening, bronchial dilatation and expiratory air trapping. Identification of bronchiectasis in patients with asthma but without allergic bronchopulmonary aspergillosis (ABPA) is plausible because bronchiectatic changes are seen at autopsy in patients who have died with long standing asthma. The true prevalence of bronchiectasis in patients with uncomplicated chronic asthma however remains unclear (LYNCH 1998). Mild cylindrical bronchiectasis based on a mild elevation of the bronchoarterial ratio may occur in patients with asthma, due to hypoxic pulmonary vasoconstriction related to localized areas of air trapping (LYNCH et al. 1993). Pulmonary arteries are reactive to changes in alveolar oxygen tension, and an increased bronchial-to-arterial diameter ratio has been observed in HRCT studies performed at altitude compared with those performed at sea level, presumably owing to the reduced ambient oxygen tension at altitude (KIM et al. 1997a). In addition, a visual illusion is that thick-walled bronchi appear larger than the adjacent vessel, even if their internal diameters are the same. So mild cylindrical bronchiectasis should be diagnosed with caution in patients with asthma and should not be the sole criterion for suggesting the diagnosis of ABPA in these patients (LYNCH 1998).

3.4.2
Quantitative Assessment of the Airways at CT

The real current challenge for CT in asthma is to visualize and quantify the lumen and wall of airways and lung attenuation to assess airway reactivity and airway wall remodeling. This will become crucial in the monitoring of current and future therapy.

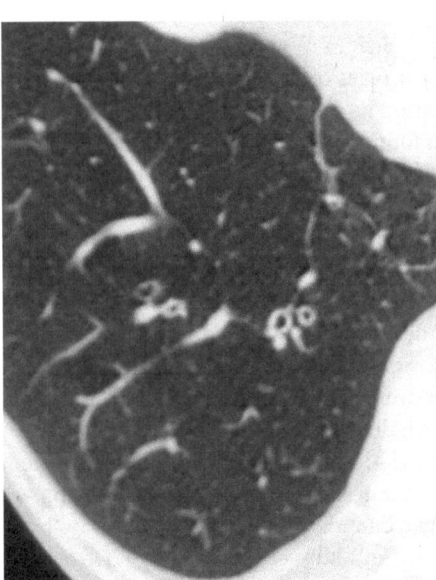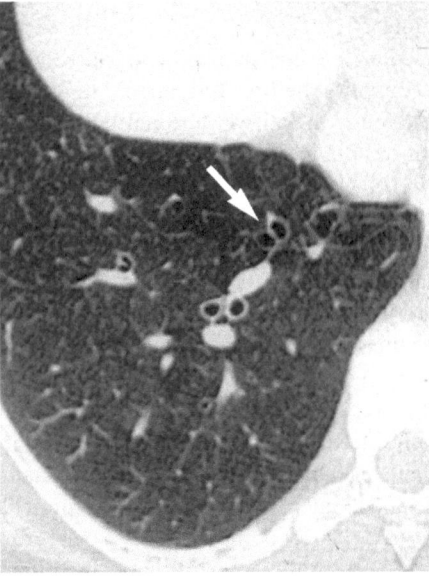

Fig. 3.6. Thin section CT scan targeted in the right lower lobe in two asthmatic patients. Left image shows bronchial wall thickening, and right image shows bronchial diameter appearing larger than the homologous pulmonary arteries (*arrow*) in a patient with severe persistent asthma

3.4.2.1
Airway Reactivity

Airway reactivity, the hallmark of asthma, refers to the ability of the airways to reversibly alter their diameters in response to stimuli. It is well documented that heightened airway reactivity can also contribute significantly to the morbidity and mortality of other airway diseases, including chronic bronchitis and cigarette smoke-induced chronic obstructive pulmonary disease.

Airway reactivity can be evaluated directly by measuring airflow changes induced by a bronchoconstrictive challenge or evaluated indirectly by measuring reversible airflow obstruction following the administration of a bronchodilator in a patient with airflow obstruction. A number of investigators have used CT to study patients in vivo with asthma to evaluate airway reactivity.

HRCT scan has been used to compare airway dimensions before and after the administration of a bronchoconstricting agonist (OKAZAWA et al. 1996). The ability to match the levels pre-intervention and post-intervention has been greatly aided by the development of volumetric helical CT (GOLDIN et al. 1998; BEIGELMAN-AUBRY et al. 2002). Data from a large portion of the lungs can be acquired during a single breath hold while the chosen collimation is thin (0.75–1.5 mm). Since lung volume influences airway wall and lumen dimensions, the use of a spirometer triggered to CT acquisition is used to ensure that lung volumes are comparable before and after a bronchoconstriction stimulus.

OKAZAWA et al. evaluated six normal patients and six mild to moderate asthmatics before and after methacholine challenge and found that airway luminal narrowing could be quantified both in normal patients and asthmatics (OKAZAWA et al. 1996). The airway wall area decreased in normal subjects who developed bronchial constriction caused by methacholine and in the asthmatics. In asthmatics this decrease in airway wall area did not occur and it was postulated that this was caused by the more edematous, stiffer airway wall being relatively uncoupled from the elastic recoil of the surrounding lung tissue. In a separate study, KEE et al. (1996) confirmed that CT produced quantified changes in the internal luminal diameter of asthmatic airways provoked by methacholine and albuterol inhalation.

BEIGELMAN-AUBRY et al. (2002) demonstrated recently that 12 patients with mild intermittent asthma present with baseline bronchoconstriction compared to 6 normal subjects when examined with CT performed at a spirometrically controlled lung volume (65% of total lung capacity after full inspiration), as they had stopped all treatment 48 h prior to study. This was confirmed by the fact that inhalation of salbutamol after methacholine challenge brought their bronchi not only back to cross section areas comparable to those of the control group, but also above their own baseline values (Fig. 3.7). This suggests that bronchoconstriction is due to an increased baseline tone or an impaired stretching in asthmatic patients. In the same study, bronchial cross section area was not influenced by methacholine challenge more in asthmatics than in normal subjects, but the presence of expiratory air trapping, induced by methacholine challenge and partly reversible after salbutamol inhalation in asthmatic patients, confirmed that the methacholine-induced bronchoconstriction involves mainly or exclusively the smallest airways (BEIGELMAN-AUBRY et al. 2002). In the same way, GOLDIN et al. (1998) showed significant leftward shifts in frequency distribution of lung parenchymal attenuation values measured by 3-mm-thick CT scans of lungs before and after methacholine challenge in 15 mild asthmatic patients. These effects returned to normal after the administration of albuterol. The same effects were not seen in the six control subjects. These studies suggest the lower attenuating areas observed on expiratory HRCT scans correlate to areas of air trapping (GOLDIN 2002).

Because lung attenuation can be quantified in a reproducible way with CT and values for normal subjects (examined at a controlled lung volume) lie within a normal range, BEIGELMAN-AUBRY et al. (2002) applied the same approach as WOLLMER et al. (1986) used to assess various stages of inflammation in smokers. The lung attenuation and anteroposterior attenuation gradient values found for intermittent asthmatic patients at a selected lung volume (65% of total lung capacity), monitored by pneumotachography were significantly higher than normal subjects (BEIGELMAN-AUBRY et al. 2002). The authors hypothesized that the well-known peribronchial and small airway inflammation occurring in asthma explains the attenuation and gradient increases. The observation that these increased attenuation values were not affected by methacholine and salbutamol challenges supports the hypothesis that bronchoconstriction played only a small or no role in attenuation changes and that distal inflammation was a more likely contender. The ability to follow up bronchial reactivity and lung attenuation by CT over time in cohorts of patients receiving different treatments can provide an independent tool to assess and monitor current and new therapy in asthmatic patients (GOLDIN et al. 1999).

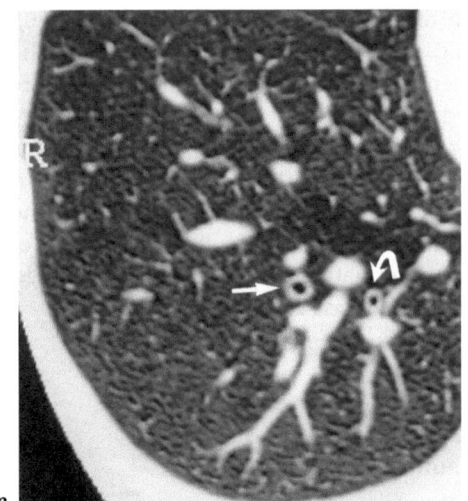

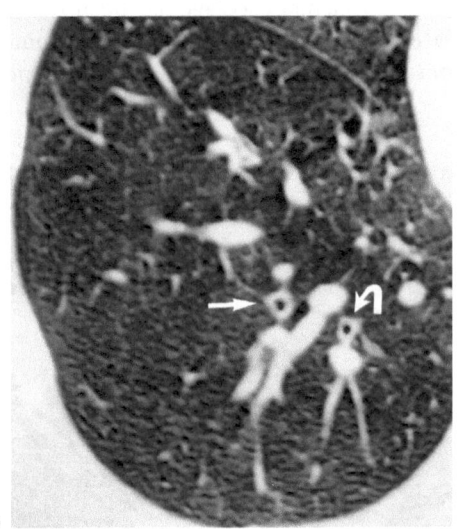

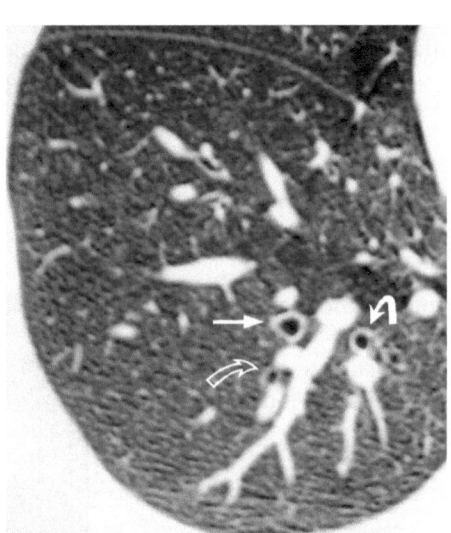

Fig. 3.7. Airway hyperresponsiveness in mild intermittent asthma. Thin section CT scans targeted on the right lower lobe at 65% of total lung capacity controlled by spirometrically triggering, at baseline (**a**), after methacholine inhalation (**b**), and 10 min after salbutamol inhalation (**c**). The cross sections of 2 thick-walled subsegmental bronchi of the posterobasal segment of the right lower lobe (*white arrows*) present with a certain degree of constriction after methacholine (**b**). After salbutamol (**c**), the bronchial lumens dilated and appear larger than before challenge. Other small bronchial lumens (*open arrows*) not previously visible have dilated sufficiently to be depicted after salbutamol. From GRENIER et al. (2002), with permission

3.4.2.2
Airway Wall Remodeling

Asthma is known to induce structural changes in the airways including subepithelial fibrosis, mucous gland and goblet cell hyperplasia and smooth muscle hypertrophy and hyperplasia. The latter change defines airway wall remodeling which is attributed to chronic inflammation. This airway wall remodeling is responsible for the faster and higher decrease related to age of forced expiratory volume per second (FEV1) in asthmatics than controls (LANGE et al. 1998). Bronchial wall thickness measured at CT has proven to be prominent in patients with more severe asthma. NIIMI et al. (2000) measured the bronchial area (mm^2) on axial CT scans obtained at full inspiration in normal volunteers and asthmatic patients. They measured only the airway wall area on the cross section on the bronchus of the superior segment of the right upper lobe. Included were 28 non-asthmatic healthy volunteers and 81 asthmatic patients (7 intermittent asthma, 13 mild persistent asthma, 33 moderate persistent asthma, 22 severe persistent asthma). The authors compared the measured values between groups and attempted to correlate the airway wall area values to the results of pulmonary function tests. They showed that the airway wall area values as well as the airway wall area values corrected by the body surface were significantly higher in patients with persistent asthma than those with intermittent asthma and in controls. These values also correlated with the duration and severity of asthma and negatively correlated with the forced expiratory volume per second (FEV1), FEV1/forced vital capacity (FVC) and midexpiratory phase of forced expiratory flow (FEF$_{25-75\%}$) (NIIMI et al. 2000). This observation supports the concept that quantitative assessment of bronchial wall area at CT could be used to assess airway wall remodeling in asthmatic patients for longitudinal studies to evaluate the effects of new therapies. Requirements should include:

1. Volumetric CT acquisition with thin collimation at spirometrically controlled lung volume
2. Selection of anatomical locations on the bronchial tree
3. Bronchial cross section images perpendicular to the bronchial central axis at the selected anatomical levels
4. Automatic segmentation of the bronchial wall contours (GRENIER et al. 2002).

Such longitudinal prospective studies are needed to be carried out to monitor changes in airway wall remodeling and potential reversibility.

3.5
Chronic Obstructive Pulmonary Disease (COPD)

Abnormalities of the trachea and major bronchi are common in COPD and involve virtually all tissue components, including the surface epithelium, tracheobronchial glands, muscularis mucosa, submucosa and cartilage.

3.5.1
Bronchial Wall Thickening and Diverticula

Bronchial wall thickening is commonly present on thin-section CT scans of patients with COPD (TAGASUKI and GODWIN 1998). This abnormality however has been assessed only subjectively. At the present time quantitative assessment of the bronchial wall area has become possible in prospective longitudinal studies.

The airway remodeling occurs in COPD as well as in asthma, but this abnormality involves essentially the small airways (KUWANO et al. 1993). The amount of smooth muscle in the bronchial wall of patients having chronic bronchitis has been found to be increased by some investigators (KUWANO et al. 1993) but to be normal by others (DUNNILL et al. 1969). The relationships between such an increase and the presence of airway responsiveness or variability in the severity of airflow obstruction in COPD is not clear.

The increase in size of tracheobronchial glands in patients with COPD is sometimes apparent by simple observation of histologic sections (DOUGLAS 1980). The openings of the bronchial gland ducts into the airway lumen may be plugged with mucous and are often dilated (WANG and YING 1976). This abnormality can be appreciated on bronchograms or MDCT scans as small depression or diverticula on the airway lumen surface (Fig. 3.8). It has also been suggested that some diverticula are related to loss of subepithelial connective tissue and herniation of airway mucosa between smooth muscle bundles (WANG and YING 1976).

3.5.2
Expiratory Bronchial Collapse and Tracheobronchomalacia

The quantity of bronchial cartilage has been found to be decreased in patients who have COPD in some investigations (MAISEL et al. 1968; TANDON and

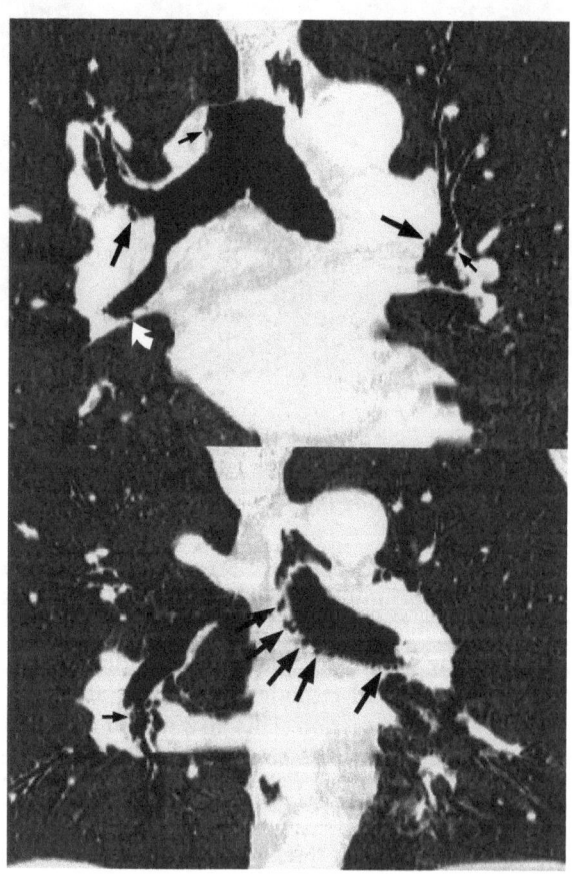

Fig. 3.8. Minimal intensity projection on a 5 -mm thick coronal slab in a patient who has COPD. Small bronchial diverticula (*arrows*) seen as outpouchings of the bronchial lumen visible along the left main bronchus, and the right upper and lower lobar bronchi

CAMPBELL 1969; THURLBECK et al. 1974). The most severe deficiency has been seen in the segmental and subsegmental bronchi, generally being more apparent in the lower than the upper lobes. In both normal individuals and COPD patients, maximum forced expiratory maneuver results in a flow-limiting collapse of segments of cartilaginous airways. Because cartilage provides an important contribution to the relative incompressibility of these airways, its deficiency may be expected to result in more prominent collapse. *Expiratory bronchial collapse* in patients who have COPD may be apparent using CT (Fig. 3.9). *Tracheobronchomalacia* is a variant of expiratory airway collapse due to abnormal flaccidity involving the trachea and main bronchi. The increase in compliance is due to the loss of integrity of the wall structural components and is particularly associated with damaged or destroyed cartilages. The coronal diameter of the trachea becomes significantly larger than the sagittal diameter, producing a lunate configuration to the trachea. The flaccidity of the trachea or bronchi is usually most apparent during coughing or forced expiration. In patients with COPD with high downstream resistance particularly high dynamic pressure gradients can be generated across the tracheal wall, and it is likely that caliber changes of more than 50% can occur at expiration with normal tracheal compliance. As a result only a decrease in cross-sectional area of the tracheal lumen greater than 70% at expiration indicates tracheomalacia (STERN et al. 1993). Dynamic expiratory multislice CT may offer a feasible alternative to bronchoscopy in patients with suspected tracheobronchomalacia. Dynamic expiratory CT may show complete collapse

or collapse of greater than 75% of the airway lumen (Fig. 3.10). Involvement of the central tracheobronchial tree may be diffuse or focal. The reduction of airway lumen may have an oval or crescent shape. The crescent form is due to the bowing of posterior membranous trachea (GILKESON et al. 2001).

3.5.3
Saber-Sheath Trachea

Saber-sheath trachea is a deformity defined as excessive coronal narrowing of the intrathoracic trachea in tandem with widening of the sagittal tracheal diameter (Fig. 3.11). In pronounced examples the tracheal index (coronal-sagittal diameters ratio) can be less than 0.6. Concurrently, the cross-sectional diameter of the trachea decreases to less than 60% of normal (STARK and NORBASH 1998). This deformity is highly characteristic of COPD, affects primarily the intrathoracic trachea and the main bronchi and usually spares the cervical trachea, which retains normal dimensions and a normal configuration. The tracheal wall is slightly thickened, rigid, smooth, or slightly corrugated with occasional ossification or calcification of the cartilaginous elements.

3.6
Summary

Multidetector row CT may result in a better morphological and functional evaluation of proximal airways in

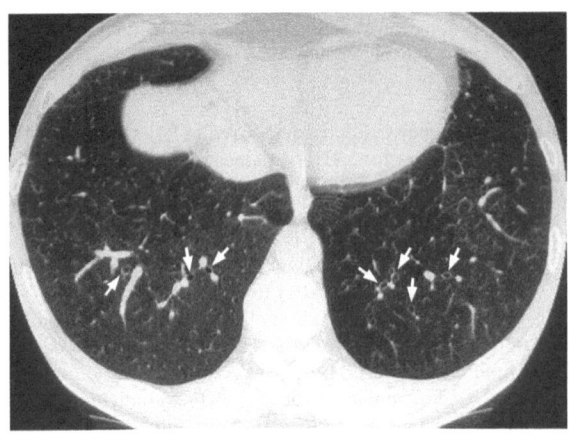 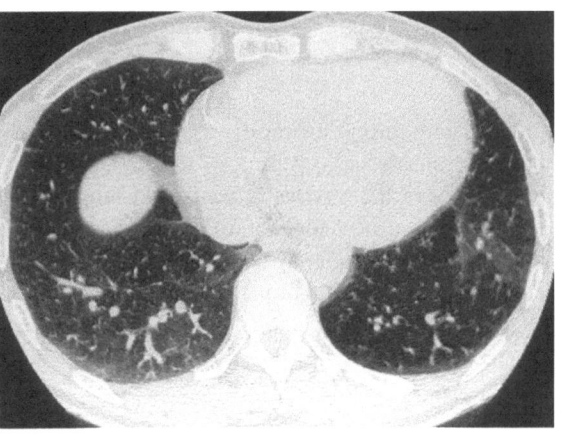

a   b

Fig. 3.9. Thin-section CT scan at full inspiration (**a**) and full expiration (**b**) in a patient with COPD. Expiratory bronchial collapse is seen on the segmental and subsegmental bronchi in the lower lobes. The bronchial collapse is complete in the posterobasal segment of the left lower lobe. Notice also the decreased lung attenuation with expiratory air trapping in the posterobasal segment of the left lower lobe

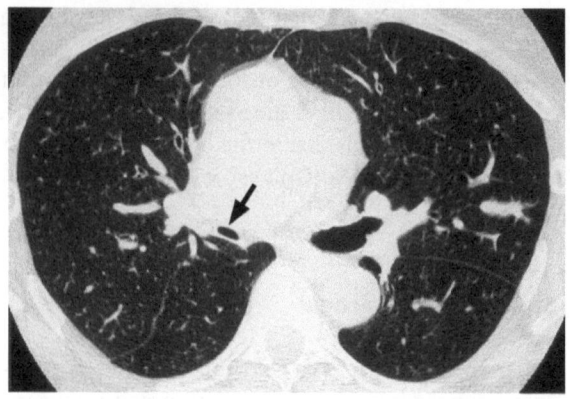
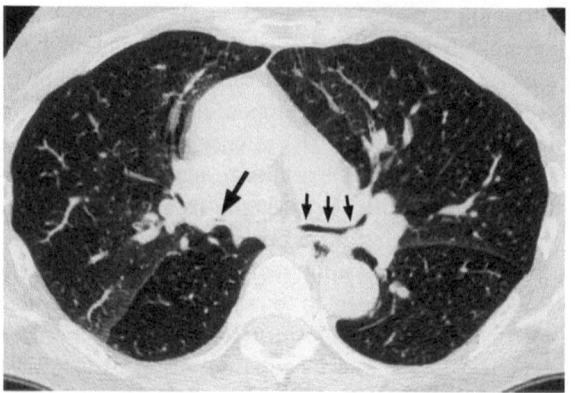

a   b

Fig. 3.10. Tracheobronchomalacia. Inspiratory (**a**) and expiratory (**b**) thin sections. On the inspiratory scan the lumen of the right intermediate bronchus is reduced (*arrows*) whereas the lumen of the main and upper lobar bronchi is normal. On expiratory scan (**b**), the lumen of the right intermediate bronchus is completely collapsed, while the reduction in caliber of the left main and upper lobar bronchi is over 70%. From GRENIER et al. (2002), with permission

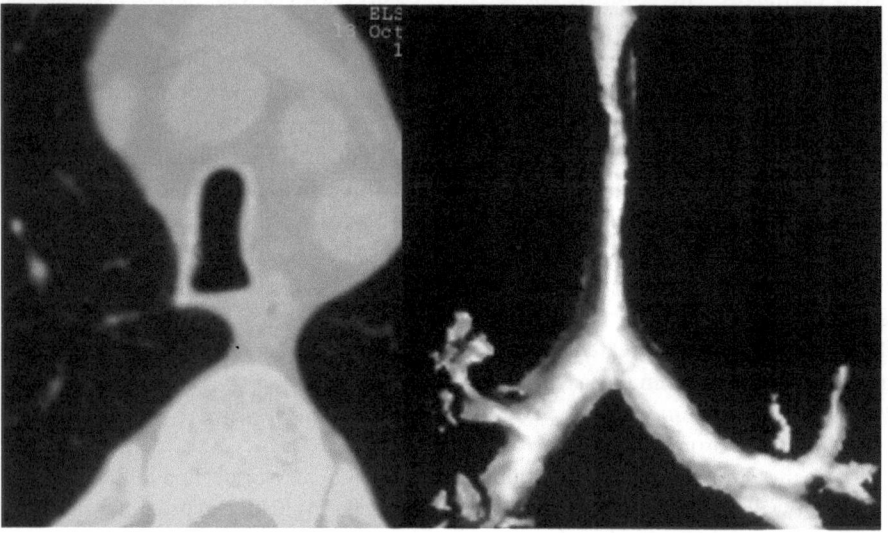

Fig. 3.11. Saber-sheath trachea in a patient with severe COPD. *Left* Thin section CT scan showing the saber-sheath trachea deformity, *right* 3D surface rendering display after multidetector row CT acquisition showing the excessive coronal narrowing of the intrathoracic trachea

patients with chronic airway disease, particularly those having bronchiectasis, asthma, or COPD. Quantitative assessment of the airways at controlled lung volume may become a new tool for monitoring changes in airways that reflect airway wall remodeling, bronchial expiratory collapse and bronchial reactivity.

References

Amirav I, Kramer SS, Grunstein M et al. (1993) Assessment of methacholine-induced airway constriction with ultrafast high-resolution computed tomography. J Appl Physiol 75: 2239–2250

Beigelman-Aubry C, Capderou A, Grenier PA et al. (2002) Mild intermittent asthma: CT assessment of bronchial cross-section area and lung attenuation at controlled lung volume. Radiology 223:181–187

Boiselle PM, Reynolds KF, Ernst A (2002) Multiplanar and three-dimensional imaging of the central airways with multidetector CT. AJR Am J Roentgenol 179:301–308

Brown R, Mitzner W, Bulut Y et al. (1997) Effect of lung inflation in vivo on airways with smooth-muscle tone or edema. J Appl Physiol 82:491–499

Brown R, Georakopoulos I, Mitzner W (1998) Individual canine airways responsiveness to aerosol histamine and methacholine in vivo. Am J Respir Crit Care Med 157:491–497

Choi YW, McAdams HP, Jeon SC et al. (2002) Low-dose spiral CT: application to surface-rendered three-dimensional imaging of central airways. J Comput Assist Tomogr 26:335–341

Douglas AN (1980) Quantitative study of bronchial mucous gland enlargement. Thorax 35:198–201

Dunnill MS, Massarella GR, Anderson JA (1969) A comparison of the quantitative anatomy of the bronchi in normal subjects, in status asthmaticus, in chronic bronchitis and in emphysema. Thorax 24:176–179

Ferretti GR, Bricault I, Coulomb M (2001) Virtual tools for imaging of the thorax. Eur Respir J 18:381–392

Fetita CI, Preteux F, Beigelman C et al. (1999) 3D CT bronchography: a new segmentation and reconstruction-based method. Radiology 213(P):197

Fraser RS, Muller NL, Coman N et al. (2000) Diagnosis of disease of the chest, vol 3, 4th edn. Sanders, Philadelphia, pp 2274–2277

Gilkeson RC, Ciancibello LM, Hejal RB et al. (2001) Tracheobronchomalacia: dynamic airway evaluation with multidetector CT. AJR Am J Roentgenol 176:205–210

Goldin JG (2002) Quantitative CT of the lung. Radiol Clin North Am 40:145–166

Goldin JG, McNitt-Gray MF, Sorenson SM et al. (1998) Airway hyperreactivity: assessment with helical thin-section CT. Radiology 208:321–329

Goldin JG, Tashkin DP, Kleerup EC et al. (1999) Comparative effects of HFA- and CPC-beclomethasone diproprionate inhalation on small airways: assessment using functional helical thin-section computed tomography. J Allergy Clin Immunol 104:S258–S267

Gotway MB, Lee ES, Reddy GP et al. (2000) Low-dose, dynamic, expiratory thin-section CT of the lungs using a spiral CT scanner. J Thorac Imaging 15:168–172

Grenier P, Maurice F, Musset D et al. (1986) Bronchiectasis: assessment by thin section CT. Radiology 161:95–99

Grenier P, Mourey-Gerosa I, Benali K et al. (1996) Abnormalities of the airways and lung parenchyma in asthmatics: CT observations in 50 patients and inter- and intraobserver variability. Eur Radiol 6:199–206

Grenier P, Beigelman-Aubry C, Fétita C et al. (2002) New frontiers in CT imaging of airway disease. Eur Radiol 12:1022–1044

Hopper KD, Iyriboz TA, Mahraj RPM et al. (1998) CT bronchoscopy: optimization of imaging parameters. Radiology 209:872–877

Kang EY, Miller RR, Muller NL (1995) Bronchiectasis: comparison of preoperative thin-section CT and pathologic findings in resected specimens. Radiology 195:649–654

Kauczor HU, Wolcke B, Fischer B et al. (1996) Three-dimensional helical CT of the tracheobronchial tree: evaluation of imaging protocols and assessment of suspected stenoses with bronchoscopic correlation. AJR Am J Roentgenol 167:419–424

Kee ST, Fahy JV, Chen DR et al. (1996) High-resolution computed tomography of airway changes after induced bronchoconstriction and bronchodilation in asthmatic volunteers. Acad Radiol 3:389–394

Kim JS, Muller NL, Park CS et al. (1997a) Bronchoarterial ratio on thin-section CT: comparison between high altitude and sea level. J Comput Assist Tomogr 21:306–311

Kim JS, Muller NL, Park CS et al. (1997b) Cylindrical bronchiectasis: diagnostic findings on thin-section CT. AJR Am J Roentgenol 168:751–754

King GG, Muller NL, Pare PD (1999) Evaluation of airways in obstructive pulmonary disease using high-resolution computed tomography. Am J Respir Crit Care Med 159:992–1004

King GG, Muller NL, Whittall KP et al. (2000) An analysis algorithm for measuring airway lumen and wall areas from high-resolution computed tomographic data. Am J Respir Crit Care Med 161:574–580

Kuwano K, Bosken CH, Pare PD et al. (1993) Small airways dimensions in asthma and in chronic obstructive pulmonary disease. Am Rev Respir Dis 148:1220–1225

Lange P, Parner J, Vestbo J et al. (1998) A 15-year follow-up study of ventilatory function in adults with asthma. N Engl J Med 339:1194–1200

Lucidarme O, Grenier PA, Coche E et al. (1996) Bronchiectasis: comparative assessment with thin-section CT and helical CT. Radiology 200:673–679

Lucidarme O, Grenier PA, Cadi M et al. (2000) Evaluation of air trapping at CT: comparison of continuous-versus suspended expiration CT techniques. Radiology 216:768–772

Lynch DA (1998) Imaging of asthma and allergic bronchopulmonary mycosis. Radiol Clin North Am 36:129–142

Lynch DA, Newell JD, Tschomper BA et al. (1993) Uncomplicated asthma in adults: comparison of CT appearance of the lungs in asthmatic and healthy subjects. Radiology 188:829–833

Maisel JC, Silvers GW, Mitchell RS et al. (1968) Bronchial atrophy and dynamic expiratory collapse. Am Rev Respir Dis 98:988–997

McAdams HP, Palmer SM, Erasmus JJ et al. (1998) Bronchial anastomotic complications in lung transplant recipients: virtual bronchoscopy for noninvasive assessment. Radiology 209:689–695

McGuinness G, Naidich DP, Leitman BS et al. (1993) Bronchiectasis: CT evaluation. AJR Am J Roentgenol 160:253–259

McNamara AE, Muller NL, Okazawa M et al. (1992) Airway narrowing in excised canine lungs measured by high-resolution computed tomography. J Appl Physiol 73:307–316

Niimi A, Matsumoto H, Amitani R et al. (2000) Airway wall thickness in asthma assessed by computed tomography. Relation to clinical indices. Am J Respir Crit Care Med 162:1518–1523

Okazawa M, Muller N, McNamara AE et al. (1996) Human airway narrowing measured using high resolution computed tomography. Am J Respir Crit Care Med 154:1557–1562

Ooi GC, Khong PL, Chan-Yeung M et al. (2002) High-resolution CT quantification of bronchiectasis: clinical and functional correlation. Radiology 225:663–672

Paganin F, Trussard V, Seneterre E et al. (1992) Chest radiography and high resolution computed tomography of the lungs in asthma. Am Rev Respir Dis 146:1084–1087

Paganin F, Seneterre E, Chanel P et al. (1996) Computed tomography of the lungs in asthma: influence of disease severity and etiology. Am J Respir Crit Care Med 153:110–114

Park CS, Muller NL, Worthy SA et al. (1997) Airway obstruction in asthmatic and healthy individuals: inspiratory and expiratory thin-section CT findings. Radiology 203:361–367

Perot V, Desberat P, Berger P et al. (2001) Nouvel algorithme d'extraction des paramètres géométriques des bronches en TDM-HR (abstract). J Radiol 82:1213

Prêteux F, Fetita CI, Capderou A et al. (1999) Modeling, segmentation, and caliber estimation of bronchi in high resolution computerized tomography. J Electron Imaging 8:36–45

Quint LE, Whyte RI, Kazerooni EA et al. (1995) Stenosis of the central airways: evaluation by using helical CT with multiplanar reconstructions. Radiology 194:871–877

Rémy J, Rémy-Jardin M, Artaud D et al. (1998) Multiplanar and three-dimensional reconstruction techniques in CT: impact on chest diseases. Eur Radiol 8:335–351

Rémy-Jardin M, Rémy J, Deschildre F et al. (1996) Obstructive lesions of the central airways: evaluation by using spiral CT with multiplanar and three-dimensional reformations. Eur Radiol 6:807–816

Rémy-Jardin M, Rémy J, Artaud D et al. (1998a) Tracheobronchial tree: assessment with volume rendering – technical aspects. Radiology 208:393–398

Rémy-Jardin M, Rémy J, Artaud D et al. (1998b) Volume rendering of the tracheobronchial tree: clinical evaluation of bronchographic images. Radiology 208:761–770

Roberts HR, Wells AU, Milne DG et al. (2000) Airflow obstruction in bronchiectasis: correlation between computed tomography features and pulmonary function tests. Thorax 55:198–204

Rubin GD (1996) Techniques of reconstruction. In: Rémy-Jardin M, Rémy J (eds) Spiral CT of the chest. Springer, Berlin Heidelberg New York, pp 101–127 (Medical radiology series)

Schoepf UJ, Becker CR, Bruening RD et al. (1999) Electrocardiographically gated thin-section CT of the lung. Radiology 212:649–654

Schaefer-Prokop C, Prokop M (1996) Spiral CT of the trachea and main bronchi. In: Rémy-Jardin M, Rémy J (eds) Spiral CT of the chest. Springer, Berlin Heidelberg New York, pp 161–183 (Medical radiology series)

Stark P, Norbash A (1998) Imaging of the trachea and upper airways in patients with chronic obstructive airway disease. Radiol Clin North Am 36:91–105

Stern EJ, Graham CM, Webb WR et al. (1993) Normal trachea during forced expiration: dynamic CT measurements. Radiology 187:27–31

Summers RM, Shaw DJ, Shelhamer JH (1998) CT virtual bronchoscopy of simulated endobronchial lesions: effect of scanning, reconstruction, and display settings and potential pitfalls. AJR Am J Roentgenol 170:947–950

Tagasuki JE, Godwin D (1998) Radiology of chronic obstructive pulmonary disease. Radiol Clin North Am 36:39–55

Tandon MK, Campbell AH (1969) Bronchial cartilage in chronic bronchitis. Thorax 24:607–612

Thurlbeck WM, Pun R, Toth J et al. (1974) Bronchial cartilage in chronic obstructive lung disease. Am Rev Respir Dis 109:73–80

Wang NS, Ying WL (1976) Morphogenesis of human bronchial diverticulum. A scanning electron microscopic study. Chest 69:201–204

Webb WR, Gamsu G, Wall SD et al. (1984) CT of a bronchial phantom: factors affecting appearance and size measurements. Invest Radiol 19:394–398

Wollmer P, Albrechtsson U, Brauer K et al. (1986) Measurement of pulmonary density by means of X-ray computerized tomography. Relation to pulmonary mechanics in normal subjects. Chest 90:387–391

Wood SA, Zerhouni EA, Hoford JD et al. (1995) Measurement of three-dimensional lung tree structures by using computed tomography. J Appl Physiol 79:1687–1697

4 Small Airways, Bronchiolitis, Air Trapping

Johny Verschakelen

CONTENTS

4.1
Introduction

Small airways disease (SAD) is defined as a pathological condition in which the small conducting airways are affected either primarily or in addition to alveolar or interstitial lung changes. Diseases affecting the small airways are often difficult to detect and to quantify by traditional diagnostic tests. The pulmonary function deterioration is indeed not only the result of small airways involvement but is also related to associated lung parenchyma changes or large airways involvement that are often present. Many histopathological subtypes have been defined. A problem is that these subtypes do not always have obvious clinical correlates, and even when there are obvious correlates, confusion can be induced by the fact that limited disease shown on pathological or radiological examinations does not necessarily result in symptoms or functional deterioration that needs to be present to make the clinical diagnosis. The increased use of high resolution computed tomography (HRCT) to study the appearances of the various pathologic subtypes is responsible for an enormous progress in the understanding of small airways disease. Although the visualization of normal small airways is limited by the spatial resolution limits of HRCT, these airways may become directly visible when inflammation of the bronchiolar wall and accompanying exudates develop. Obstruction of the bronchioles may be detected indirectly because regional under-ventilation results in reduced perfusion, which can be seen on HRCT. Some of these features can be quantified and can be used for functional imaging. The HRCT features of SAD depend on the bronchiolar branches that are involved, on the type of reaction to the bronchiolar injury, and on the extent and the distribution of this reaction in the lungs.

J. VERSCHAKELEN, MD
Professor, Department of Radiology, U.Z. Gasthuisberg, Herestraat 49, 3000 Leuven, Belgium

4.2
Bronchiolar Anatomy and Pathology

Pathologists consider those airways that do not contain cartilage as small airways and use the term "bronchiole" to describe these small airways. The internal diameter of 2 mm is another often used division between "small" and "large" airways.

Although both definitions do not correspond because cartilage may be found in some peripheral airways less than 1 mm in diameter, the latter definition is more practical and most frequently used. Distinguishing small airways from large airways is indeed very useful because small and large airways often have very distinctive clinical and physiologically abnormalities (Green and Turton 1982). The most distal bronchioles are classified as terminal (membranous) bronchioles and respiratory bronchioles (Wright et al. 1992). The terminal bronchioles are found between the 6th and 23rd generations of branching (Plopper and Ten Have-Opbroek 1994), have an internal diameter of approximately 0.6 mm and their length varies from 0.8 to 2.5 mm (Hansen et al. 1975; Plopper and Ten Have-Opbroek 1994). These terminal membranous bronchioles have a complete fibromuscular wall and no alveoli arise directly from them. Downstream from these airways are the respiratory bronchioles. The major difference between terminal (membranous) and respiratory bronchioles is that the latter have gas exchanging alveoli arising from their walls while terminal bronchioles are purely air conducting. Further downstream these respiratory bronchioles communicate through alveolar ducts with numerous alveolar sacs (Plopper and Ten Have-Opbroek 1994).

The secondary pulmonary lobule is defined as the smallest portion of the lung that is surrounded by connective tissue septa (Miller 1947). It is irregularly polyhedral and has a diameter between 0.5 and 3 cm. An important feature of this secondary pulmonary lobule is the "core structure" comprising the supplying terminal bronchiole and the accompanying pulmonary artery branch, surrounded by lymph vessels. The secondary pulmonary lobules are demarcated by the "interlobular septa" containing pulmonary veins, lymphatics and connective tissue stroma. Between the "core" structures and the interlobular septa numerous alveolar sacs, respiratory bronchioles and capillaries are found.

Since normal airways with a diameter below approximately 1.5 mm cannot be identified on HRCT, the majority of the normal bronchioles including the centrilobular terminal bronchioles are not visible. Diseased bronchioles, however, are often readily identifiable (Gruden et al. 1994; Itoh et al. 1993; Murata et al. 1986). In general, small airways disease can be considered as a reaction to an injury. This injury can be focal or multifocal or can affect the whole lung inducing diffuse lung disease. The injury can have a known cause but can also be the result of an unknown trigger. Schematically, the bronchiolar reactions resulting from this trigger can be divided into four categories. The first group is characterized by inflammatory thickening of the bronchiolar wall and a bronchiolar lumen filled with mucus and exudates. In the second group bronchiolar wall changes are limited but the characteristic abnormality is the proliferation of granulation tissue in the lumen. The third group is characterized by the development of submucosal (mural) fibrosis resulting in bronchiolar narrowing and distortion. Finally bronchioles can also be involved indirectly: changes adjacent to the bronchioles such as lung fibrosis and abnormalities in the bronchiolar wall such as granulomas can also cause distortion of the bronchiole and narrowing of the lumen (Fig. 4.1).

4.3
Classification of Small Airways Disease

Classification of small airways disease is usually based on pathologic subtypes and on clinical criteria. However, both the histopathological schemes and the clinical classifications are very often unsatisfactory. This is true for an important part related to the fact that histopathological subtypes do not always have obvious clinical correlates. In addition, the number of causes of small airways disease and diseases associated with small airways involvement described in the literature is continuously increasing. Myers and Colby (1993) published one of the first histologic classifications that served as a basis for other classification schemes. Their classification includes the following subtypes: 1) acute bronchiolitis (infectious bronchiolitis), 2) cryptogenic organizing pneumonia (COP) or bronchiolitis obliterans organizing pneumonia (BOOP), 3) constrictive bronchiolitis (obliterative bronchiolitis, bronchiolitis obliterans), 4) adult bronchiolitis, 5) respiratory bronchiolitis (smokers bronchiolitis) respiratory bronchiolitis associated interstitial lung disease (RBILD), 6) mineral dust airways disease, 7) follicular bronchiolitis, and 8) diffuse panbronchiolitis. Others proposed different versions of this pathological classification (Hwang et al. 1997; Worthy and Muller 1998) and also classifications based on a combination of clinical and histopathological criteria were suggested (POLETTI et al. 1999). MÜLLER and MILLER (1995) were the first to use a classification system based on radiological features. They noticed that the various forms of bronchiolitis generally resulted in one of three predominant CT patterns: 1) nodules and branching lines (Fig. 4.2), 2) ground-glass attenuation and consolidation (Fig. 4.6), or 3) low attenuation and mosaic perfusion (Fig. 4.3). Their first pattern is predominantly seen in acute infectious bronchiolitis, in diffuse panbronchiolitis and is also often found in patients with chronic inflammation of the bronchioles such as asthma, chronic bronchitis, and bronchiectasis. Bronchiolitis obliterans organizing pneumonia (BOOP) and respiratory bronchiolitis interstitial lung disease (RBILD) were classified in the second group because ground-glass opacities and consolidation are their pre-

Fig. 4.1. Types of bronchiolar reactions to injuries

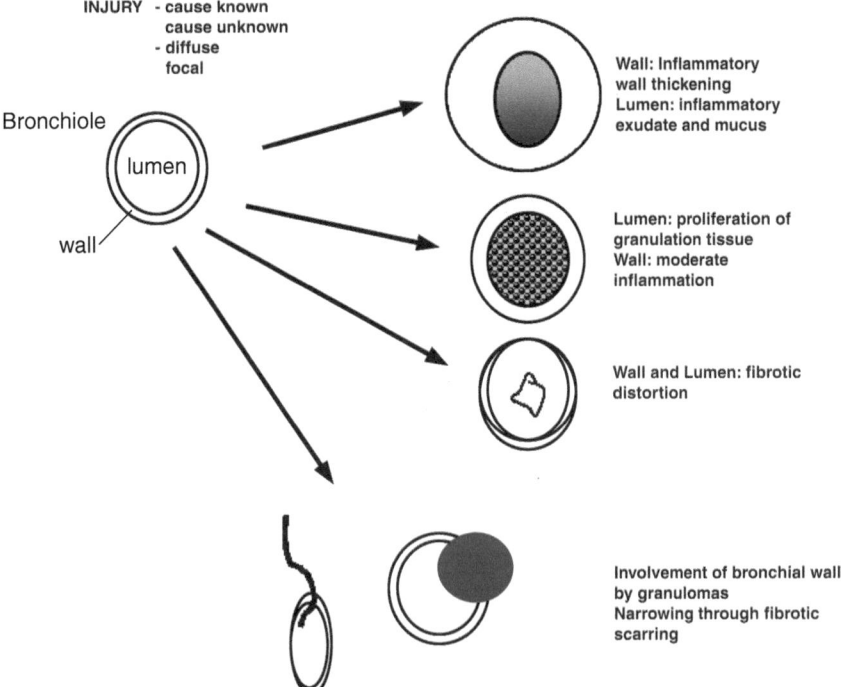

Fig. 4.2a, b. Direct signs of small airways disease include: 1) poorly defined centrilobular opacities (**a** *arrowhead*); 2) increased prominence or size of centrilobular branching structures (**a** *arrow*) and 3) visibility and dilatation of centrilobular bronchioles (**b** *arrow*). In combination these signs can produce the "tree in bud" sign (**b**)

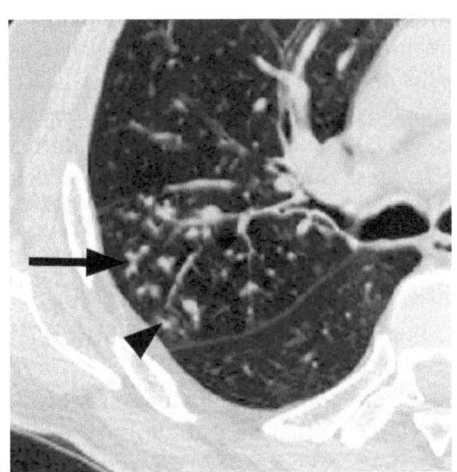

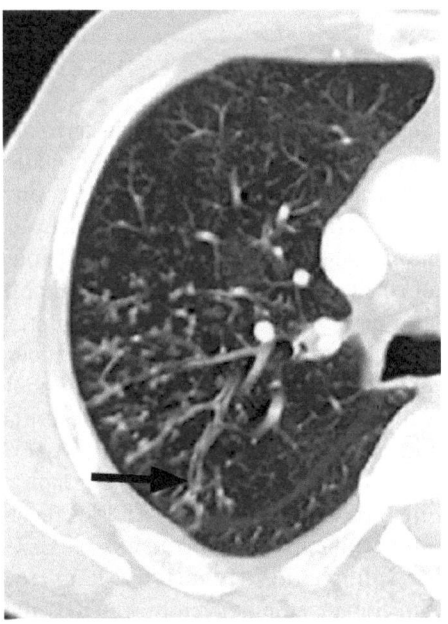

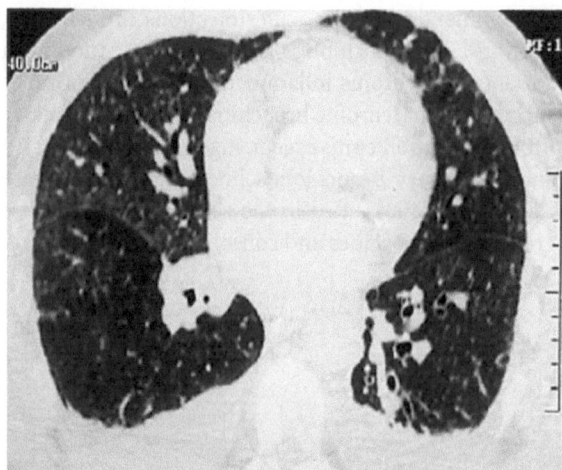

Fig. 4.3a, b. Indirect signs of small airways disease include: a widespread area of decreased lung attenuation in the right lower lobe. Notice also the reduction in caliber of the macroscopic pulmonary vessels in this area; b and c multiple areas of decreased attenuation, surrounded by areas of increased attenuation. Notice the small decrease in size of the vessels in the areas of low attenuation when compared with the surrounding "high" attenuation areas: "mosaic perfusion" (*arrows*)

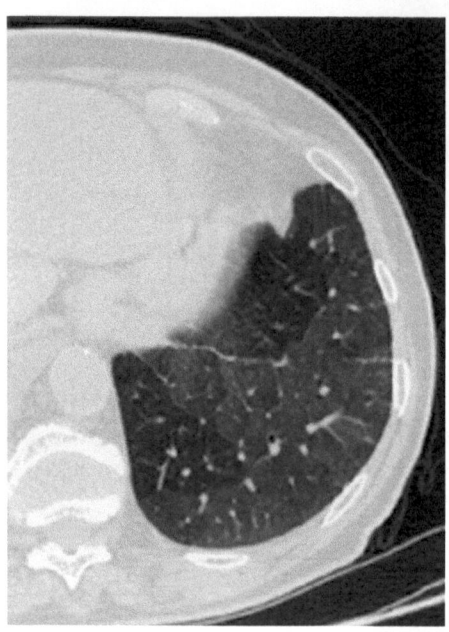

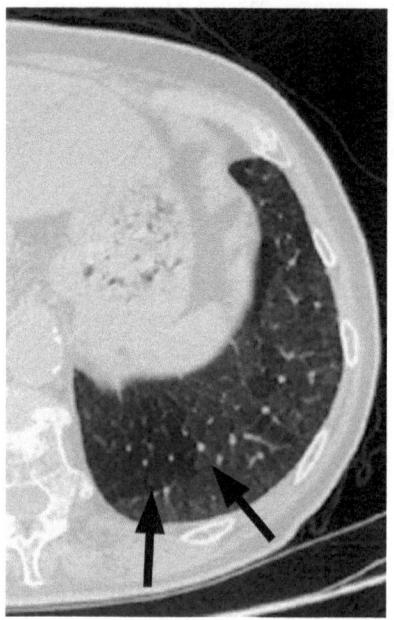

dominant CT features. Finally the form of bronchiolitis characteristically associated with low attenuation and mosaic perfusion on CT scans is constrictive bronchiolitis (also known as obliterative bronchiolitis and bronchiolitis obliterans, see paragraph on terminology of small airways disease). MÜLLER and MILLER (1995) also described a group of chronic infiltrative diseases that are often associated with bronchiolocentric infiltrates and in which one or more of the abovementioned patterns can be recognized. Examples are allergic alveolitis and sarcoidosis.

As explained in the next paragraph according to the recently developed ATS/ERS INTERNATIONAL MULTIDISCIPLINARY CONSENSUS CLASSIFICATION OF IDIOPATHIC INTERSTITIAL PNEUMONIAS (2002), the clinicopathologic entities BOOP and RBILD are now considered as quite distinct from bronchiolitis and classified as idiopathic interstitial pneumonias rather than as small airways diseases. That is why more recent radiological classifications can be based on the fundamental differences between the nodular and branching lines pattern and the low attenuation and mosaic perfusion pattern.

4.4
Terminology in Small Airways Disease

Until recently the term bronchiolitis obliterans was often used to describe different clinicopathological types of SAD. Together with its use in the context of bronchiolitis obliterans organizing pneumonia, this has induced a lot of confusion in the literature. Historically, Wilhelm Lange used this term first in 1902 in a report that described an idiopathic respiratory illness that resulted in the death of two patients (LANGE 1901). Microscopically the lumina of the bronchioles in these patients were irregularly narrowed because they were filled with spindle cells and young granulation tissue plugs. Similar plugs were seen in some alveoli adjacent to these obliterated bronchioles. Since that moment many reports have used this term to describe similar findings, but also totally different clinical and pathological changes were associated with this term. For example, BAAR and GALINDO (1966) described the case of a 56-year-old man who died from an illness of unknown cause 9 months after onset of symptoms. In this patient concentric rings of submucosal and peribronchiolar fibrosis were replacing the bronchiolar wall while the interstitium was spared. They called this entity "bronchiolitis fibrosa obliterans" because fibrosis was the predominant finding while intraluminal granulation tissue plugs were not observed.

In order to reduce the confusion in terminology GOSINK et al. (1973) and GARG et al. (1994) proposed to divide bronchiolitis obliterans into two subgroups based on the pathological findings: *proliferative bronchiolitis* and *constrictive bronchiolitis*. This approach had the advantage that their two subgroups not only have distinct pathological findings but also that the radiological and clinical features are different. The major pathological feature in the proliferative type is the presence of granulation tissue plugs in the bronchiolar lumen. However, these plugs are not only found in the bronchiolar lumen but also in the alveoli. That is how the term bronchiolitis obliterans organizing pneumonia (BOOP) was introduced in the literature (EPLER et al. 1985). Very soon it turned out that in most cases the alveolar involvement was the most predominant finding. For this reason it has been suggested to discard the bronchiolitis obliterans (subtype proliferative bronchiolitis) part altogether (SULAVIK 1989) and to use a better term to describe this pathological entity: cryptogenic organizing pneumonia (COP) (DAVISON et al. 1983; GEDDES 1991) and to classify this entity as an idiopathic interstitial pneumonia rather than as small airways disease. The major pathological feature of the constrictive type is the development of submucosal fibrosis resulting in bronchiolar narrowing and distortion. The term obliterative bronchiolitis is also used to describe this "constrictive" subtype, while the term bronchiolitis obliterans is nowadays considered

as a synonym of the terms constrictive and obliterative bronchiolitis (HANSELL et al. 1997). These terms, indeed, express better the clinical abnormalities found in patients with enough lung involvement to cause symptoms. In contrast to COP where pulmonary function changes suggest interstitial lung disease (abnormal VC and diffusion), lung function deterioration in constrictive bronchiolitis is obstructive (abnormal FEV1 and FEV1/VC).

Besides constrictive bronchiolitis there is another presentation of SAD that has caused less confusion in terminology because the term bronchiolitis obliterans has never been used to describe it. This type of SAD is, however, less well defined and has as main characteristic the presence of a cellular inflammation of the bronchiolar wall together with inflammatory exudates and mucus in the lumen. This type also has a very distinct radiological presentation and is often described as *cellular or exudative bronchiolitis*.

4.5
High Resolution CT Signs of Small Airways Disease

In general two different groups of CT signs can be found in patients with small airways diseases. On the one hand there are direct signs that are caused by the fact that the bronchiolar changes themselves become visible. On the other hand, bronchiolar changes can be too small to be visible directly but can cause indirect signs that suggest small airways involvement.

4.5.1
Direct Signs

The main direct signs suggesting small airways disease are: 1) poorly defined opacities that are adjacent to, surround or obscure centrilobular arteries and that are centered or clustered 5–10 mm from the lobular periphery or pleural surface (MURATA et al. 1986), 2) increased prominence or size of centrilobular branching structures or loss of their definition, and 3) visibility and dilatation of centrilobular bronchioles. These signs do not need to be present simultaneously but in combination they can produce a sign that has been aptly described as "tree in bud" (Fig. 4.2). The pathological correlates of these HRCT signs are inflammatory thickening of the bronchiolar wall, bronchiolar dilatation, intraluminal mucus, fluid or pus and secondary involvement of the peribronchovascular interstitium or centrilobular alveoli (GRUDEN et al. 1994).

It is easy to understand that these radiological findings are not only found in processes that predominantly affect the bronchioles but that they can also result from processes that affect the pulmonary arteries or lymphatic vessels with secondary involvement of the peribronchovascular interstitium or centrilobular alveoli. Because lymph vessels are also present in the centrilobular region surrounding the terminal bronchioles and arterioles, lymphangitic spread of tumor and sarcoidosis – two diseases that typically spread along the lymph vessels – can also produce centrilobular nodules (COLLINS 2001; LYNCH et al. 1989; STEIN et al. 1987). However, differential diagnosis with bronchiolar disease is usually easy because of other signs of lymphangitic spread of disease such as the presence of thickened interlobular septa, subpleural nodular thickening and irregular interface between vessels and lung create the typical lymphangitic pattern. Diseases of the arterial walls or perivascular tissues can also result in centrilobular abnormality. Although bronchiolectasis is absent, inflammatory responses within the perivascular interstitium can cause apparent bronchiolar wall thickening and ill-defined centrilobular nodules. HRCT can depict these abnormalities in mild cases of pulmonary edema, in vasculitis syndromes or in reactions to substances injected iv. Differential diagnosis with bronchiolar disease is based on associated findings such as pleural effusion in edema or the distribution pattern (CONNOLLY et al. 1996; FRAZIER et al. 1998).

Bronchial filling with tumor cells as seen in bronchoalveolar cell carcinoma can also produce centrilobular nodules and branching lines. Differential diagnosis with bronchiolitis can be difficult but in bronchoalveolar cell carcinoma the changes are usually slowly progressive and associated with areas of ground-glass attenuation and consolidation (KUSHIHASHI et al. 1994).

A special group of diseases are those that affect either alone or in combination the bronchiolar wall, the bronchiolar lumen and peribronchiolar interstitium and alveoli not as the predominant finding but as a small or more important part of the syndrome. A typical example is bronchiolitis obliterans organizing pneumonia (BOOP) or as earlier mentioned better described as cryptogenic organizing pneumonia (COP). The predominant finding is filling of the airspace lumen with young granulation tissue causing areas of ground-glass opacity and consolidation (ALASALY et al. 1995; LEE et al. 1994; MÜLLER et al. 1990) (Fig. 4.6a). However, because these granulation tissue polyps are also growing into the bronchiole areas with centrilobular ill-defined nodules, branching lines and even "tree in bud" can be found indicating the small airway part of the disease (Fig. 4.6b). This small airway part has been the cause of a lot of confusion in the terminology because it introduced the descriptive term bronchiolitis obliterans into this pathological entity (see chapter small airways disease and terminology). Other diseases associated with inflammatory small airway involvement resulting in ill-defined centrilobular opacities are hypersensitivity pneumonitis, follicular bronchiolitis, respiratory bronchiolitis and respiratory bronchiolitis interstitial lung disease and the sinobronchial (panbronchiolitis) syndrome (HANSELL et al. 1996; HARTMAN et al. 1997; HOWLING et al. 1999; NISHIMURA et al. 1992; REMY-JARDIN et al. 1993c).

4.5.2
Indirect Signs

Indirect signs of small airways disease, in contrast to direct signs that correspond with the bronchiolar abnormalities themselves, reflect the consequences of this bronchiolar disease and are the result of one specific bronchiolar change: narrowing of the lumen. This narrowing should be important enough but also firm enough to cause disturbance in airflow.

Indirect signs are the hallmark of fibrotic narrowing of the bronchioles seen in constrictive bronchiolitis.

Although fibrosis is usually also present and part of the pathological changes, other processes that affect the bronchiolar wall and lumen can also cause bronchiolar narrowing. Granulomatous lesions involving the bronchiolar wall are believed to be responsible for airflow obstruction resulting in indirect HRCT signs in sarcoidosis and hypersensitivity pneumonitis (DAVIES et al. 2000; GLEESON et al. 1996; HANSELL et al. 1996).

These HRCT signs comprise areas of reduced density of the lungs, constriction of the pulmonary vessels within these areas of decreased lung density and lack of change of the of the cross section of the affected parts of the lung on scans obtained at end-expiration. Often there are also associated bronchial abnormalities such as bronchial wall thickening and bronchiectasis (MARTI-BONMATI et al. 1989; PADLEY et al. 1993; HARTMAN et al. 1994; STERN and FRANK 1994; MÜLLER and MILLER 1995; HANSELL et al. 1997; HWANG et al. 1997).

4.5.2.1
Areas of Decreased Density of the Lung Parenchyma

Bronchiolar obstruction causes hypoxic vasoconstriction resulting in areas of decreased lung attenuation

(Fig. 4.3). These areas can be patchy (Fig. 4.3b) or can be widespread (Fig. 4.3a) and are usually poorly defined but can have a well-demarcated geographical outline as well. Redistribution of blood flow to the normal surrounding areas causes increased density of these areas. In this way, especially when bronchiolar involvement is patchy, a patchwork of regions of varied attenuation develops. These density differences between different adjacent lung areas gave rise to the term "mosaic attenuation" (Austin et al. 1996).

As we will see further, not only a difference in lung attenuation between the affected and non-affected lung areas but also a difference in vessel size in these areas are major features of mosaic attenuation (Fig. 4.3b): that is why the term "mosaic perfusion" is also used to describe this entity. Although theoretically correct here, the term is not always correct. It should not be used in those cases where the inhomogeneous attenuation of the lung parenchyma is caused by a patchy distribution of the lung disease for example when multiple areas of ground-glass opacity can be recognized in the lung. The differential diagnosis between the "real" mosaic perfusion and this mosaic attenuation is usually easy. When the mosaic attenuation is caused by a patchy distribution of the lung disease the vessels have the same (normal) caliber in both the high-density and the low-density areas. Moreover in this latter condition the low attenuation areas usually correspond with normal lung tissue. In the "real" mosaic perfusion both low attenuation and high attenuation areas represent pathologic lung tissue with smaller than normal vessel size in the low and larger than normal vessel size in the high attenuation areas.

Mosaic attenuation, however, is not always the result of bronchiolar disease or patchy distribution of lung disease but can also be caused by direct vascular obstruction. The obstructed small vessels are then responsible for the low attenuation areas while redistribution of blood to surrounding normal lung causes increased attenuation. Worthy et al. (1997b) showed that infiltrative lung disease and airway disease are usually differentiated reliably as the cause of mosaic attenuation whereas vascular disease is often misinterpreted as infiltrative lung disease or airway disease. In most cases of vascular disease leading to mosaic perfusion on HRCT, the cause is chronic thromboembolic pulmonary artery hypertension. However, the pattern has also been described in patients with primary pulmonary artery hypertension, pulmonary capillary hemangiomatosis (Primack et al. 1994), pulmonary veno-occlusive disease (Mandel et al. 2000), polyarteritis nodosa (Worthy et al. 1997b), scleroderma

(Sherrick et al. 1997), and intimal sarcoma of the pulmonary arteries (Dennie et al. 2002).

4.5.2.2
Reduction in Caliber of the Macroscopic Pulmonary Vessels

Reduction of the lung attenuation is an indirect sign of decrease of perfusion in the small vessels and capillaries that are beyond the resolution of HRCT. The reduction of the caliber of the larger vessels in the low attenuation areas is often directly visible (Fig. 4.3). In acute bronchiolar obstruction this represents a physiologic reflex of hypoxic vasoconstriction (Guckel et al. 1999) but in the chronic state there is vascular remodeling and the reduced caliber becomes irreversible. Because of redistribution of flow to the normal areas, vessel caliber in these areas is typically increased (Fig. 4.3b). A very important differential diagnostic feature with emphysema, as mentioned earlier, is the fact that the vessels are not distorted. Reduced lung perfusion and hence reduced attenuation of the lung parenchyma and reduced caliber of the vessels is also an important feature in occlusive vascular disease such as chronic pulmonary embolism and chronic pulmonary hypertension. The differential diagnosis between a bronchiolar and a vascular cause is sometimes possible and based on the presence or absence of air trapping on expiratory CT, respectively.

4.5.2.3
Air Trapping at Expiratory CT

Air trapping at expiratory CT is defined as "retention of excess gas (air) in all or part of the lung, especially during expiration, either as a result of complete or partial airway obstruction or as a result of local abnormalities in pulmonary compliance" (Austin et al. 1996). The air is trapped and the cross-sectional area of the affected parts of the lung do not decrease in size on expiratory CT (Fig. 4.4). Usually the regional inhomogeneity of the lung density (mosaic attenuation) seen at end-inspiration HRCT scans is accentuated on sections obtained at end, or during, expiration because the high attenuation areas increase in density while the low attenuation areas remain unchanged (Fig. 4.12). Sometimes however, the areas of decreased attenuation are not visible on inspiratory CT but are only detectable on expiratory CT when normal areas decrease in size and increase in density and the affected areas remain more or less unchanged (Stern and Frank 1994;

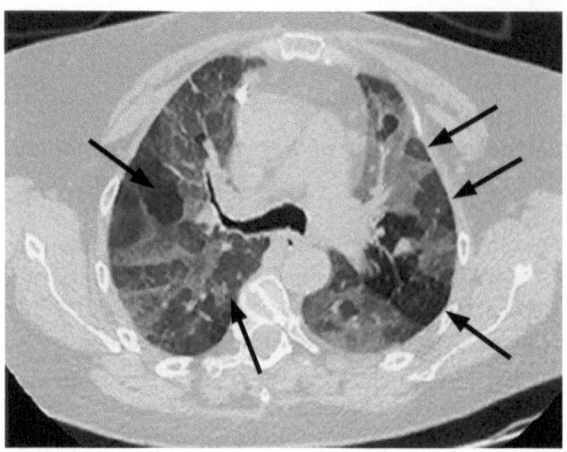

Fig. 4.4. Expiratory HRCT showing multiple areas of air trapping (*arrows*)

DESAI and HANSELL 1997; ARAKAWA and WEBB 1998; LUCIDARME et al. 1998; VERSCHAKELEN et al. 1998). As mentioned earlier, looking for air trapping on expiratory CT may be helpful to differentiate between occlusive vascular diseases and bronchiolar disease as a cause of mosaic attenuation. When it is caused by bronchiolar obstruction, mosaic perfusion is accentuated on expiratory CT because the low attenuation areas show air trapping. In case of an occlusive vascular disease air trapping usually does not occur (STERN et al. 1995b; ARAKAWA et al. 1998b). However, although the presence of mosaic perfusion is considered uncommon in acute pulmonary embolism (STERN and FRANK 1994; STERN et al. 1995b; DESAI and HANSELL 1997; ARAKAWA and WEBB 1998; ARAKAWA et al. 1998b; COCHE et al. 1998; LUCIDARME et al. 1998; VERSCHAKELEN et al. 1998), air trapping may be seen on expiratory high resolution CT in these patients (WORTHY et al. 1997b; ARAKAWA et al. 2002). ARAKAWA et al. (2002) found in a series of 41 patients with acute pulmonary embolism 1 or more areas of air trapping in 71.9% of the patients. This air trapping was seen not only in areas with pulmonary embolism but also in areas without embolism. The proposed mechanism of bronchoconstriction in acute pulmonary embolism includes bronchoactive amines released from platelet aggregations in the thrombus or a change in parasympathetic nervous system tension, which centralize the bronchial smooth muscle tension.

Finally an important caveat is that in patients with widespread small airways disease, inspiratory and expiratory scans may look very similar because of the severity of air trapping.

4.5.2.4
Involvement of the Macroscopic Airways

Most patients with bronchiolitis have some degree of bronchial thickening and dilatation (HANSELL et al. 1997). This can be related to the continuity of the bronchial tree where disease affects both the large and small airways such as in infectious bronchiolitis or because the disease directly involves the large and small airways. In immunologically mediated constrictive bronchiolitis marked dilatation of the bronchi is a frequent finding (REMY-JARDIN et al. 1994; LOUBEYRE et al. 1995).

4.6
Diseases Associated with Small Airways Involvement

The small airways are involved in a wide variety of diffuse interstitial and airway diseases. In some diseases bronchiolitis is the major component of the pathological changes, has an apparent cause and corresponds with a clinically recognizable entity. In other diseases bronchiolitis can be the major component of the pathology giving even recognizable symptoms and/or radiological changes but without a clear clue towards the cause of this small airways involvement. Finally, in some diseases involvement of the small airways is not the predominant pathological abnormality but part of a more widespread involvement of the lung parenchyma although it can be important enough to cause radiological and functional changes. This paragraph discusses briefly the most important airway and interstitial diseases that can have a major or minor component of small airway involvement that is directly or indirectly visible on HRCT scan.

4.6.1
Small Airways Involvement Characterized by Direct HRCT Signs

4.6.1.1 Acute (Infectious) Bronchiolitis

A variety of infectious agents may result in bronchiolar inflammation (Table 4.1). In acute cellular bronchiolitis the luminal exudate is rich in neutrophils. Acute bronchiolitis is not only seen in acute viral infections, mycoplasma pneumonia, acute aspergillus infection, post-primary tuberculosis pneumonia (Fig. 4.5), but is

Table 4.1. Causes of acute bronchiolitis

-Acute viral infections in children and adults (adenovirus)
-*Mycoplasma pneumoniae* pneumonia
-Acute aspergillus pneumonia (immunocompromised)
-Bacterial infection (immunocompromised)
-Bronchogenic spread of post-primary TB
 (immunocompromised)
-Bronchiectasis
-Asthma, chronic bronchitis
-Aspiration
-Acute fume or toxic exposures
-Cystic fibrosis

also found in acute fumes or toxin exposure, asthma, bronchiectasis and chronic bronchitis (LOGAN et al. 1994; McGUINNESS et al. 1994; PRIMACK et al. 1995) (Table 4.1). HRCT features are caused by the inflammatory changes in the centrilobular bronchioli and in the surrounding alveoli and consist of ill-defined centrilobular nodules and branching lines ("tree in bud" appearance (direct signs) (AQUINO et al. 1996). Often also bronchial wall thickening and areas of ground-glass opacity and lung consolidation are found corresponding with bronchitis and alveolar filling.

4.6.1.2
Diffuse Panbronchiolitis (Sino-Bronchial Syndrome)

In diffuse panbronchiolitis a lymphocytic infiltration of the bronchioles causing wall thickening is seen together with bronchiolectasis with secretions and foamy macrophages filling these airways and the immediately adjacent alveoli. Initially, it was thought to be a disease confined to Asian countries, but it has been reported worldwide (NISHIMURA et al. 1992; FITZGERALD et al. 1996). HRCT findings reflect the pathological changes: centrilobular nodules and branching lines ("tree in bud" pattern) usually asso-

ciated with mild cylindrical bronchiectasis. Mosaic perfusion and air trapping are only rarely seen.

4.6.1.3
Follicular Bronchiolitis

Follicular bronchiolitis is defined as lymphoid hyperplasia of the bronchus associated lymphoid tissue and is characterized by the presence of hyperplastic lymphoid follicles with reactive germinal centers distributed along the bronchioles and to a lesser extent along the bronchi. Most cases are associated with collagen vascular disease but are also seen during immunodeficiency or hypersensitivity reactions. In a recent report HOWLING et al. (1999) have shown that the cardinal HRCT finding consists of centrilobular nodules variably associated with peribronchial nodules and patchy areas of ground-glass opacity (direct signs). Less common CT findings are bronchial dilatation and bronchial wall thickening.

4.6.1.4
Cryptogenic Organizing Pneumonia

In cryptogenic organizing pneumonia young granulation tissue is found in the alveoli. This proliferation of granulation tissue extends into the lumina of the bronchioli and the alveolar ducts whose walls show a mild to moderate inflammatory infiltrate. The extent of disease both in the bronchiolar and in the alveolar lumen was responsible for the earlier mentioned confusion in terminology. In 1985 EPLER et al. described this pathological entity as bronchiolitis obliterans with organizing pneumonia (BOOP). A few years earlier, however, DAVISON et al. (1983) already described the same pathological entity and this author used the term cryptogenic organizing pneumonia (COP). Both terms are synonyms and have been used since then. Although

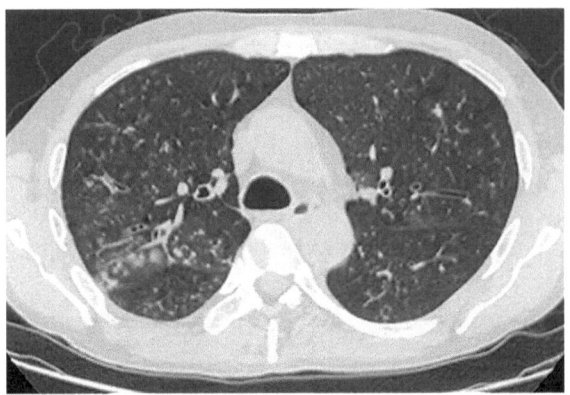
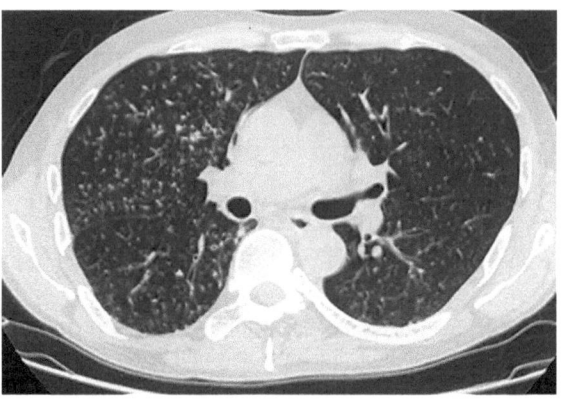

a   b

Fig. 4.5a, b. Immunocompromised patient with acute aspergillus infection predominantly affecting the right lung

the term BOOP is more popular, the term COP is probably better because organizing pneumonia is often the dominating finding. It was also decided to consider this disease not as small airways disease but as an idiopathic interstitial pneumonia (ATS/ERS INTERNATIONAL MULTIDISCIPLINARY CONSENSUS CLASSIFICATION OF IDIOPATHIC INTERSTITIAL PNEUMONIAS 2002).

The bronchiolar involvement can be recognized on HRCT as small ill-defined nodules and branching lines (Fig. 4.6b). The alveolar filling presents as predominantly bronchocentric and subpleural areas of lung consolidation or ground-glass opacity (MÜLLER et al. 1990; LEE et al. 1994; ALASALY et al. 1995) (Fig. 4.6a). However, BOOP/COP presenting as lines (MURPHY et al. 1999) or nodules has also been described. BOOP/COP can be idiopathic but can also be associated with several diseases (BOOP-reaction) (LOGAN et al. 1995; LOHR et al. 1997).

4.6.1.5
Respiratory Bronchiolitis (RB) and Respiratory Bronchiolitis Interstitial Lung Disease (RBILD)

RB is characterized by an abundance of macrophages within the lumina of the respiratory bronchioles often associated with profusion of macrophages in the alveolar lumen. Cigarette smoking causes it and the pathological changes in the respiratory bronchioles are sufficiently distinct to establish the diagnosis. However the majority of patients with RB are asymptomatic and RB is usually an incidental finding during pathological examination of lung resection specimens or during autopsy in heavy smokers (HARTMAN et al. 1997; MOON et al. 1999). HRCT can be normal but can also show centrilobular ill-defined nodules and ground-glass opacity. Today, RB-interstitial lung disease (RBILD) is considered to be a more severe reaction of the bronchioles and the lung parenchyma to cigarette smoke (HARTMAN et al. 1997). In RBILD besides ground-glass opacity and ill-defined nodules also signs of lung fibrosis can be seen (HEYNEMAN et al. 1999). Because of the overlap in pathological and radiological findings, desquamative interstitial pneumonia is also considered to be a (mild) reaction to cigarette smoke (HEYNEMAN et al. 1999).

4.6.2
Small Airways Involvement Characterized by Indirect HRCT Signs

4.6.2.1
Constrictive Bronchiolitis

In constrictive (obliterative) bronchiolitis the bronchiolar injury is probably older or chronic. Indeed, in this type of bronchiolitis submucosal and peribronchiolar fibrosis, accompanied by bronchiolar narrowing, bronchiolar distortion and mucostasis, are the most dominant findings.

Constrictive bronchiolitis can result from many causes of bronchiolar injury. The most common causes

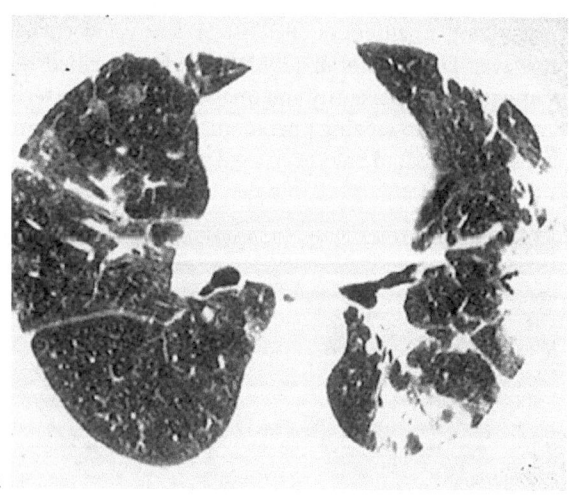

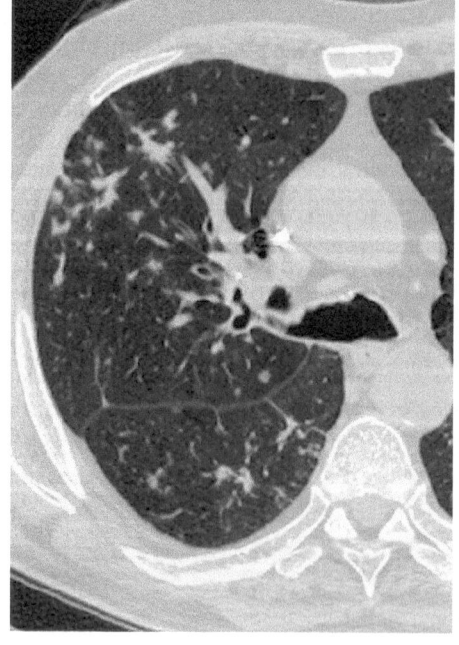

Fig. 4.6a, b. Cryptogenic organizing pneumonia (COP) typically presents as bronchocentric and subpleural areas of lung consolidation or ground-glass opacity (**a**). However the bronchiolar involvement can often be recognized on HRCT as centrilobular small ill-defined nodules and branching lines (**b**)

are listed in Table 4.2 (PADLEY et al. 1993; WORTHY et al. 1997a,c; PEREZ et al. 1998).

Direct signs of bronchiolitis (ill-defined nodules, branching lines) are rare. Typical features of constrictive bronchiolitis are indirect: 1) widespread or patchy areas of decreased lung attenuation and reduced perfusion (mosaic perfusion) (STERN et al. 1995a) and 2) widespread or patchy areas of air trapping on expiratory CT (STERN et al. 1995b; DESAI and HANSELL 1997; WORTHY et al. 1997b; ARAKAWA et al. 1998b; WORTHY and MULLER 1998). In some cases associated abnormalities of the macroscopic airways are found.

Table 4.2. Causes of constrictive bronchiolitis

-Idiopathic (rare)
-Healed infections (especially viral and mycoplasma)
-Component of chronic bronchitis, cystic fibrosis, bronchiectasis
-Inhalation of toxins or fumes
-Connective tissue diseases (rheumatoid arthritis, Sjögren)
-Transplant-associated airway injury (bone marrow, heart-lung, lung)
-Drug reaction (penicillamine)
-Other conditions (inflammatory bowel disease)

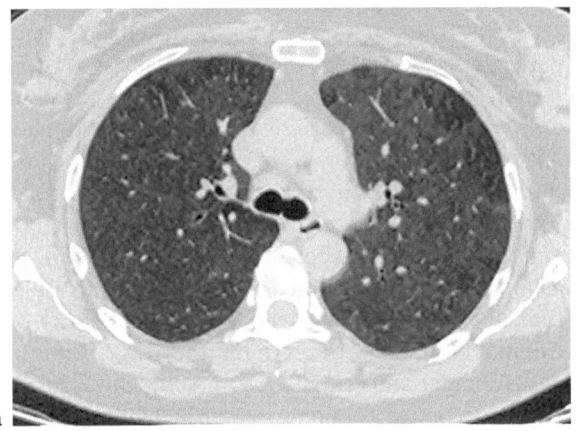

a

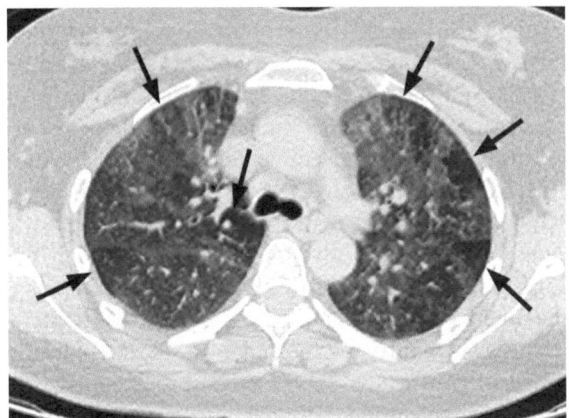

b

4.6.2.2
Hypersensitivity Pneumonitis

Hypersensitivity pneumonitis, also called extrinsic allergic alveolitis (EAA), is an allergic lung disease caused by the inhalation or organic dusts. In susceptible individuals the deposition of these organic dusts in the terminal and respiratory bronchioles causes an inflammatory granulomatous bronchiolitis of variable degrees. More than 50 different substances have been associated with EAA. Farmer's lung caused by the inhalation of thermophilic actinomycetes and bird fancier's lung resulting from the inhalation of avian protein in feces and feathers are the most common types (SELMAN-LAMA and PEREZ-PADILLA 1993). The clinical features are classically divided into three syndromes: acute, subacute and chronic. However, there is often significant overlap between the syndromes. Characteristic histopathologic features of these three syndromes include cellular bronchiolitis, diffuse lymphocytic interstitial infiltration and non-caseating granulomas (SOLEMAN and COLBY 1988). Interstitial fibrosis and end-stage lung disease may result from chronic exposure (MATAR et al. 2000). The HRCT findings reflect these pathological changes.

In the acute phase the HRCT features are characterized by the presence of bilateral air space consolidation and ill-defined opacities. In the subacute phase the HRCT findings include extensive ground-glass opacity, small ill-defined centrilobular nodules, areas of decreased attenuation, mosaic perfusion and air trapping (SILVER et al. 1989; HANSELL and MOSKOVIC 1991; REMY-JARDIN et al. 1993c; HANSELL et al. 1996; SMALL et al. 1996) (Fig. 4.7).

These signs of hypoperfusion and air trapping are frequent findings in patients with hypersensitivity pneumonitis. SMALL et al. (1996) scanned 20 patients with proven subacute extrinsic allergic alveolitis and demonstrated areas of decreased attenuation consistent with small airways disease in 15/20 patients which were confirmed to be areas of air trapping in 11/12 patients on expiratory scans. Additional areas of air trapping were identified in 5 patients on expiratory scans. The CT signs in the chronic phase of the disease include small nodules, irregular linear opacities, traction bronchiectasis, architectural distortion and honeycombing (ADLER et al. 1992; LYNCH et al. 1995).

Fig. 4.7a, b. Subacute hypersensitivity pneumonitis: inspiratory HRCT (a) and expiratory HRCT (b). On inspiratory HRCT (a) extensive ground-glass opacity can be seen involving both lungs together with a few areas of decreased lung attenuation. Expiratory HRCT (b) shows multiple areas of air trapping (*arrows*)

4.6.2.3
Sarcoidosis

Sarcoidosis is a systemic granulomatous disease of unknown etiology, characterized by multiorgan involvement and non-caseating epithelioid granulomas. In the lungs, which are affected in about 90% of the patients, the granulomas are typically located in a perilymphatic distribution. That is why HRCT signs can be very typical especially when nodular opacities are seen in the areas where lymphatics are found: nodular thickening of the interlobular septa and of the bronchovascular bundles and the presence of subpleural and perifissural nodules (Fig. 4.13a). Conglomeration of granulomatous lesions can induce larger mass-like opacities (HANSELL et al. 1998). By virtue of this perilymphatic distribution, sarcoid granulomas are also concentrated around the small airways and may be responsible for airflow obstruction presenting as mosaic perfusion and patchy air trapping on HRCT (Fig. 4.13b). Support for this theory comes from CARRINGTON (1976) who demonstrated bronchiolar and peribronchiolar granulomata histologically. In fact air trapping on expiratory CT is a frequent finding in patients with pulmonary sarcoidosis occurring in 95% of the patients in one study (DAVIES et al. 2000). Air trapping may involve any lung zone and may even be the only HRCT feature of pulmonary sarcoidosis (MAGKANAS et al. 2001).

4.7
Radiological-Functional Correlation in Small Airways Disease

4.7.1
Introduction

Many attempts have been made to quantify morphological changes seen on CT in patients with small airways disease. Despite the fact that many of these attempts were successful and that many reports have suggested that computed tomography can be helpful to predict lung function deterioration in SAD, these techniques never reached the daily clinical practice. The investigation of the structure-function relationship is indeed difficult.

1. The pathological changes in SAD are often very complex and result in a mixture of abnormally increased and decreased lung densities on CT. An automated objective quantification of these density changes can be elusive because high and low density areas can extinguish each other. Other automated density measurements such as density thresholding depend on a clear difference between normal and abnormal densities. Such a difference exists in emphysema where low attenuation areas can be easily recognized from normal lung regions in most cases but is less obvious in the heterogeneous lung changes of SAD. Also simple visual scoring systems have been applied to quantify SAD. Although interobserver variation can be a problem, good interobserver correlations are found in most studies. However, semi quantitative estimation of lung changes is laborious and time-consuming.

2. Pulmonary function tests measure global lung function while CT can show focal or multifocal abnormalities that are perhaps not widespread enough to cause pulmonary function decrease. Although SAD is accompanied by bronchiolar narrowing and in this way induces obstructive pulmonary function changes, some diseases of the small airways are also responsible for restrictive pulmonary function changes (KING 1993). In addition specific tests such as MEF 25% should be used to measure small airways function because more general indices of airway obstruction can be negative.

Airflow obstruction can be the result of several parenchymal and airway abnormalities. These include bronchitis and bronchiectasis, asthma, emphysema and constrictive (obliterative) bronchiolitis. It has been shown that HRCT is very helpful in the differential diagnosis of these various causes. The diagnosis of bronchiectasis is usually not difficult because of the typical HRCT features. Signs of asthma, emphysema and constrictive (obliterative) bronchiolitis are often overlapping. In a study by COPLEY et al. (2002) examining the HRCT features of panlobular and centrilobular emphysema, asthma and constrictive (obliterative) bronchiolitis, it was shown that HRCT had a very high accuracy in the diagnosis of constrictive (obliterative) bronchiolitis and centrilobular emphysema. In this study the most significant differences in HRCT findings between asthmatics and normal subjects were the presence of bronchial wall thickening and vascular attenuation (i.e. thinning of the vessels). Also decreased parenchymal attenuation as part of a mosaic pattern was found more frequently in asthmatics although this difference was not statistically significant. Comparing asthmatics and patients with constrictive (obliterative) bronchiolitis the most significant differences in favor of constrictive (obliterative) bronchiolitis were bronchial dilatation, vascular attenuation (i.e. thinning of the vessels) and decreased attenuation of the lung parenchyma as part of the mosaic pattern, which was seen in all patients with constrictive (obliterative) bronchiolitis. Finally, in this study the authors were also able in most cases to differentiate between emphysema and constrictive (obliterative) bronchiolitis. The absence of parenchymal destruction always indicated cases of obliterative bronchiolitis.

Overinflation, a sign of chronic obstructive airways disease can be determined and quantified using CT. ARAKAWA et al. (1998a) correlated pulmonary function tests with several CT parameters. These authors found significant correlations between FEV1/FVC and the tracheal index (transverse/anteroposterior diameter), the thoracic cage ratios (anteroposterior/transverse diameters) at the tracheal carina and 5 cm below and the presence of intercostal lung bulging. Correlations were significant but weak between FEV1/FVC and the sternoaortic distance and the depth of the azygoesophageal recess.

Although the contribution and roles of emphysema and small airways disease in causing expiratory airflow limitation in patients with COPD is somewhat controversial, it appears that emphysema is not primarily responsible for the severe expiratory airflow limitation in these patients but that airflow limitation is more related to narrowing of the small airways. This was suggested in a study of pulmonary function by GELB et al. (1996) who found that patients could have a loss of lung elastic recoil causing hyperinflation with increased TLC and decreased diffusing capacity and expiratory airflow – all physiologic hallmarks of emphysema – with no or trivial emphysema on CT but with severe small airways disease (GELB et al. 1998). Many studies examining the CT signs of constrictive (obliterative) bronchiolitis have confirmed the strong correlation between the presence and extent of these and the depression of pulmonary function tests of the small airways. HANSELL et al. (1997) correlated the individual features of small airways disease depicted at CT with functional indices in patients with constrictive (obliterative) bronchiolitis. The CT scans were scored semi-quantitatively by two observers for extent of decreased attenuation of the lung parenchyma, end-expiration CT signs of air trapping and bronchial dilation, wall thickening, and mucous plugging. Correlations of the extent of decreased attenuation and measures of airflow obstruction were strongest between decreased attenuation at end-expiration and airflow at low lung volumes. These authors concluded that decreased attenuation is the cardinal sign for further quantitative studies of constrictive (obliterative) bronchiolitis. However, not all studies could show significant correlations between the extent of CT abnormalities in constrictive (obliterative) bronchiolitis and functional impairment (PADLEY et al. 1993). This is very likely related to the fact that CT can depict lung parenchyma changes long before they are important enough to induce pulmonary function changes. Also LUCIDARME et al. (1998) concluded in a study where findings on expiratory CT were correlated with pulmonary function tests in patients with chronic airway disease, that air trapping might permit detection of airway obstruction even when these pulmonary function tests were normal.

Most studies on CT functional correlations have concentrated on constrictive (obliterative) bronchiolitis and on small airway involvement causing airway narrowing. It is generally accepted that this small airways narrowing is reflected in the "black lung" component of the mosaic pattern seen on HRCT. For this reason quantifying small airways disease corresponds in most studies with an objective measurement or a subjective estimation of these areas of low lung attenuation. But also bronchial lumen and wall changes (KING et al. 2000; NIIMI et al. 2000; BEIGELMAN-AUBRY et al. 2002) and cross-sectional area of the lungs (MITCHELL et al. 1996) have been used to quantify small airways disease.

4.7.2
Air Trapping

Expiratory air trapping is the key finding for depicting small airways obstruction (ARAKAWA et al. 1998b). In the normal lung during expiration lung attenuation increases as a result of the reduction in gas volume. Due to the influence of gravity, dependent lung areas have both on inspiratory and on expiratory scans a higher density than non-dependent lung regions. However, during expiration dependent areas show a greater increase in lung density than non-dependent areas. As a result the anteroposterior density gradient is significantly greater on expiratory scans than on inspiratory scans (VERSCHAKELEN et al. 1993). Often, especially on expiratory scans, the posterior aspect of the upper lobe anterior to the major fissure appears denser than the anterior aspect of the lower lobe immediately behind this major fissure (GRENIER et al. 2002). In about 50–80% of the normal individuals, with normal pulmonary function tests one or more areas of air trapping often limited to one or a few adjacent secondary pulmonary lobules can be seen (WEBB et al. 1993; VERSCHAKELEN et al.

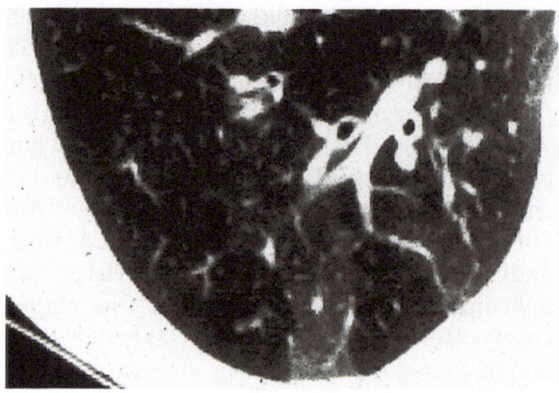

Fig. 4.8. Expiratory HRCT of a normal individual with normal pulmonary function tests. Visualization of a few areas of air trapping limited to one or a few secondary pulmonary lobules

1998) (Fig. 4.8). These areas of air trapping are predominantly seen in the dependent parts of the lower lobes (K.W. LEE et al. 2000). The frequency increases with age while severity increases with age and smoking (K.W. LEE et al. 2000; VERSCHAKELEN et al. 1998). Air trapping is usually considered abnormal when it affects a volume of lung equal to or greater than a pulmonary segment (GRENIER et al. 2002).

Air trapping can be found in patients with normal findings on inspiratory scans (ARAKAWA and WEBB 1998) and should alert the observer for the presence of small airways disease. Pathological air trapping is also more frequently found in the inner segment of the lung than in the outer segment, and the contribution of the inner segment to the pulmonary function deterioration may be greater than the outer segment (NAKANO et al. 1999). In some patients it may difficult to determine air trapping because they are unable to suspend respiration during the time required to obtain a scan, because they are coughing, especially at end exhalation, or because they have an inadequate rate of exhalation. In these patients lateral decubitus CT could be a useful adjunct to depict air trapping (FRANQUET et al. 2000). However, determining the state of inflation of the lungs on inspiratory and expiratory CT is largely subjective. One can look at the invagination of the posterior membrane of the trachea on the expiratory CT and at the normal decrease in cross-sectional area of the lungs that should be approximately 55% at end-expiration but these signs are, in practice, not so easy to gauge (MITCHELL et al. 1996). Performing complementary continuous-expiration CT can be another useful adjunct to depict air trapping when suspended-end-expiration CT images are ambiguous. LUCIDARME et al. (2000) showed that the extent of and relative attenuation decrease in air trapping areas increased significantly in scans obtained with continuous-expiration CT compared with those obtained with suspended-end-expiration CT.

The extent of air trapping can be quantified using objective measurements or (subjective) semi-quantitative scoring systems that estimate the percentage of air trapping on each scan. Objective density measurement of the lung can be expressed in several ways. It can be expressed as a mean density of the voxels included in a chosen region of interest (ROI) (GODDARD et al. 1982), as a histogram that shows the distribution of attenuation values within the ROI (WEGENER et al. 1978), as a density mask that highlights (MÜLLER et al. 1988; ADAMS et al. 1991) or as a calculation that summarizes the pixels with a density above or below a certain critical value (Fig. 4.9). This latter method has been used successfully to quantify air trapping in patients with asthma. NEWMAN et al. (1994) calculated the percentage of pixels below –900 Hounsfield units in a group of asthmatics and compared this pixel index with the pixel index found in a normal control group. The mean pixel index was significantly higher in asthmatic subjects compared with normal individuals indicating more areas of low attenuation in asthmatics. In addition, the percentage of abnormal lung in asthmatics as determined by CT had a significant correlation with the PFTs that reflect air trapping.

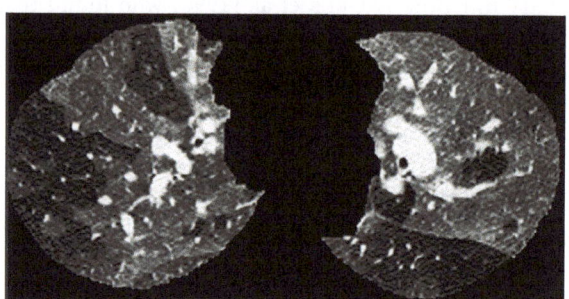

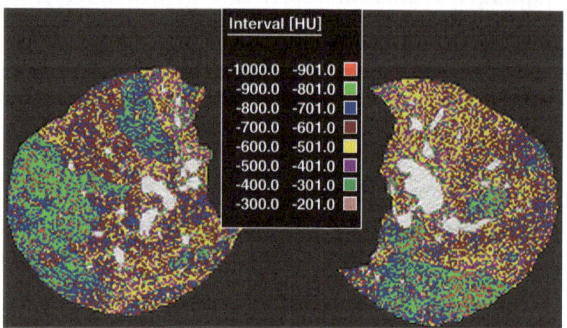

Fig. 4.9a, b. One way to quantify air trapping on expiratory CT (**a**) is to use a density mask that highlights pixels with a certain predefined density range (**b**). This density mask can help to detect and quantify low attenuation lung areas

Expressing lung density on a histogram has the advantage that changes in the distribution of attenuation values are detectable when mean attenuation is unchanged. Mean lung density, however, is easier to obtain and compare. Using this method, density changes between full inspiration and full expiration can be compared and expiratory/inspiratory ratios can be calculated (KUBO et al. 1999). The density mask has the advantage that it combines density measurement with the visual assessment of pathology so that is possible to allow for regional abnormalities such as hyperinflation and fibrosis that influence mean lung attenuation. Variable degrees of inspiration can be a problem in quantitative CT scanning and spirometrically controlled scanning (BEINERT et al. 1995; KOHZ et al. 1995) can solve this. KOHZ et al. (1995) obtained highly reproducible lung density measurements using spirometrically controlled CT.

Several semi-quantitative scoring systems have been used to estimate the percentage of abnormal lung on CT scans. However, very often the scoring system is based on the scoring system proposed by STERN et al. (1994). In this scoring system air trapping is estimated at each level and for each lung on a 5-point scale: 0=no air trapping, 1=1–25%, 2=26–50%, 3=51–75% and 4=76–100% of cross-sectional areas of the lung affected. The air trapping score is the summation of these numbers for the different levels studied.

The visual assessment of air trapping is based on the comparison of paired inspiratory and expiratory scans and on the detection of focal areas of decreased attenuation on the expiratory scan, i.e. areas that do not increase in density as compared with the surrounding normal lung regions. In patients with diffuse air trapping with only limited increase in lung density, depicting focal air trapping can be a problem. KAUCZOR et al. (2000) found in a mixed group of patients with and without pulmonary disease that the visual assessment of focal – not diffuse – air trapping at expiratory high-resolution CT did not correlate with physiological evidence of obstruction as derived from pulmonary function tests and concluded that the perception of focal air trapping requires an adequate expiratory increase in lung density which is decreased or absent in patients with diffuse air trapping. DUPONT et al. (1999) showed in a group of asthmatic patients that more focal air trapping was found after bronchodilatation than before which is probably also related to the fact that bronchodilatation induces an expiratory density decrease of the normal lung.

Despite these potential difficulties in recognizing focal areas of air trapping NG et al. (1999) have shown

that visual quantification is reliable and allows a good interobserver and intraobserver agreement although observer variation increased when a finer scoring system was used (nearest 5% instead of a coarser 5-point scale). Using a grid superimposed on the CT images can also reduce interobserver variation (K.W. LEE et al. 2000).

In general, focal air trapping shows good correlation with static lung volumes and dynamic lung parameters of airway obstruction (KAUCZOR et al. 2000). On the other hand, results of pulmonary function tests in patients with air trapping and normal findings on inspiratory scans were intermediate, falling between those of patients with normal findings on inspiratory and expiratory HRCT scans and those of patients with air trapping and abnormal findings on inspiratory scans (ARAKAWA and WEBB 1998). The development of spiral CT and multidetector-row CT has opened new perspectives in the study of focal air trapping that are not fully explored yet. Using multidetector-row CT with 1 mm slice thickness over the lungs performed at full expiration, an exhaustive assessment of the volume of air trapping and even a 3D visualization of its distribution can be obtained (KALENDER et al. 1990). In the same way precise quantitative data on lung cross section surfaces and lung volumes at deep inspiration and expiration can be calculated (KAUCZOR et al. 1998). Several studies have shown that spiral CT minimum intensity projection images (MinIPs) improve the detection of air trapping and are associated with an increased observer confidence and agreement as compared with HRCT alone (BHALLA et al. 1996; REMY-JARDIN et al. 1996; FOTHERINGHAM et al. 1999; WITTRAM et al. 2002).

4.7.3
Bronchial Lumen and Wall Changes

Several techniques that measure bronchial lumen and bronchial wall changes have been used with success to study the magnitude and distribution of airway narrowing. However, these studies were usually performed in a research environment. Measurement of bronchial lumen and bronchial wall areas is indeed very critical and depends on the lung volume and angle between the airway central axis and the plane of section. Volumetric acquisition at controlled lung volume is required in order to precisely match the airways of an individual on repeated studies (GRENIER et al. 2002). Some methods are based on visual analysis of photographed images, others on automated computerized algorithms to measure airway lumen area, airway wall area, and in some studies even airway angle of orientation (WEBB et al. 1984; WOOD et al. 1995; PRETEUX et al. 1999; KING et al. 2000). All these analysis algorithms have been validated using data from phantom studies and excised animal lungs and have been used to quantify airway narrowing in animal lungs in vivo as well in normal and asthmatic subjects. In this way NIIMI et al. (2000) found that airway wall thickening occurs in patients with asthma and is not limited to those with severe disease. In addition airway wall thickening may relate to the duration and severity of disease and the degree of airflow obstruction. BEIGELMAN-AUBRY et al. (2002) found that in patients with asthma, bronchial cross-sectional areas were significantly smaller than in healthy volunteers except after the inhalation of a bronchodilator. It can be expected that these tests, with continued refinement, will one day become of benefit in the clinical practice of radiology (KING et al. 1999).

4.7.4
Quantitative CT Assessment of Lung Parenchymal Changes in some Small Airways Diseases

This paragraph will review CT-functional correlations in different diseases and conditions in which the small airways are involved.

4.7.4.1
Lung Parenchymal Changes Secondary to Cigarette Smoking

The relationship between cigarette smoking and obstructive airways disease is well known but the effects of cigarette smoking on the development of interstitial lung diseases are poorly understood. Nonetheless, available epidemiological data suggest that cigarette smoking is causally related to the development of certain interstitial lung diseases, including respiratory bronchiolitis interstitial lung disease (RBILD), desquamative interstitial pneumonia (DIP) and pulmonary Langerhans' cell histiocytosis (PLCH) (RYU et al. 2001). The significant overlap between the CT findings of respiratory bronchiolitis, RBILD and DIP suggests the concept that they represent different degrees of severity of small airways and parenchymal reaction to cigarette smoke (HEYNEMAN et al. 1999).

In a prospective study of 175 healthy adult volunteers (current smokers, ex-smokers and non-smokers) with overall normal function tests REMY-JARDIN et al. (1993a) detected lung parenchymal changes in current smokers and ex-smokers that were not or to a lesser degree present in non-smokers. About 20% of the smokers showed multiple areas of ground-glass attenuation, which was significantly more than in ex-smokers (4%) and non-smokers (0%). About one-quarter of the current smokers had small micronodules again significantly more than in ex-smokers (4%) and non-smokers (0%). Emphysema was another feature significantly more seen in smokers than the other two groups. Differences between the 3 groups for other findings such as the presence of bronchial wall thickening (smokers 33%, ex-smokers 16%, non-smokers 18%), dependent areas of attenuation (smokers 34%, ex-smokers 43%, non-smokers 12%), septal lines (smokers 6%, ex-smokers 8%, non-smokers 8%) and subpleural micronodules (smokers 38%, ex-smokers 39%, non-smokers 22%) were not significant. The areas of ground-glass attenuation were corresponding either with an accumulation of pigmented macrophages and mucus in the alveolar spaces, associated with mild interstitial inflammation and/or fibrosis, a thickening of inflammatory walls with inflammatory cells with normal alveolar spaces or an organizing alveolitis. The parenchymal micronodules correspond to bronchiolectasis with peribronchiolar fibrosis (REMY-JARDIN et al. 1993b). These abnormalities are predominantly seen in the upper lung zones. In this study the presence of emphysema and the abnormal bronchial wall thickening were the only HRCT signs with significantly lower values of functional parameters.

Since the introduction of expiratory HRCT another CT sign of lung and more specific of small airway involvement in cigarette smokers became evident: air trapping. VERSCHAKELEN et al. (1998) examined 30 healthy subjects (11 non-smokers, 7 ex-smokers for >2 years, 12 current smokers; age range 35–55 years) with a smoking history between 0 and 28.5 pack-years) and normal pulmonary function tests and correlated HRCT findings at suspended deep inspiration and deep expiration with pulmonary function. In 24 subjects (7 non-smokers, 7 ex-smokers, 10 current smokers) areas of focal air trapping were found. Scores of focal air trapping were not significantly different between smokers and ex-smokers, but were significantly lower ($p<0.05$) in non-smokers and showed a significant ($p<0.0005$) correlation with pack-years. The degree of air trapping was also associated with several lung function tests, especially RV, DLCO, FRC, FEV1 and FEV1/VC.

Others confirmed these findings. Comparing smokers and non-smokers K.W. LEE et al. (2000) showed that the frequency of air trapping tended to be higher in smokers although there was no statistical significance. Although pulmonary function tests were within normal ranges in all subjects, regardless of the presence of air trapping, a statistically significant difference was found in the mean values of FVC and FEV1/FVC between the air trapping group and the non-air trapping group, while the grade of air trapping had a significant correlation with FEV1/FVC. This grade of air trapping was higher in heavy smokers. Also MASTORA et al. (2001) found that among smokers and ex-smokers air trapping more often involved a lung segment and lobe than in non-smokers where air trapping was limited to one or a few lobules in most cases. In addition in subjects who had air trapping, significantly more micronodules, dependent densities and heterogeneous lung attenuation were found. Among smokers with air trapping a significantly higher amount of micronodules were seen compared to the subjects without air trapping. They did not find any significant difference between the mean values of functional parameters in subjects who had air trapping, as compared with subjects who did not have air trapping. Also no relationship was found between CT findings at expiration and functional indexes of small airways disease.

From these studies it is obvious that air trapping is related to smoking and intensity of smoking and can be observed before pulmonary function deteriorates. However the pattern and not simply the presence of air trapping seems important.

Some investigators studied overall lung density changes in smokers during inspiration and expiration, compared these with density changes in non-smokers and correlated them with pulmonary function tests and measurements of cross-sectional area changes. PELINKOVIC et al. (1997) found that healthy asymptomatic male smokers had a significantly higher lung density at all inspiratory states compared to healthy asymptomatic non-smokers and although they had a higher vital capacity they showed a smaller cross-sectional area increase of the lung during inspiration. According to the authors this could be related to the decreasing compliance of the smoker's lung as a result of small airways disease and hypoxic vasoconstriction. Comparing HRCT density at full expiration/full inspiration (E/I ratio) of smokers with normal lung function, smokers with functional characteristics of COPD and smokers with functional characteristics of emphysema, KUBO et al. (1999) saw a significant elevation of the E/I ratio in the COPD and the emphysema group indicating that this index reflects hyperinflation and airway obstruction regardless of the functional characteristics of emphysema.

A few studies followed a group of healthy smokers, ex-smokers and non-smokers during several years and examined sequential changes in structural and functional abnormalities of the lung (SOEJIMA et al. 2000; REMY-JARDIN et al. 2002). Significant differences in CT findings between the initial CT and a CT performed after a minimum follow-up period of 4 years, were seen in only the group of persistent current smokers, who showed a significantly higher frequency of emphysema and ground-glass attenuation. In the persistent current smokers with micronodules at the initial CT no changes were seen, in another one-third micronodules showed a higher profusion, while in the remaining one-third micronodules were replaced with emphysema. Subjects with emphysema and/or areas of ground-glass attenuation at the initial CT had a significantly more rapid decline in lung function than did those with a normal CT scan (REMY-JARDIN et al. 2002). SOEJIMA et al. (2000) performed an annual inspiratory and expiratory HRCT for 5 years in a group of current smokers, ex-smokers and non-smokers and calculated the mean lung density, the CT value with the most frequent appearance and the relative area of low attenuation with CT values less than –912 HU. Except for FEV1, pulmonary function tests did not change annually. In non-smokers, only the percentage of the relative area of low attenuation with CT values less than –912 HU in the middle or lower lung fields exhibited an annual increase. In current smokers, the percentage of the relative area of low attenuation with CT values less than –912 HU in the upper lung field was augmented, while inspiratory mean lung density and the CT value with the most frequent appearance in the middle and lower lung field became more positive, confirming the results of the study performed by PELINKOVIC et al. (1997). In past smokers, the percentage of the relative area of low attenuation with CT values less than –912 HU in any lung field examined increased. The annual change in the percentage of the relative area of low attenuation with CT values less than –912 HU in the upper lung field was larger for past smokers than non-smokers, with little difference between past and current smokers. Expiratory CT parameters showed few annual changes in all groups.

4.7.4.2
Lung and Heart-Lung Transplantation

Many studies have investigated the CT changes in patients with constrictive (obliterative) bronchiolitis after lung transplantation and have correlated these findings with pulmonary function deterioration

(Leung et al. 1998; E.S. Lee et al. 2000; Bankier et al. 2001; Siegel et al. 2001). These studies provided evidence that air trapping on expiratory CT is an accurate indicator of constrictive (obliterative) bronchiolitis (Fig. 4.10). Constrictive (obliterative) bronchiolitis is indeed a major factor that limits the survival of lung transplant recipients and is found in up to 70% of patients who survive transplantation for 5 years (Arcasoy and Kotloff 1999).

The diagnosis of constrictive (obliterative) bronchiolitis in these patients is based on histological findings, but because of the patchy distribution of the small airways involvement, bronchial biopsy cannot provide material for histologic proof in a substantial number of patients (Arcasoy and Kotloff 1999). That is why the International Society for Heart and Lung Transplantation has proposed to use a spirometric definition for the clinical diagnosis of

chronic allograft dysfunction (Cooper et al. 1993). "Bronchiolitis obliterans syndrome" (BOS) is used for lung transplant patients with an otherwise unexplained and sustained decline of the forced expiratory volume in 1 s (FEV1) to a level of 80% or less of the best postoperative value. Leung et al. (1998) obtained inspiratory and expiratory HRCT images at 5 different levels and spirometry in 21 lung transplant recipients. On inspiratory images, bronchiectasis and mosaic patterns of lung attenuation were present in 4 (36%) and 7 (64%) out of 11 patients with pathologic and functional evidence of constrictive (obliterative) bronchiolitis, and 2 (20%) and 1 (10%) out of 10 patients without constrictive (obliterative) bronchiolitis ($p>0.05$ and $p<0.05$, respectively). The sensitivity, specificity, and accuracy of bronchiectasis and mosaic pattern for constrictive (obliterative) bronchiolitis were 36%, 80%, and 57%, and 64%, 90%,

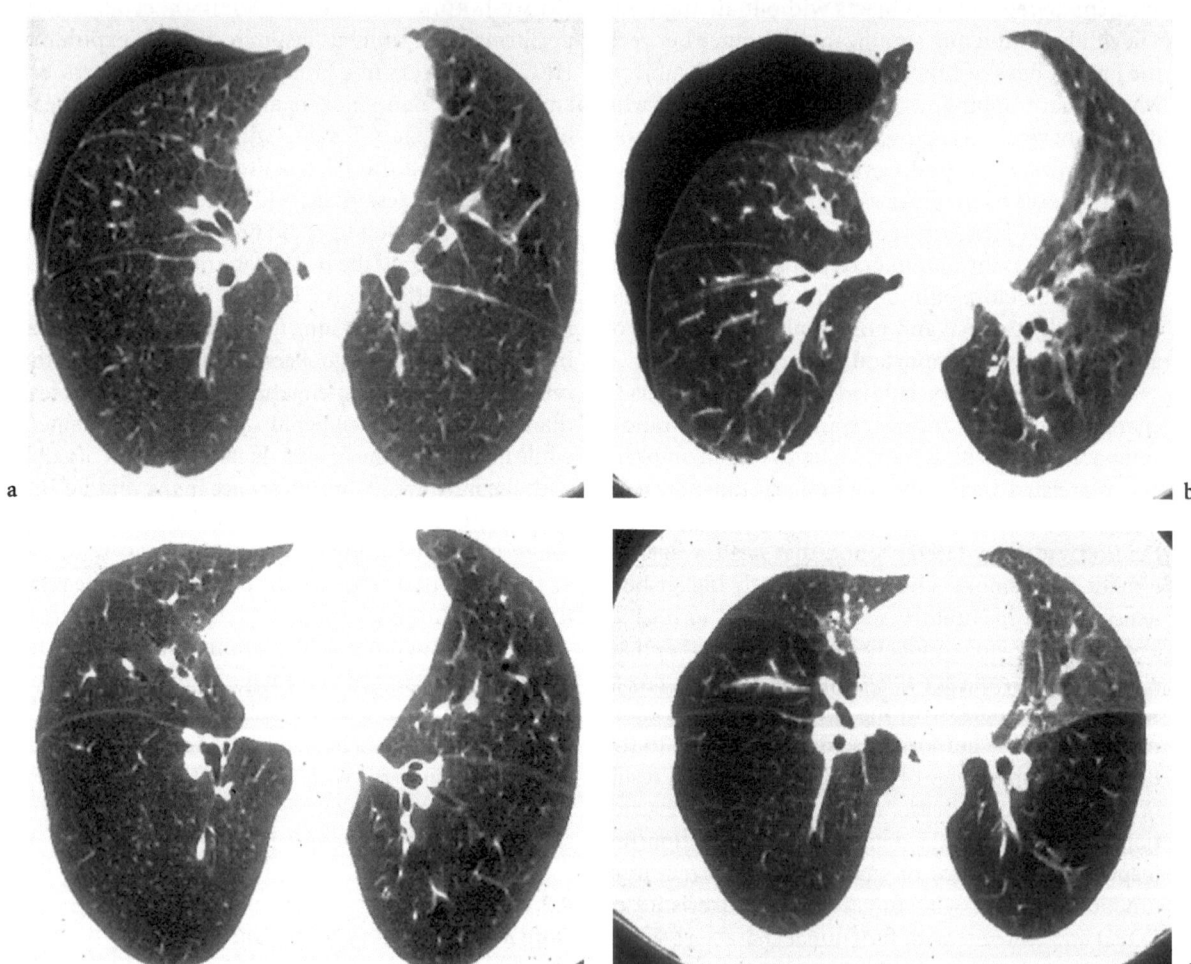

Fig. 4.10a–d. Patient with bilateral lung transplantation. A first CT scan shows a post-biopsy pneumothorax visible on both the inspiratory (**a**) and expiratory (**b**) images. Lung attenuation is normal. A second CT scan performed 1 month later shows decreased lung attenuation of the right lower lobe on inspiratory scan (**c**) corresponding with a large area with air trapping on the expiratory scan (**d**) caused by constrictive (obliterative) bronchiolitis (BOS)

and 70%, respectively. On expiratory images, air trapping was found in 10 out of 11 (91%) patients with constrictive (obliterative) bronchiolitis compared to 2 out of 10 (20%) patients without constrictive (obliterative) bronchiolitis ($p<0.002$). Air trapping was found to have a sensitivity of 91%, a specificity of 80%, and an accuracy of 86% for constrictive (obliterative) bronchiolitis. BANKIER et al. (2001) examined a larger group of patients and performed a visual quantification of the air trapping. They found that the extent of air trapping increased with BOS severity ($p=0.001$). In addition a threshold of 32% of air trapping turned out to be optimal for distinguishing between patients with and those without BOS and provided a sensitivity of 83%, a specificity of 89%, and an accuracy of 88%. The prevalence of BOS and positive predictive value of air trapping increased with postoperative time, but the negative predictive value of air trapping remained high throughout the study. Patients without BOS who had air trapping exceeding 32% of the parenchyma were at significantly increased risk of developing BOS ($p=0.004$). The usefulness of expiratory HRCT in detecting airway obstruction and air trapping was also assessed in pediatric lung transplant recipients (SIEGEL et al. 2001). In a population of 21 pediatric lung transplant recipients, the sensitivity of expiratory CT for enabling the diagnosis of BOS was 100%; the specificity 71%; the positive predictive value 64%; and the negative predictive value 100%. Expiratory CT scores correlated strongly (rho=0.75, $p<0.01$) with pulmonary function test-based scores. Dilatation of the lower lobe bronchi is another sign of constrictive (obliterative) bronchiolitis in lung transplantation patients. It has been shown that the percentage of dilated bronchi increases with increasing pulmonary dysfunction (LENTZ et al. 1992).

4.7.4.3
Bronchiectasis

The association between the presence of bronchiectasis and obstructive pulmonary function deterioration is well known, but before the advent of CT this paradoxical combination of bronchial dilatation and an obstructive functional defect was not well understood.

HRCT has shown that areas of decreased attenuation visible as mosaic attenuation and air trapping on expiratory HRCT and reflecting constrictive (obliterative) bronchiolitis are very common in patients with severe bronchiectasis and can even precede the development of bronchiectasis (HANSELL et al. 1994)

(Fig. 4.11). IM et al. (1996) examined 48 consecutive patients with lobular low attenuation areas of the lung on HRCT and found that the majority (85%) had bronchiectasis. Small airways narrowing causing these areas of low attenuation is to a large extent very likely responsible for the obstructive pulmonary function changes. This was suggested in a study performed by ROBERTS et al. (2000). These authors examined the inspiratory and expiratory features on the CT scan of 100 patients with bronchiectasis undergoing concurrent lung function tests and found that bronchial wall thickness and decreased attenuation were consistently the strongest independent determinants of airflow obstruction. Endobronchial secretions seen on CT scanning had no functional significance; the severity of bronchial dilatation was negatively associated with airflow obstruction after adjustment for other morphological features.

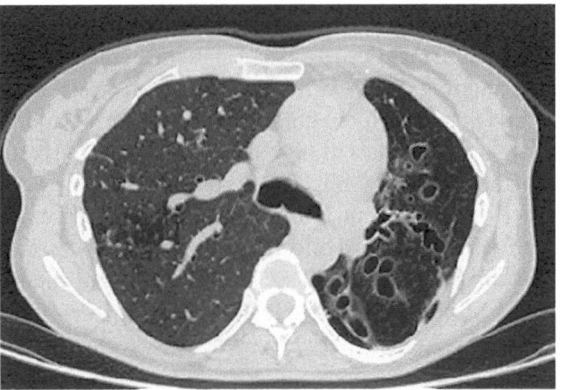

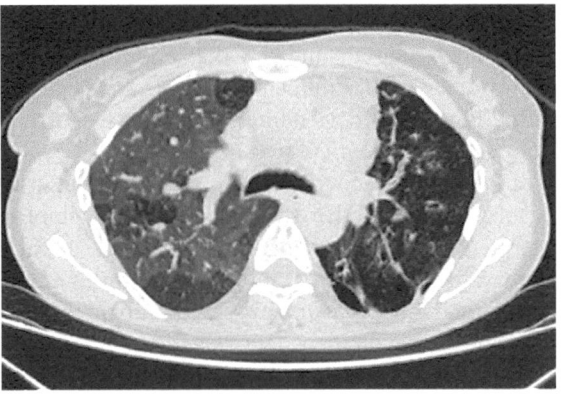

Fig. 4.11a, b. Patient with bronchiectasis in the left lung. On inspiratory CT (**a**) a large area of decreased attenuation is seen on the left side. Expiratory CT (**b**) shows areas of air trapping not only in the left lung but also in the right lung suggesting that air trapping can precede the development of bronchiectasis

4.7.4.4
Rheumatoid Arthritis and Sjögren Syndrome

Pulmonary involvement is a frequent extraarticular manifestation in rheumatoid arthritis (RA). Interstitial lung disease and subclinical alveolitis have been found in up to 40% of RA patients (FRANK et al. 1973; PEREZ et al. 1989). RA is also strongly associated with airway disease. Bronchiectasis can be recognized in up to 35% of patients with RA (MCDONAGH et al. 1994; REMY-JARDIN et al. 1994) while constrictive (obliterative) bronchiolitis is also present in many patients and can even be the dominant presenting feature. In a CT study of 84 patients, 30% showed features of bronchiectasis/bronchiolectasis presumed to reflect small airways involvement (REMY-JARDIN et al. 1994). PEREZ et al. (1989) prospectively evaluated, with high-resolution computed tomography (HRCT) and pulmonary function tests (PFTs), 50 patients with RA including 39 non-smokers and 11 smokers without radiographic evidence of RA-related lung changes. PFTs demonstrated airway obstruction in 18% and small airways disease in 8% of patients. However, HRCT demonstrated bronchial and/or lung abnormalities in 70% of patients, consisting of air trapping (32%), cylindrical bronchiectasis 30%), mild heterogeneity in lung attenuation (20%), and/or centrilobular areas of high attenuation (6%). Airway obstruction and SAD were correlated with the presence of bronchiectasis and bronchial wall thickening and with bronchial infection but were unrelated to rheumatologic data. FEF (25–75) was reduced and the slope of phase III by single breath nitrogen washout was increased in patients with airway changes on HRCT scans, whereas no PFT abnormalities were found in the majority of patients with normal HRCT scans. Because of the high number of non-smokers in this study group it is unlikely that smoking-induced airways disease has influenced the results. Penicillamine, however, has been incriminated as a contributory causative agent (WOLFE et al. 1983).

In the Sjögren syndrome, unlike rheumatoid arthritis, constrictive (obliterative) bronchiolitis is rarely the dominant presenting feature. In most cases signs of interstitial lung fibrosis and lymphocytic interstitial pneumonia predominate but air trapping on expiratory HRCT has been found in 32% of patients in one study (FRANQUET et al. 1999) (Fig. 4.12). However, in this study no significant correlation was found between the extent of air trapping and pulmonary function tests suggesting again that HRCT is more sensitive than lung function tests to measure early subclinical bronchiolar changes. In

a recent study TAOULI et al. (2002) found signs of large and/or small airways disease in 54% of a group 35 patients with primary Sjögren syndrome. These authors also found a significant correlation between the extent of air trapping and FEV1.

4.7.4.5
Hypersensitivity Pneumonitis

Lung areas with decreased attenuation and air trapping are together with ground-glass opacification the most common CT patterns in subacute hypersensitivity pneumonitis followed by the presence of a micronodular pattern and reticulation (SMALL et al. 1996) (Fig. 4.7). It has been shown that the extent of these low attenuation areas correlates significantly with the severity of pulmonary obstruction measured by the increase of the residual volume whereas ground-glass opacification and reticulation

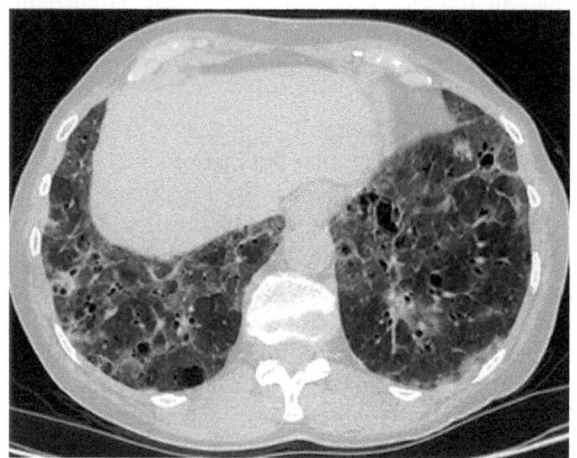

a

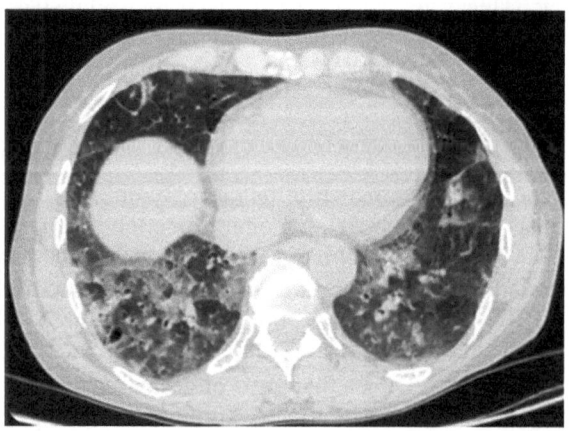

b

Fig. 4.12a, b. Patient with Sjögren syndrome. Both inspiratory (a) and expiratory CT (b) show ground-glass opacities and areas of decreased lung attenuation which suggest small airways disease

correlate with restrictive lung function changes (HANSELL et al. 1996). In a study by REMY-JARDIN et al. (1993c) all 21 patients with subacute and all 24 patients with chronic bird breeder hypersensitivity pneumonitis had an abnormal HRCT scan. In this study a micronodular pattern observed as an isolated finding or concurrently with ground-glass attenuation was the most frequently occurring CT abnormality in the subgroup with subacute disease. In this group emphysematous changes were present in 3 patients and air trapping in 1 patient. The main CT pattern observed in the group with chronic hypersensitivity pneumonitis was honeycombing (50%) variably associated with ground-glass attenuation, parenchymal micronodules and/or emphysema. In the absence of honeycombing the most frequent finding was emphysema diversely associated with ground-glass attenuation and micronodules. These authors concluded that emphysema must be considered as an integrated part of bird breeder lung and proposed the theory that this emphysema could be the result of constrictive (obliterative) bronchiolitis. On follow-up studies after cessation of exposure, CT showed a return to normal or dramatic improvement in the group with subacute disease. In the group with chronic hypersensitivity pneumonitis a considerable reduction in ground-glass attenuation and micronodules was seen. No fibrotic or emphysematous changes were depicted during follow-up.

4.7.4.6
Sarcoidosis

Air trapping on expiratory CT is a frequent finding in patients with pulmonary sarcoidosis occurring in 95% of the patients in 1 study (DAVIES et al. 2000) (Fig. 4.13). Air trapping may involve any lung zone, can occur at the level of the secondary pulmonary lobule as well as in distributions suggesting sublobular, subsegmental, and segmental involvements, is not specific for a given stage of disease and may even be the only HRCT feature of pulmonary sarcoidosis (BARTZ and STERN 2000; MAGKANAS et al. 2001). Significant correlations between air trapping extent and pulmonary function tests indicating airway obstructions were found (MAGKANAS et al. 2001). DAVIES et al. (2000) examined 22 patients with pulmonary sarcoidosis and compared inspiratory and expiratory HRCT features with pulmonary function tests. Air trapping was found in 95% of the patients and the extent correlated significantly with residual volume (RV)/total lung capacity (TLC) and maximal mid-expiratory flow rate between 25 and 75% of the vital capacity. Other signs such as nodules,

septal thickening, traction bronchiectasis, lung distortion and ground-glass opacification were present in 86%, 86%, 67%, 57% and 19% of the patients, respectively but did not correlate with pulmonary function. However, the results of studies correlating radiographic features of sarcoidosis with clinical and physiological parameters of abnormal lung function have been inconsistent. MÜLLER et al. (1989) demonstrated that nodular opacities on CT correlated with less severe dyspnea and larger lung volumes. Another study showed that the presence and extent of a reticular pattern was the main determinant of functional impairment particularly airflow obstruction (HANSELL et al. 1998) and not the CT features of small airways disease but in this study the proportion of subjects having the fibrotic stage of the disease was larger.

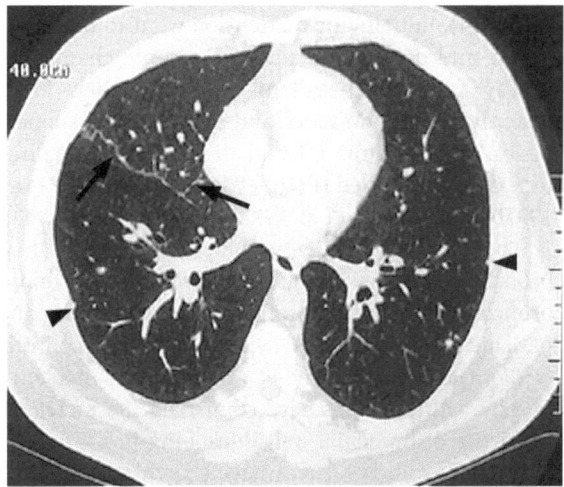
a

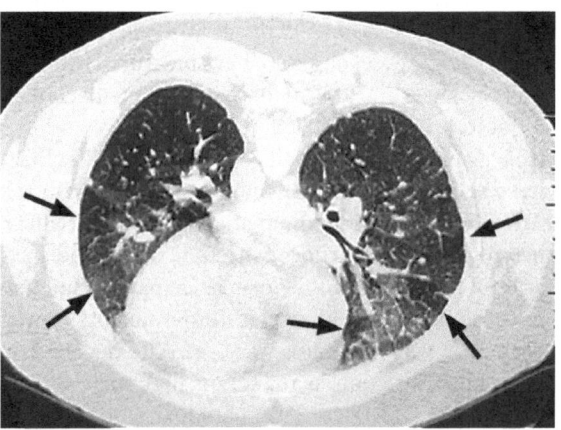
b

Fig. 4.13a, b. Patient with pulmonary sarcoidosis. Inspiratory scan (**a**) shows nodular thickening of the interlobular septa (*arrows*) and subpleural nodules (*arrowheads*). Expiratory scan (in prone body position) (**b**) shows multiple areas of air trapping (*arrows*)

4.7.4.7
Asthma

Asthma is characterized by the reversible obstruction of airways due to bronchial hyperresponsiveness associated with airway inflammation involving both proximal and distal airways. Despite the fact that many studies have focused on the CT visualization and quantification of airway and lung parenchymal changes in patients with acute and chronic asthma, high resolution CT is still rarely used in everyday practice in these patients. This is probably related to the fact that the HRCT changes are variable and often very subtle. HRCT changes include airway wall thickening, bronchiectasis, centrilobular opacities, mucoid impactions, emphysema, mosaic perfusion and air trapping (LYNCH et al. 1993; GEVENOIS et al. 1996; GRENIER et al. 1996; PAGANIN et al. 1996; PARK et al. 1997; CARR et al. 1998; LUCIDARME et al. 1998) (Fig. 4.14). Initially most attention went to the study of site and degree of acute and chronic changes in airway wall thickness and airway caliber. More recently also the presence and extent of mosaic perfusion on inspiratory CT and air trapping on expiratory CT were studied (PARK et al. 1997; WORTHY et al. 1997b; LUCIDARME et al. 1998; LAURENT et al. 2000).

The majority of patients with chronic asthma have bronchial wall thickening. LYNCH et al. (1993) found bronchial wall thickening in 42 out of 44 asthmatic patients (92%) compared with 5 out of 27 controls (19%). GRENIER et al. (1996) reported an incidence of 82%. Other studies have confirmed the increased incidence of bronchial wall thickening in asthmatics (PAGANIN et al. 1992) and have shown that patients with severe disease have thicker airways than patients with milder disease (AWADH et al. 1998; NIIMI et al. 2000). OKAZAWA et al. (1996) have measured bronchial wall thickness and looked for the site of methacholine-induced bronchoconstriction in asthmatics and controls using HRCT. The walls of the asthmatic subjects were significantly thicker than those of the controls. After administration of methacholine a significant heterogeneous bronchoconstriction was found in all sizes of airways with the greatest change being found in those 2–4 mm in diameter. In the normal airways bronchial wall thickness was seen to decrease on bronchoconstriction, a reduction that was not seen in asthmatic airways. This paradoxical decrease in bronchial wall thickness during bronchoconstriction was also seen in animal experiments (MCNAMARA et al. 1992) and may be due to a reduction in bronchial wall blood volume and may be a contributing factor to hyperresponsiveness in asthma.

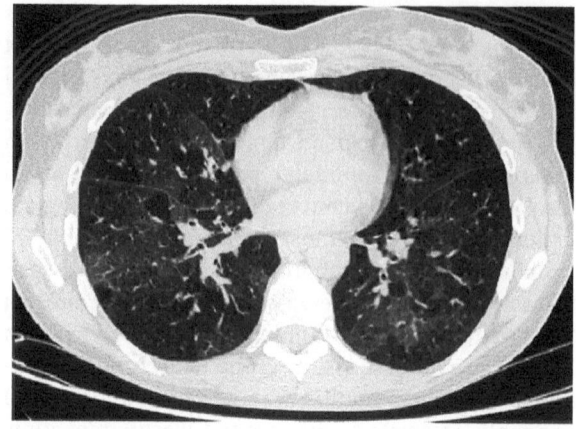

Fig. 4.14. Patient with asthma. Expiratory CT scan shows multiple areas of air trapping

The airway involvement in asthmatic patients is also reflected in a higher incidence of indirect HRCT signs of airway obstruction such as decreased lung density, mosaic perfusion and air trapping. There is uncertainty over the nature of similar hyperlucent areas seen on CT scans of asthmatic patients and whether these represent either emphysema with alveolar wall destruction or simply non-destructive hyperinflation. Estimations of prevalence of emphysema in asthmatic lungs using CT or HRCT scanning range from 0–80% (KINSELLA et al. 1988; PAGANIN et al. 1992, 1996; LYNCH et al. 1993; ANGUS et al. 1994). The high incidence of emphysema in some studies can be explained by the fact that an important part of the studied asthmatics were smokers (LYNCH et al. 1993). However also the type of asthma seems to be important. PAGANIN et al. (1996) used HRCT scanning to examine 126 asthmatics who had never smoked. They found a significantly greater frequency of emphysema in patients with non-allergic asthma compared to the patients with allergic asthma. BIERNACKI et al. (1997) compared lung density measured in patients with chronic obstructive pulmonary disease (COPD) and in asthmatics with lung density of normal subjects. They found that some patients with chronic asthma developed a reduction in computed tomography lung density similar to that in patients with emphysema. However, in a recent study BEIGELMAN-AUBRY et al. (2002) measured lung density in 12 patients with mild intermittent asthma who were non-smokers at 65% of TLC and found, in comparison with normal volunteers measured in the same circumstances, an increase in lung attenuation instead of a decrease. They speculated that this increase was related to inflammatory changes or to increased lung perfusion related to inflammation.

Besides a global change in lung density some asthmatics also show patchy areas of low lung attenuation presenting as mosaic perfusion and air trapping. GRENIER et al. (1996) found areas of low lung attenuation in 31% of patients with asthma. LAURENT et al. (2000) compared the inspiratory and expiratory HRCT findings of 22 patients classified as moderate asthma with 22 healthy volunteers (10 smokers and 12 non-smokers). Mosaic perfusion was found in 22.7% of the asthmatics, in 5% of the smokers and 0% of the non-smokers. Air trapping scores were significantly higher in the asthmatics than in non-smoking control subjects, but not in smokers. This difference was ascribed to non-dependent zones of the lung for which air trapping scores were also higher in asthmatic subjects. As mentioned in the section "Airways involvement related to cigarette smoking" air trapping is a frequent finding in smokers especially in the dependent parts of the lungs and can probably not be used to differentiate asthmatics from smokers. The relationship between the extent of CT diagnosed air trapping and pulmonary function tests of asthmatic patients remains unclear. LAURENT et al. (2000) found only a significant correlation between the air trapping score and FEV1. Others found significant correlations with FRC and RV (NEWMAN et al. 1994) or with VC and FEF 25, too (LUCIDARME et al. 1998). These differences in correlation are probably predominantly related to the selection of the study population but other factors may be important. CT may depict air trapping and small airways disease before lung function deteriorates. On the other hand, before air trapping becomes visible on expiratory scans, the surrounding normal lung should increase in lung density. In case of diffuse airway obstruction and hyperinflation of the entire lung, expiratory density increase could be insufficient to show air trapping (DUPONT et al. 1997) An interesting observation in the study of LAURENT et al. (2000) was the fact that after inhalation of a short acting bronchodilator (salbutamol) no changes could be found in terms of air trapping scores suggesting that the air trapping is reflecting permanent changes due to small airways remodeling. This was also suggested by their observation that air trapping was negatively correlated with FEF25–75 reversibility after bronchodilatation. However, other studies found a partial reduction of the air trapping after bronchodilatation probably related to a difference of reversibility in the population studied or to a difference of diffusion of the inhaled bronchodilator to the distal small airways. BEIGELMAN-AUBRY et al. (2002) found a reduction of air trapping induced by methacholine after salbutamol inhalation in patients with asthma. Also GOLDIN et al. (1998) performed HRCT scans in patients with mild asthma and in a control population before and after bronchial provocation with methacholine chloride and after reversal of provocation with albuterol and measured lung attenuation. They found that at baseline, lung attenuation frequency distribution curves were similar between the control and asthma groups. After methacholine, control subjects showed a decrease of less than 10% in the FEV1 and no significant differences in lung attenuation curves. Patients with asthma showed a 20%–36% decrease in FEV1, with significant decreases in the median and lowest 10th percentile regions of the attenuation curves. After albuterol, control subjects showed no change in spirometric measurements, lung attenuation, or bronchial size, whereas all such parameters returned to baseline levels in patients with asthma.

4.8
Conclusion

Many studies have shown good correlations between morphological changes of small airways disease on computed tomography and lung function changes and have suggested that computed tomography may be helpful to predict this lung function deterioration.

Quantifying small airways disease corresponds in most studies with an objective measurement or a subjective estimation of areas of low attenuation visible on inspiratory and/or expiratory HRCT.

The fact that the CT quantification of small airways disease has not reached the daily clinical practice is related to the complexity of the measurements and to the high sensitivity of computed tomography depicting abnormalities that do not cause any functional deterioration. It can be expected that with the introduction of faster acquisition systems making "real time" functional imaging possible and computer-assisted diagnosis (CAD) tools, these techniques will become even more important and available for clinical routine.

References

Adams H, Bernard MS, McConnochie K (1991) An appraisal of CT pulmonary density mapping in normal subjects. Clin Radiol 43:238–242

Adler BD, Padley SP, Muller NL et al. (1992) Chronic hypersensitivity pneumonitis: high-resolution CT and radiographic features in 16 patients. Radiology 185:91–95

Alasaly K, Muller N, Ostrow DN et al. (1995) Cryptogenic organizing pneumonia. A report of 25 cases and a review of the literature. Medicine (Baltimore) 74:201–211

Angus RM, Davies M-I, Cowan MD et al. (1994) Computed tomographic scanning of the lung in patients with allergic bronchopulmonary aspergillosis and in asthmatic patients with a positive skin test to *Aspergillus fumigatus*. Thorax 49:586–589

Aquino SL, Gamsu G, Webb WR et al. (1996) Tree-in-bud pattern: frequency and significance on thin section CT. J Comput Assist Tomogr 20:594–599

Arakawa H, Webb WR (1998) Air trapping on expiratory high-resolution CT scans in the absence of inspiratory scan abnormalities: correlation with pulmonary function tests and differential diagnosis. AJR Am J Roentgenol 170:1349–1353

Arakawa H, Kurihara Y, Nakajima Y et al. (1998a) Computed tomography measurements of overinflation in chronic obstructive pulmonary disease: evaluation of various radiographic signs. J Thorac Imaging 13:188–192

Arakawa H, Webb WR, McCowin M et al. (1998b) Inhomogeneous lung attenuation at thin-section CT: diagnostic value of expiratory scans. Radiology 206:89–94

Arakawa H, Kurihara Y, Sasaka K et al. (2002) Air trapping on CT of patients with pulmonary embolism. AJR Am J Roentgenol 178:1201–1207

Arcasoy SM, Kotloff RM (1999) Lung transplantation. N Engl J Med 340:1081–1091

ATS/ERS (2002) ATS/ERS International Multidisciplinary Consensus Classification of Idiopathic Interstitial Pneumonias. Am J Respir Crit Care Med 165:277–304

Austin JH, Muller NL, Friedman PJ et al. (1996) Glossary of terms for CT of the lungs: recommendations of the Nomenclature Committee of the Fleischner Society. Radiology 200:327–331

Awadh N, Müller NL, Park CS et al. (1998) Airway wall thickness in patients with near fatal asthma and control groups: assessment with high resolution computed tomography. Thorax 53:248–253

Baar HS, Galindo J (1966) Bronchiolitis fibrosa obliterans. Thorax 21:209–214

Bankier AA, Van Muylem A, Knoop C et al. (2001) Bronchiolitis obliterans syndrome in heart-lung transplant recipients: diagnosis with expiratory CT. Radiology 218:533–539

Bartz RR, Stern EJ (2000) Airways obstruction in patients with sarcoidosis: expiratory CT scan findings. J Thorac Imaging 15:285–289

Beigelman-Aubry C, Capderou A, Grenier PA et al. (2002) Mild intermittent asthma: CT assessment of bronchial cross-sectional area and lung attenuation at controlled lung volume. Radiology 223:181–187

Beinert T, Behr J, Mehnert F et al. (1995) Spirometrically controlled quantitative CT for assessing diffuse parenchymal lung disease. J Comput Assist Tomogr 19:924–931

Bhalla M, Naidich DP, McGuinness G et al. (1996) Diffuse lung disease: assessment with helical CT – preliminary observations of the role of maximum and minimum intensity projection images. Radiology 200:341–347

Biernacki W, Redpath AT, Best JJ et al. (1997) Measurement of CT lung density in patients with chronic asthma. Eur Respir J 10:2455–2459

Carr DH, Hibon S, Rubens M et al. (1998) Peripheral airways obstruction on high-resolution computed tomography in chronic severe asthma. Respir Med 92:448–453

Carrington CB (1976) Structure and function in sarcoidosis. Ann NY Acad Sci 278:265–283

Coche EE, Muller NL, Kim KI et al. (1998) Acute pulmonary embolism: ancillary findings at spiral CT. Radiology 207:753–758

Collins J (2001) CT signs and patterns of lung disease. Radiol Clin North Am 39:1115–1135

Connolly S, Manson D, Eberhard A et al. (1996) CT appearance of pulmonary vasculitis in children. AJR Am J Roentgenol 167:901–904

Cooper JD, Billingham M, Egan T et al. (1993) A working formulation for the standardization of nomenclature and for clinical staging of chronic dysfunction in lung allografts. International Society for Heart and Lung Transplantation. J Heart Lung Transplant 12:713–716

Copley SJ, Wells AU, Müller NL et al. (2002) Thin-section CT in obstructive pulmonary disease: discriminatory value. Radiology 223:812–819

Davies CWH, Tasker AD, Padley SPG et al. (2000) Air trapping in sarcoidosis on computed tomography: correlation with lung function. Clin Radiol 55:217–221

Davison AG, Heard BE, McAllister WA et al. (1983) Cryptogenic organizing pneumonitis. Q J Med 52:382–394

Dennie CJ, Veinot JP, McCormack DG et al. (2002) Intimal sarcoma of the pulmonary arteries seen as a mosaic pattern of lung attenuation on high-resolution CT. AJR Am J Roentgenol 178:1208–1210

Desai SR, Hansell DM (1997) Small airways disease: expiratory computed tomography comes of age. Clin Radiol 52:332–337

Dupont LJ, Pype JL, Demedts MG et al. (1997) Bronchodilator-induced changes in inspiratory and expiratory lung densitometry by high resolution computed tomography in patients with COPD and asthma. Eur Respir J 10:28S

Dupont LJ, Trap K, Verschakelen JA et al. (1999) Detection of patchy air trapping on expiratory HRCT scans in patients with asthma and COPD. Am J Respir Crit Care Med 159:A653

Epler GR, Colby TV, McLoud TC et al. (1985) Bronchiolitis obliterans organizing pneumonia. N Engl J Med 312:152–158

Fitzgerald JE, King TE Jr, Lynch DA et al. (1996) Diffuse panbronchiolitis in the United States. Am J Respir Crit Care Med 154:497–503

Fotheringham T, Chabat F, Hansell DM et al. (1999) A comparison of methods for enhancing the detection of areas of decreased attenuation on CT caused by airways disease. J Comput Assist Tomogr 23:385–389

Frank ST, Weg JG, Harkleroad LE et al. (1973) Pulmonary dysfunction in rheumatoid disease. Chest 63:27–34

Franquet T, Diaz C, Domingo P et al. (1999) Air trapping in primary Sjogren syndrome: correlation of expiratory CT with pulmonary function tests. J Comput Assist Tomogr 23:169–173

Franquet T, Stern EJ, Gimenez A et al. (2000) Lateral decubitus CT: a useful adjunct to standard inspiratory-expiratory CT for the detection of air-trapping. AJR Am J Roentgenol 174:528–530

Frazier AA, Rosado-de-Christenson ML, Galvin JR et al. (1998) Pulmonary angiitis and granulomatosis: radiologic-pathologic correlation. Radiographics 18:687–710

Garg K, Lynch DA, Newell JD et al. (1994) Proliferative and constrictive bronchiolitis: classification and radiologic features. AJR Am J Roentgenol 162:803–808

Geddes DM (1991) BOOP and COP. Thorax 46:545–547

Gelb AF, Hogg JC, Muller NL et al. (1996) Contribution of emphysema and small airways in COPD. Chest 109:353–359

Gelb AF, Zamel N, Hogg JC et al. (1998) Pseudophysiologic emphysema resulting from severe small-airways disease. Am J Respir Crit Care Med 158:815–819

Gevenois PA, Scillia P, Maertelaer V de et al. (1996) The effects of age, sex, lung size, and hyperinflation on CT lung densitometry. AJR Am J Roentgenol 167:1169–1173

Gleeson FV, Traill ZC, Hansell DM (1996) Expiratory CT evidence of small airways obstruction in sarcoidosis. AJR Am J Roentgenol 166:1052–1054

Goddard PR, Nicholson EM, Laszlo G et al. (1982) Computed tomography in pulmonary emphysema. Clin Radiol 33:379–387

Goldin JG, McNitt-Gray MF, Sorensen SM et al. (1998) Airway hyperreactivity: assessment with helical thin-section CT. Radiology 208:321–329

Gosink BB, Friedman PJ, Liebow AA (1973) Bronchiolitis obliterans. Roentgenologic-pathologic correlation. Am J Roentgenol Radium Ther Nucl Med 117:816–832

Green M, Turton CW (1982) Bronchiolitis and its manifestations. Eur J Respir Dis 63:36–42

Grenier P, Mourey-Gerosa I, Benali K et al. (1996) Abnormalities of the airways and lung parenchyma in asthmatics: CT observations in 50 patients and inter- and intra-observer variability. Eur Radiol 6:199–206

Grenier PA, Beigelman-Aubry C, Fetita C et al. (2002) New frontiers in CT imaging of airway disease. Eur Radiol 12:1022–1044

Gruden JF, Webb WR, Warnock M (1994) Centrilobular opacities in the lung on high-resolution CT: diagnostic considerations and pathologic correlation. AJR Am J Roentgenol 162:569–574

Guckel C, Wells AU, Taylor DA et al. (1999) Mechanism of mosaic attenuation of the lungs on computed tomography in induced bronchospasm. J Appl Physiol 86:701–708

Hansell DM, Moskovic E (1991) High-resolution computed tomography in extrinsic allergic alveolitis. Clin Radiol 43:8–12

Hansell DM, Wells AU, Rubens MB et al. (1994) Bronchiectasis: functional significance of areas of decreased attenuation at expiratory CT. Radiology 193:369–374

Hansell DM, Wells AU, Padley SP et al. (1996) Hypersensitivity pneumonitis: correlation of individual CT patterns with functional abnormalities. Radiology 199:123–128

Hansell DM, Rubens MB, Padley SPG et al. (1997) Obliterative bronchiolitis: individual CT signs of small airways disease and functional correlation. Radiology 203:721–726

Hansell DM, Milne DG, Wilsher ML et al. (1998) Pulmonary sarcoidosis: morphologic associations of airflow obstruction at thin-section CT. Radiology 209:697–704

Hansen JE, Ampaya PE, Bryant GH et al. (1975) Branching pattern of airways and airspaces of single human terminal bronchiole. J Appl Physiol 38:983–989

Hartman TE, Swensen SJ, Müller NL (1994) Bronchiolar diseases: computed tomography. In: Epler GR (ed) Diseases of the bronchioles. Raven, New York, pp 43–58

Hartman TE, Tazelaar HD, Swensen SJ et al. (1997) Cigarette smoking: CT and pathologic findings of associated pulmonary diseases. Radiographics 17:377–390

Heyneman LE, Ward S, Lynch DA et al. (1999) Respiratory bronchiolitis, respiratory bronchiolitis-associated interstitial lung disease, and desquamative interstitial pneumonia: different entities or part of the spectrum of the same disease process? AJR Am J Roentgenol 173:1617–1622

Howling SJ, Hansell DM, Wells AU et al. (1999) Follicular bronchiolitis: thin-section CT and histologic findings. Radiology 212:637–642

Hwang JH, Kim TS, Lee KS et al. (1997) Bronchiolitis in adults: pathology and imaging. J Comput Assist Tomogr 21:913–919

Im JG, Kim SH, Chung MJ et al. (1996) Lobular low attenuation of the lung parenchyma on CT: evaluation of forty-eight patients. J Comput Assist Tomogr 20:756–762

Itoh H, Murata K, Konishi J et al. (1993) Diffuse lung disease: pathologic basis for the high-resolution computed tomography findings. J Thorac Imaging 8:176–188

Kalender WA, Rienmuller R, Seissler W et al. (1990) Measurement of pulmonary parenchymal attenuation: use of spirometric gating with quantitative CT. Radiology 175:265–268

Kauczor HU, Heussel CP, Fischer B et al. (1998) Assessment of lung volumes using helical CT at inspiration and expiration: comparison with pulmonary function tests. AJR Am J Roentgenol 171:1091–1095

Kauczor HU, Hast J, Heussel CP et al. (2000) Focal airtrapping at expiratory high-resolution CT: comparison with pulmonary function tests. Eur Radiol 10:1539–1546

King GG, Müller NL, Paré PD (1999) Evaluation of airways in obstructive pulmonary disease using high-resolution computed tomography. Am J Respir Crit Care Med 159:992–1004

King GG, Muller NL, Whittall KP et al. (2000) An analysis algorithm for measuring airway lumen and wall areas from high-resolution computed tomographic data. Am J Respir Crit Care Med 161:574–580

King TE Jr (1993) Overview of bronchiolitis. Clin Chest Med 14:607–610

Kinsella M, Muller NL, Staples C et al. (1988) Hyperinflation in asthma and emphysema. Assessment by pulmonary function testing and computed tomography. Chest 94:286–289

Kohz P, Stabler A, Beinert T et al. (1995) Reproducibility of quantitative, spirometrically controlled CT. Radiology 197:539–542

Kubo K, Eda S, Yamamoto H et al. (1999) Expiratory and inspiratory chest computed tomography and pulmonary function tests in cigarette smokers. Eur Respir J 13:252–256

Kushihashi T, Munechika H, Ri K et al. (1994) Bronchioalveolar adenoma of the lung: CT-pathologic correlation. Radiology 193:789–793

Lange W (1901) Ueber eine eigenthumliche Erkrankung der kleinen Bronchien und Bronchiolen (bronchitis et bronchiolitis obliterans). Dtsch Arch Klin Med 70:342–364

Laurent F, Latrabe V, Raherison C et al. (2000) Functional significance of air trapping detected in moderate asthma. Eur Radiol 10:1404–1410

Lee ES, Gotway MB, Reddy GP et al. (2000) Early bronchiolitis obliterans following lung transplantation: accuracy of expiratory thin-section CT for diagnosis. Radiology 216:472–477

Lee KS, Kullnig P, Hartman TE et al. (1994) Cryptogenic organizing pneumonia: CT findings in 43 patients. AJR Am J Roentgenol 162:543–546

Lee KW, Chung SY, Yang I et al. (2000) Correlation of aging and smoking with air trapping at thin-section CT of the lung in asymptomatic subjects. Radiology 214:831–833

Lentz D, Bergin CJ, Berry GJ et al. (1992) Diagnosis of bronchiolitis obliterans in heart-lung transplantation patients: importance of bronchial dilatation on CT. AJR Am J Roentgenol 159:463–467

Leung AN, Fisher K, Valentine V et al. (1998) Bronchiolitis obliterans after lung transplantation: detection using expiratory HRCT. Chest 113:365–370

Logan PM, Primack SL, Miller RR et al. (1994) Invasive aspergillosis of the airways: radiographic, CT, and pathologic findings. Radiology 193:383–388

Logan PM, Miller RR, Muller NL (1995) Cryptogenic organizing pneumonia in the immunocompromised patient: radiologic findings and follow-up in 12 patients. Can Assoc Radiol J 46:272–279

Lohr RH, Boland BJ, Douglas WW et al. (1997) Organizing pneumonia. Features and prognosis of cryptogenic, secondary, and focal variants. Arch Intern Med 157:1323–1329

Loubeyre P, Revel D, Delignette A et al. (1995) Bronchiectasis detected with thin-section CT as a predictor of chronic lung allograft rejection. Radiology 194:213–216

Lucidarme O, Coche E, Cluzel P et al. (1998) Expiratory CT scans for chronic airway disease: correlation with pulmonary function test results. AJR Am J Roentgenol 170:301–307

Lucidarme O, Grenier PA, Cadi M et al. (2000) Evaluation of air trapping at CT: comparison of continuous – versus suspended – expiration CT techniques. Radiology 216:768–772

Lynch DA, Webb WR, Gamsu G et al. (1989) Computed tomography in pulmonary sarcoidosis. J Comput Assist Tomogr 13:405–410

Lynch DA, Newell JD, Tschomper BA et al. (1993) Uncomplicated asthma in adults: comparison of CT appearance of the lungs in asthmatic and healthy subjects. Radiology 188:829–833

Lynch DA, Newell JD, Logan PM et al. (1995) Can CT distinguish hypersensitivity pneumonitis from idiopathic pulmonary fibrosis? AJR Am J Roentgenol 165:807–811

Magkanas E, Voloudaki A, Bouros D et al. (2001) Correlation of expiratory high-resolution CT findings with inspiratory patterns and pulmonary function tests. Acta Radiol 42:494–501

Mandel J, Mark EJ, Hales CA (2000) Pulmonary veno-occlusive disease. Am J Respir Crit Care Med 162:1964–1973

Marti-Bonmati L, Ruiz Perales F, Catala F et al. (1989) CT findings in Swyer-James syndrome. Radiology 172:477–480

Mastora I, Remy-Jardin M, Sobaszek A et al. (2001) Thin-section CT finding in 250 volunteers: assessment of the relationship of CT findings with smoking history and pulmonary function test results. Radiology 218:695–702

Matar LD, McAdams HP, Spron TA (2000) Hypersensitivity pneumonitis. AJR Am J Roentgenol 174:1061–1066

McDonagh J, Greaves M, Wright AR et al. (1994) High resolution computed tomography of the lungs in patients with rheumatoid arthritis and interstitial lung disease. Br J Rheumatol 33:118–122

McGuinness G, Scholes JV, Garay SM et al. (1994) Cytomegalovirus pneumonitis: spectrum of parenchymal CT findings with pathologic correlation in 21 AIDS patients. Radiology 192:451–459

McNamara AE, Muller NL, Okazawa M et al. (1992) Airway narrowing in excised canine lungs measured by high-resolution computed tomography. J Appl Physiol 73:307–316

Miller WS (1947) The lung, 2nd edn. Thomas, Springfield, pp 203–205

Mitchell AW, Wells AU, Hansell DM (1996) Changes in cross-sectional area of the lungs on end expiratory computed tomography in normal individuals. Clin Radiol 51:804–806

Moon J, duBois RM, Colby TV et al. (1999) Clinical significance of respiratory bronchiolitis on open lung biopsy and its relationship to smoking related interstitial lung disease. Thorax 54:1009–1014

Müller NL, Miller RR (1995) Diseases of the bronchioles: CT and histopathologic findings. Radiology 196:3–12

Müller NL, Staples CA, Miller RR et al. (1988) "Density mask". An objective method to quantitate emphysema using computed tomography. Chest 94:782–787

Müller NL, Mawson JB, Mathieson JR et al. (1989) Sarcoidosis: correlation of extent of disease at CT with clinical, functional, and radiographic findings. Radiology 171:613–618

Müller NL, Staples CA, Miller RR (1990) Bronchiolitis obliterans organizing pneumonia: CT features in 14 patients. AJR Am J Roentgenol 154:983–987

Murata K, Itoh H, Todo G et al. (1986) Centrilobular lesions of the lung: demonstration by high-resolution CT and pathologic correlation. Radiology 161:641–645

Murphy JM, Schnyder P, Verschakelen J et al. (1999) Linear opacities on HRCT in bronchiolitis obliterans organising pneumonia. Eur Radiol 9:1813–1817

Myers JL, Colby TV (1993) Pathologic manifestations of bronchiolitis, constrictive bronchiolitis, cryptogenic organizing pneumonia, and diffuse panbronchiolitis. Clin Chest Med 14:611–622

Nakano Y, Sakai H, Muro S et al. (1999) Comparison of low attenuation areas on computed tomographic scans between inner and outer segments of the lung in patients with chronic obstructive pulmonary disease: incidence and contribution to lung function. Thorax 54:384–389

Newman KB, Lynch DA, Newman LS et al. (1994) Quantitative computed tomography detects air trapping due to asthma. Chest 106:105–109

Ng CS, Desai SR, Rubens MB et al. (1999) Visual quantitation and observer variation of signs of small airways disease at inspiratory and expiratory CT. J Thorac Imaging 14:279–285

Niimi A, Matsumoto H, Amitani R et al. (2000) Airway wall thickness in asthma assessed by computed tomography. Am J Respir Crit Care Med 162:1518–1523

Nishimura K, Kitaichi M, Izumi T et al. (1992) Diffuse panbronchiolitis: correlation of high-resolution CT and patho logic findings. Radiology 184:779–785

Okazawa M, Muller N, McNamara AE et al. (1996) Human airway narrowing measured using high resolution computed tomography. Am J Respir Crit Care Med 154:1557–1562

Padley SP, Adler BD, Hansell DM et al. (1993) Bronchiolitis obliterans: high resolution CT findings and correlation with pulmonary function tests. Clin Radiol 47:236–240

Paganin F, Trussard V, Seneterre E et al. (1992) Chest radiography and high resolution computed tomography of the lungs in asthma. Am Rev Respir Dis 146:1084–1087

Paganin F, Seneterre E, Chanez P et al. (1996) Computed tomography of the lungs in asthma: influence of disease severity and etiology. Am J Respir Crit Care Med 153:110–114

Park CS, Muller NL, Worthy SA et al. (1997) Airway obstruction in asthmatic and healthy individuals: inspiratory and expiratory thin-section CT findings. Radiology 203: 361–367

Pelinkovic D, Lorcher U, Chow KU et al. (1997) Spirometric gated quantitative computed tomography of the lung in healthy smokers and nonsmokers. Invest Radiol 32: 335–343

Perez T, Farre JM, Gosset P et al. (1989) Subclinical alveolar inflammation in rheumatoid arthritis: superoxide anion, neutrophil chemotactic activity and fibronectin generation by alveolar macrophages. Eur Respir J 2:7–13

Perez T, Remy-Jardin M, Cortet B (1998) Airways involvement in rheumatoid arthritis: clinical, functional, and HRCT findings. Am J Respir Crit Care Med 157:1658–1665

Plopper CG, Ten Have-Opbroek AAW (1994) Anatomical and histological classification of the bronchioles. In: Epler GR (ed) Diseases of the bronchioles. Raven, New York, pp 15–25

Poletti V, Zompatori M, Cancellieri A (1999) Clinical spectrum of adult chronic bronchiolitis. Sarcoidosis Vasc Diffuse Lung Dis 16:183–196

Prêteux F, Fetita CI, Capderou A et al. (1999) Modeling, segmentation, and caliber estimation of bronchi in high resolution computerized tomography. J Electron Imaging 8:36–45

Primack SL, Muller NL, Mayo JR et al. (1994) Pulmonary parenchymal abnormalities of vascular origin: high-resolution CT findings. Radiographics 14:739–746

Primack SL, Logan PM, Hartman TE et al. (1995) Pulmonary tuberculosis and *Mycobacterium avium-intracellulare*: a comparison of CT findings. Radiology 194:413–417

Remy-Jardin M, Remy J, Boulenguez C et al. (1993a) Morphologic effects of cigarette smoking on airways and pulmonary parenchyma in healthy adult volunteers: CT evaluation and correlation with pulmonary function tests. Radiology 186:107–115

Remy-Jardin M, Remy J, Gosselin B et al. (1993b) Lung parenchymal changes secondary to cigarette smoking: pathologic-CT correlations. Radiology 186:643–651

Remy-Jardin M, Remy J, Wallaert B et al. (1993c) Subacute and chronic bird breeder hypersensitivity pneumonitis: sequential evaluation with CT and correlation with lung function tests and bronchoalveolar lavage. Radiology 189: 111–118

Remy-Jardin M, Remy J, Cortet B et al. (1994) Lung changes in rheumatoid arthritis: CT findings. Radiology 193:375–382

Remy-Jardin M, Remy J, Gosselin B et al. (1996) Sliding thin slab, minimum intensity projection technique in the diagnosis of emphysema: histopathologic-CT correlation. Radiology 200:665–671

Remy-Jardin M, Edme J-L, Boulenguez C et al. (2002) Longitudinal follow-up study of smoker's lung with thin-section CT in correlation with pulmonary function tests. Radiology 222:261–270

Roberts HR, Wells AU, Milne DG et al. (2000) Airflow obstruction in bronchiectasis: correlation between computed tomography features and pulmonary function tests. Thorax 55:198–204

Ryu JH, Colby TV, Hartman TE et al. (2001) Smoking-related interstitial lung diseases: a concise review. Eur Respir J 17: 122–132

Selman-Lama M, Perez-Padilla R (1993) Airflow obstruction and airway lesions in hypersensitivity pneumonitis. Clin Chest Med 14:699–714

Sherrick AD, Swensen SJ, Hartman TE (1997) Mosaic pattern of lung attenuation on CT scans: frequency among patients with pulmonary artery hypertension of different causes. AJR Am J Roentgenol 169:79–82

Siegel MJ, Bhalla S, Gutierrez FR et al. (2001) Post-lung transplantation bronchiolitis obliterans syndrome: usefulness of expiratory thin-section CT for diagnosis. Radiology 220:455–462

Silver SF, Müller NL, Miller RR et al. (1989) Hypersensitivity pneumonitis. Evaluation with CT. Radiology 173:441–445

Small JH, Flower CD, Traill ZC et al. (1996) Air-trapping in extrinsic allergic alveolitis. Clin Radiol 51:684–688

Soejima K, Yamaguchi K, Kohda E et al. (2000) Longitudinal follow-up study of smoking-induced lung density changes by high-resolution computed tomography. Am J Respir Crit Care Med 161:1264–1273

Soleman A, Colby TV (1988) Histologic diagnosis of extrinsic allergic alveolitis. Am J Surg Pathol 12:514–518

Stein MG, Mayo J, Muller N et al. (1987) Pulmonary lymphangitic spread of carcinoma: appearance on CT scans. Radiology 162:371–375

Stern EJ, Frank MS (1994) Small-airways disease of the lungs: findings at expiratory CT. AJR Am J Roentgenol 163:37–41

Stern EJ, Webb WR, Gamsu G (1994) Dynamic quantitative computed tomography. A predictor of pulmonary function in obstructive lung diseases. Invest Radiol 29:564–569

Stern EJ, Muller NL, Swensen SJ et al. (1995a) CT mosaic pattern of lung attenuation: etiologies and terminology. J Thorac Imaging 10:294–297

Stern EJ, Swensen SJ, Hartman TE et al. (1995b) CT mosaic pattern of lung attenuation: distinguishing different causes. AJR Am J Roentgenol 165:813–816

Sulavik SS (1989) The concept of „organizing pneumonia". Chest 96:967–968

Taouli B, Brauner MW, Mourey I et al. (2002) Thin-section chest CT findings of primary Sjogren's syndrome: correlation with pulmonary function. Eur Radiol 12:1504–1511

Verschakelen J, Fraeyenhoven L van, Laureys G et al. (1993) Differences in CT density between dependent and nondependent portions of the lung: influence of lung volume. AJR Am J Roentgenol 161:713–717

Verschakelen JA, Scheinbaum K, Bogaert J et al. (1998) Expiratory CT in cigarette smokers: correlation between areas of decreased lung attenuation, pulmonary function tests and smoking history. Eur Radiol 8:1391–1399

Webb WR, Gamsu G, Wall SD et al. (1984) CT of a bronchial phantom. Factors affecting appearance and size measurements. Invest Radiol 19:394–398

Webb WR, Stern EJ, Kanth N et al. (1993) Dynamic pulmonary CT: findings in healthy adult men. Radiology 186:117–124

Wegener OH, Koeppe P, Oeser H (1978) Measurement of lung density by computed tomography. J Comput Assist Tomogr 2:263–273

Wittram C, Batt J, Rappaport DC et al. (2002) Inspiratory and expiratory helical CT of normal adults: comparison of thin section scans and minimum intensity projection images. J Thorac Imaging 17:47–52

Wolfe F, Schurle DRE, Lin JJ et al. (1983) Upper and lower airway disease in penicillamine treated patients with rheumatoid arthritis. J Rheumatol 10:406–410

Wood SA, Zerhouni EA, Hoford JD et al. (1995) Measurement

of three-dimensional lung tree structures by using computed tomography. J Appl Physiol 79:1687–1697

Worthy SA, Muller NL (1998) Small airway diseases. Radiol Clin North Am 36:163–173

Worthy SA, Flint JD, Muller NL (1997a) Pulmonary complications after bone marrow transplantation: high-resolution CT and pathologic findings. Radiographics 17:1359–1371

Worthy SA, Muller NL, Hartman TE et al. (1997b) Mosaic attenuation pattern on thin-section CT scans of the lung: differentiation among infiltrative lung, airway, and vascular diseases as a cause. Radiology 205:465–470

Worthy SA, Park CS, Kim JS et al. (1997c) Bronchiolitis obliterans after lung transplantation: high-resolution CT findings in 15 patients. AJR Am J Roentgenol 169:673–677

Wright JL, Cagle P, Churg A et al. (1992) Diseases of the small airways. Am Rev Respir Dis 146:240–262

5 CT and MRI of Pulmonary Emphysema: Assessment of Lung Structure and Function

Alexander A. Bankier and Pierre Alain Gevenois

CONTENTS

5.1 Computed Tomography

5.1.1 Introduction

Computed tomography (CT) is a radiological modality that provides transverse anatomical images. In these

A. A. Bankier, MD
Department of Radiology, University of Vienna, Währinger Gürtel 18-20, 1090 Vienna, Austria
P. A. Gevenois, MD
Université Libre des Bruxelles – Hopital Erasme, Route de Lennik 808, 1070 Bruxelles, Belgium

images, the value of each picture element (pixel) corresponds to the X-ray attenuation of a defined volume of tissue (voxel). The X-ray attenuation values for each set of projections (slice) are registered by a computer and organized in the form of a matrix. The number of calculated pixels not only determines the image matrix size, but also impacts on the image resolution, and should therefore be as high as possible. In clinical practice, the matrix size is 512×512 pixels. The X-ray attenuation, that is sometimes termed "tissue density", is numerically expressed in Hounsfield units (HU). The scale of attenuation values ranges from –1000 HU, corresponding to the attenuation value of air, to 3000 HU, with 0 HU corresponding to the attenuation value of water. The thousands of pixels and respective attenuation values included in one scan make CT the most precise modality for the in vivo assessment of pulmonary parenchyma (Hoffman and McLennan 1997).

In addition to providing overall anatomical information as to lung tissue destruction, the major advantage of CT in patients with emphysema is that it also identifies the specific pulmonary locations where the alveolar surface has been destroyed (Fig. 5.1). The ability to estimate the extent and severity of pulmonary emphysema in vivo appears important for several reasons:

1. Early accurate detection of lung destruction and mapping of its progression are required to understand the natural history of emphysema.
2. The treatment of advanced emphysema by lung volume reduction surgery (LVRS) requires knowledge of the location of emphysematous lesions and objective methods for evaluating post-surgical results (Gierada et al. 1997).
3. CT could be a sensitive modality for quantifying the progression of emphysema when determining the efficacy of replacement therapy in patients with a_1-antitrypsin deficiency (Dirksen et al. 1999).
4. Studies suggesting that alveolar number and surface-to-volume ratio can be restored by drugs in rats with elastase-induced emphysema imply the future need for measurements that can accurately assess the therapeutic effect of such therapies (Stockley 2000; Tepper et al. 2000).

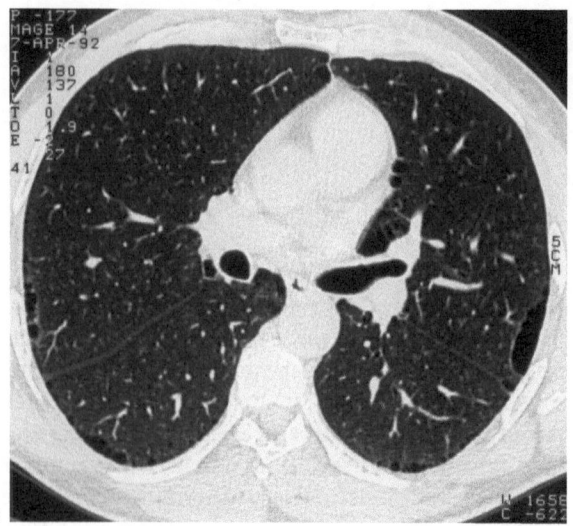

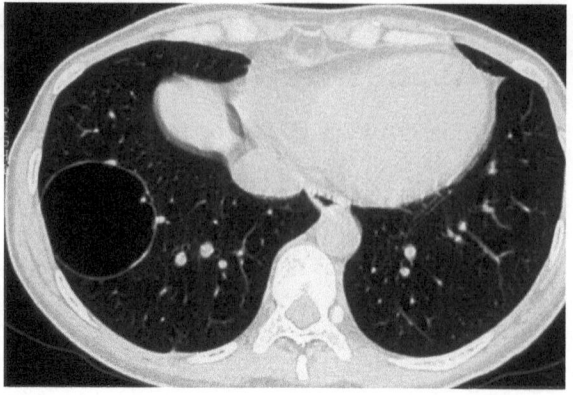

Fig. 5.1. Thin-section CT in a patient with emphysema. Both sections show typical features of the disease. Beside a decreased overall attenuation of the lung parenchyma, **a** shows subpleural emphysematous lesions. **b** shows a large bulla in the right lower lung

5. The detection of early emphysema may prevent the occurrence of obstructive ventilatory impairment by smoking cessation or other medical interventions (MORGAN 1992).

Numerous studies have addressed the capability of CT to accurately quantify the extent and the severity of pulmonary emphysema (STERN and FRANCK 1994; THURLBECK and MÜLLER 1994). To verify whether CT is adequately validated and to suggest possible directions for future research, this article provides an overview over previously published studies, often based on widely varying models. Studies have indeed been based on subjective visual grading or on objective indexes derived from attenuation values, on two-dimensional or on three-dimensional approaches, and on CT scans obtained at either full suspended inspiration or full suspended expiration.

Pulmonary emphysema is defined as "abnormal permanent enlargement of the air spaces distal to the terminal bronchioles, accompanied by destruction of the alveolar walls, and without obvious fibrosis". Because this definition is based on pathology, new modalities for diagnosis and quantification of emphysema must be validated by comparisons with this standard of reference (SNIDER et al. 1985). However, the presence and extent of emphysema can be determined by both macroscopic and microscopic assessment of lung specimens. Before discussing the CT-related issues in the quantification of emphysema, we will thus first briefly review the most widely used macroscopic and microscopic methods used in emphysema quantification.

5.1.2
Histopathological Quantification of Pulmonary Emphysema

5.1.2.1
Macroscopic Methods

Two methods were traditionally used to macroscopically quantify the severity of emphysema: point counting developed by DUNNILL (1962b), and panel grading proposed by THURLBECK et al. (1970). Point counting calculates the proportion occupied by emphysematous spaces expressed as a percentage of a lung section by using a transparent plastic sheet with a grid drawn on it and placed on the lung section. The points of this grid lie 1 cm apart and are situated at the angles of equilateral triangles with 1 cm sides. The percentage of the lung involved by emphysema is given by the number of points superimposed on emphysematous spaces multiplied by 100 and divided by the number of points on the whole lung section. This method is truly quantitative and can be performed on several sections obtained throughout a lung specimen, but it is tedious and time-consuming. Panel grading is based on the comparison of paper-mounted sagittal lung sections with a set of standards that score emphysema from 0 to 100 at intervals of 5 or 10. Scores of 5–25 indicate mild emphysema, scores ranging from 30–50 indicate moderate emphysema and scores of 60 or more correspond to severe emphysema (THURLBECK et al. 1970). Panel grading is relatively quick. However, rather than being truly quantitative, it is a method for ranking emphysema according to predefined categories of severity. Moreover, this technique tends to underestimate the extent of panlobular emphysema and does not permit combined grading of several sections from the same lung specimen. This is important because it has been shown that adequate assessment of emphysema can hardly be made from a single lung slice alone (TURNER and WHIMSTER 1981).

With respect to the drawbacks of the two above described methods, a third method has recently been developed to measure the extent of emphysema on numerous paper-mounted lung sections. This computer-assisted method follows the principles of point counting and calculates the relative area of lung macroscopically occupied by emphysema expressed as a percentage. This method has the advantages of being quick, precise, highly reproducible and permits the combination of data from several slices obtained throughout a lung specimen (GEVENOIS et al. 1995a).

5.1.2.2
Microscopic Methods

Several methods have been developed to microscopically quantify emphysema. The mean linear intercept (Lm) is defined as the ratio of the length of a test line placed on a microscopic lung sample, divided by the number of intercepts of this test line with alveolar walls (DUNNILL 1962b). The airspace wall per unit volume (AWUV) is a measurement expressing the alveolar surface area per unit of lung volume and is derived from Lm. As pointed out by THURLBECK and MÜLLER (1994), neither the loss of alveolar surface nor Lm and AWUV are sensitive methods for recognizing emphysema. Indeed, Lm is normal in 32% of emphysematous patients and AWUV is abnormal only in 26% of surgically resected patients with severe macroscopic emphysema (THURLBECK 1967b; GILLOOLY and LAMB 1993b). More recently, a computer-based method to measure the distance between alveolar walls in the lung parenchyma was used (GEVENOIS et al. 1996a). As commented on by MÜLLER and THURLBECK (1996), "this is a measurement of the average transection distance between walls of alveoli, alveolar ducts, and alveolar sacs considered together; it is not the average alveolar diameter", and this term is less ambiguous than Lm.

The destructive index (DI), which is defined as the percentage of destroyed alveolar and alveolar duct space, was introduced by SAETTA et al. (1985) as an objective criterion for alveolar wall destruction. DI has three components: breaks in alveolar walls (DI_b), metaplasia of type II cells and some degree of

fibrosis in alveolar wall (DI_f), and the so-called classic emphysema (DI_e). In the study by SAETTA et al. (1985) DI_b was increased in smokers in whom the size of the air space was still normal. Therefore, this parameter could be an early indicator of lung destruction. Nevertheless, an increased DI_b could also be related to an increased number and size of fenestrae in the normal parenchyma adjacent to emphysema, or to abnormal properties of elastic tissue in smokers without macroscopic emphysema (THURLBECK and MÜLLER 1994). BOREN (1962) reported the presence of holes in alveolar walls of normal lung specimens and suggested that holes larger than 20 μm in diameter were abnormal. NAGAI et al. (1994) measured the size of these holes and found that only 0.2% of normal subjects have fenestrae larger than 20 μm in diameter. Alveolar destruction could thus be defined as the presence of holes, also called fenestrae, larger than this diameter, that probably represent the earliest pathological evidence of emphysema (COSIO et al. 1986).

5.1.3
CT Quantification of Pulmonary Emphysema

5.1.3.1
Subjective CT Quantification

The subjective CT quantification of emphysema is based on the visual assessment of areas of vascular disruption and of decreased attenuation without clear margins, as compared to the contiguous normal parenchyma (Fig. 5.1) (BERGIN et al. 1986a). In 1986, BERGIN et al. visually estimated the percentage area that demonstrated changes suggestive of emphysema on contiguous 10 -mm thick CT sections on a study group of 32 patients (BERGIN et al. 1986b). On the basis of statistically significant correlations between visual CT scores and macroscopic emphysema graded with a system adapted from THURLBECK et al. (1970) on midsagittal sections, the authors concluded that CT is a useful adjunct in assessing the presence and extent of emphysema. In 1987, by comparing the CT scores of 1 -mm thick high resolution CT (HRCT) slices performed at 5 levels of 20 postmortem inflated lung specimens and the pathologic scores obtained at the same anatomic levels, HRUBAN et al. (1987) demonstrated that HRCT is able to distinguish emphysematous lungs from normal lungs even in the mildest degrees of emphysema. By applying the same methods in patients with mild emphysema, KUWANO et al. (1990) found statistically significant correlations between the scores established on the HRCT slices and macroscopic grading, as well as between the CT scores and the microscopic destructive index (DI). MILLER et al. (1989) assessed the extent of emphysema by superimposing a grid with squares corresponding to 1 cm² on CT images and determined the percentage of squares containing emphysema. Comparing these results obtained on 10 -mm and 1.5 -mm thick CT sections to the grading panel of parasagittal standards established by THURLBECK et al. (1970), the authors found that CT had a poor sensitivity for detecting early emphysematous lesions (MILLER et al. 1989).

In all abovementioned studies, "mental adjustments" were required in order to apply the top to bottom grading panel to transverse CT images (BERGIN et al 1986b; HRUBAN et al. 1987; MILLER et al. 1989; KUWANO et al. 1990). In addition, the grading system is not really quantitative, but rather ranks emphysema according to categories of severity (THURLBECK et al. 1970). To overcome these limitations, another group applied a quantitative computer-assisted method to horizontal paper-mounted lung sections that provided results on a continuous scale (GEVENOIS et al. 1995a). All pioneer studies have indeed shown that the extent of emphysema visually scored on CT scans does significantly correlate with the extent of emphysema scored on macroscopic lung sections obtained from resected lung specimens. Most studies, however, were focused on centrilobular emphysema, and did not provide a truly objective CT quantification as applied to HRCT scans (GEVENOIS and YERNAULT 1995).

5.1.3.2
Objective CT Quantification

5.1.3.2.1
Attenuation Measurements

To objectively quantify pulmonary emphysema, several lung attenuation parameters based on the histogram analysis of the frequency distribution of the attenuation value of the lung, have been developed (Fig. 5.2) (STOEL et al. 1999). The most commonly used methods are based on the measurement of the mean lung attenuation, the areas of lung occupied by attenuation values lower than predetermined thresholds (MÜLLER et al. 1988; GEVENOIS et al. 1995b, 1996a; BAE et al. 1997; MISHIMA et al. 1999a; NAKANO et al. 1999), and a predetermined percentile of the lung attenuation distribution curve (GOULD et al. 1988; DIRKSEN et al. 1999).

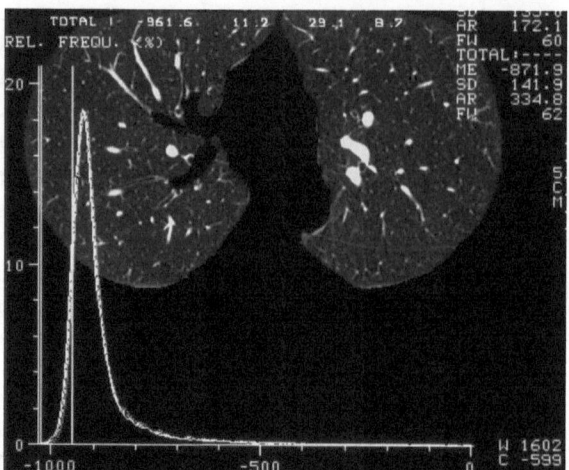

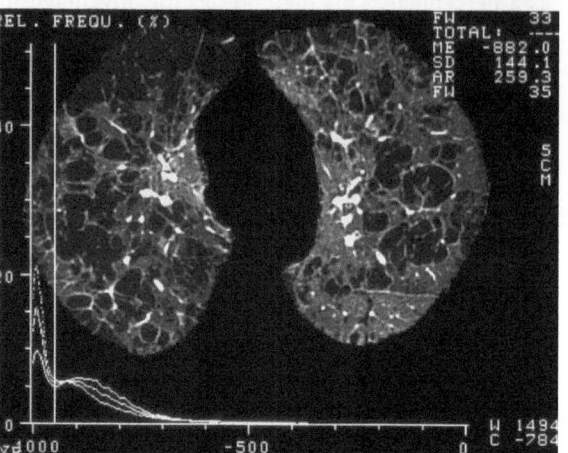

Fig. 5.2. CT-derived histogram analyses. **a** (normal volunteer) shows normal height and shape of the histogram curve. In a patient with severe emphysema (**b**), the curve has a lower peak and is clearly displaced towards the left, i.e., towards lower attenuation values

In the first pathologic-CT comparative study using numbers for attenuations, HAYHURST et al. showed that the distribution curve of these densities was significantly shifted toward lowest attenuation values in patients with emphysema compared to normal individuals (Fig. 5.2) (HAYHURST et al. 1984). In a CT-pathologic correlation study based on microscopic measurements, GOULD et al. (1988) showed that the lowest fifth percentile of the histogram of attenuation values was significantly correlated with AWUV. The lowest fifth percentile depends not only on the extent of emphysema but is also influenced by the relative amount of higher attenuation values, corresponding to airway walls, blood vessels, and potential infiltrates that tend to displace the histogram to the right (RIENMÜLLER et al. 1991; HARTLEY et al. 1994). Consequently, if emphysema is associated with other pulmonary disorders, the lowest fifth percentile will underestimate the extent of emphysema. To overcome this drawback, an absolute threshold should be used and the relative area of lung occupied by attenuation values lower than this threshold should be measured (GEVENOIS et al. 1995b). A commercially available CT program called "Density Mask" (General Electric Medical Systems, Milwaukee, Wis.) highlights pixels within a given attenuation range and automatically calculates the area of highlighted pixels (MÜLLER et al. 1988; KINSELLA et al. 1990). MÜLLER et al. (1988) compared the relative area highlighted on a single 1-cm thick CT scan – after injection of contrast material – with the corresponding macroscopic section of the fixed lung cut in the same plane as the CT scan and graded using a modification of the picture-grading system from THURLBECK et al. (1970). The highest correlation was observed at attenuation values lower than –910 HU, and as a consequence, this threshold was recommended for the identification of emphysema. Nevertheless, statistically significant correlations only indicate that CT and pathological scores are statistically linked. This, however, does not imply that the percentage areas obtained by CT quantifications are equal to the percentage areas occupied by emphysema on the pathological specimen. Moreover, the proposed threshold might have been influenced by the administration of contrast material in the original study by MÜLLER et al. Furthermore, the grading panel does not represent the extent of lung involved by emphysema and underestimates panlobular emphysema, notably in initial stages.

In an attempt to determine the best attenuation threshold for the recognition of emphysema, another group applied on 1-mm thick CT sections a program that automatically recognizes the lungs, traces the lung contours, determines histograms of attenuation values, and measures the lung area occupied by pixels included in the predetermined range of attenuation values (KALENDER et al. 1991). On thin CT sections obtained from the lung apices to the bases with 1-cm intervals, these authors calculated the relative area of lung, expressed in percentage, occupied by attenuation values lower than various thresholds ranging from –900 HU to –970 HU. On a first study based on 63 patients with lung resection, they compared the CT data with the macroscopic extent of emphysema measured, on horizontal paper-mounted whole lung sections (GEVENOIS et al. 1995b), by a previously validated computer-assisted method (GEVENOIS et al. 1995a). The authors showed that the only threshold for which there was no significant difference between the distribution of the CT measurements and the distri-

bution of macroscopic measurements was –950 HU. Thresholds lower than –950 HU underestimated emphysema, and thresholds above –950 HU overestimated emphysema. The case-by-case comparisons between the relative area occupied by attenuation values lower than –950 HU (RA_{950}) and the relative area of lung macroscopically occupied by emphysema were not identical in every patient: the mean of the absolute values of the differences between the RA_{950} and the relative area of lung macroscopically occupied by emphysema was 4.9% and ranged from 0.1 to 19.9%. These data thus suggested that the relative area of lung occupied by attenuation values lower than –950 HU calculated on thin section CT scans obtained at full inspiration, allowed an objective quantification of macroscopic emphysema in vivo and with an acceptable error margin (GEVENOIS et al. 1995b).

Because MCLEAN et al. (1992) recommended that pulmonary emphysema should be measured microscopically rather than macroscopically, comparisons between CT and morphometry should include microscopic measurements. Using AWUV as a microscopic measurement of the alveolar wall surface in 28 subjects referred for surgical resection of lung tumors, GOULD et al. (1988) reported statistically significant correlations between AWUV and the lowest fifth percentile of the frequency distribution curve of attenuation values ($r=-0.77$, $p<0.001$) calculated on 13-mm thick CT sections. In a more recent study based on 38 patients also referred for lung resection, other authors measured mean interwall distance (MIWD) and mean perimeter (MP) and compared the percentage surface area of lung occupied by attenuation values lower than thresholds ranging from –900 to –970 HU to the microscopic indexes. These authors showed that the highest correlation was obtained with –950 HU ($r=0.70$). Thus, both the macroscopic as well as the microscopic studies suggested that RA_{950} provides a valuable measurement of the extent of pulmonary emphysema (GEVENOIS et al. 1996a).

To predict the lung surface-to-volume ratio from CT attenuation values, COXSON et al. (1999) considered –910 HU as a threshold and compared CT measurements with histologic estimates of surface area. The lung volume was calculated by summing the voxel dimensions in each slice, and the lung weight was estimated by multiplying the mean lung attenuation value by the lung volume. From these measurements, COXSON et al. (1999) derived the regional lung inflation expressed in ml/g. A comparison of the amount of emphysema detected in the same lobe by both CT and point counting of the resected specimen showed that the volume fraction of lesions greater than 5 mm in diameter measured by morphometry was similar to the fraction of lung inflated above 10.2 ml/g. It also showed that lesions less than 5 mm in diameter corresponded to the fraction of the lung inflated between 6.0 and 10.2 ml/g and that regions inflated below 6.0 ml/g are morphologically normal. This method appeared more accurate than the surface area occupied by emphysema because these authors observed a reduced surface-to-volume ratio in mild emphysema, whereas surface area and tissue weight were decreased only in severe cases of emphysema.

5.1.3.3
Comparison between Objective and Subjective CT Quantification

An advantage of computer-assisted quantification is the reproducibility of the technique across readers of varying expertise and experience, and across institutions. This allows for a more accurate comparison of results among different centers (KAZEROONI 1999). On the other hand, the advantages of a subjective scoring system are ease of application and the lack of expensive dedicated software packages. In a series of 62 patients who underwent thin section CT prior to surgical lung resection, BANKIER et al. (1999) compared subjective visual grading of pulmonary emphysema with macroscopic morphometry and objective CT quantification. Three readers of varying degrees of expertise subjectively graded emphysema in two reading sessions. All three readers systematically overestimated emphysema and the interobserver agreement with weighted kappa values ranged from 0.43 to 0.59. Independent of the level of expertise of the individual reader, the correlation between subjective scores and macroscopic results was weaker than the correlation between objective CT quantification and macroscopic morphologic measurements. This study thus suggests that subjective visual grading should be supplemented with more reliable objective methods whenever a precise and reader-independent quantification of emphysema is required.

5.1.3.4
Fractal Analysis and Tissue Characterization

Quantifications of pulmonary emphysema by computer-assisted methods are based on mathematic approaches, named metrics, that may be used to describe the heterogeneity of the spatial distribution of attenuation values within the reconstructed image.

These metrics include very simple parameters such as the mean lung density to areas of low attenuation based on single or a range of densities (HOFFMAN and MCLENNAN 1997). Fractal dimensions are more complex metrics. The concept of fractals, first introduced by MANDELBROT, is used for a structure with a non-integer number of dimensions. Fractal geometry has found widespread applications in physical sciences because it is a suitable model for objective quantification of spatial heterogeneity (MCNAMEE 1991). Unlike Euclidean shape, a fractal cannot be represented by closed-form algebraic relationships and unlike Euclidean shapes, a fractal has continued detail at higher magnifications (MCNAMEE 1991). A more common definition of a fractal is a shape composed with smaller parts that, when enlarged, are similar to the whole shape. Thus, copies of itself can be found at any scale. In summary, the properties of fractals are scale invariance, self-similarity, and fractional dimension.

To differentiate both normal from emphysematous lungs and normal from emphysematous regions within one lung, UPPALURI et al. (1997) developed an adaptive multiple features method (AMFM) based on fractal analysis. With an electron beam CT scanner, they acquired two-dimensional slices of the whole lung with 3 mm collimation at maximal inspiration. Adjacent pixels with small differences of gray level were merged. Then, first-order and second-order statistical measurements were separately computed for each slice. First-order measurements were mean, variance, and skewness of the attenuation distribution curve. Second-order measurements included co-occurrence entropy, contrast and angular second moment. These authors observed that second-order statistics and fractal dimension were sensitive to the gray level and spatial relationships between pixels in a region. This suggests that these parameters can be used for tissue characterization. These authors compared AMFM, mean lung density, and the lowest fifth percentile of the distribution histogram of attenuation values to discriminate normal from emphysematous lungs. The accuracy of AMFM, mean lung density, and the lowest fifth percentile was 100%, 95%, and 97%, respectively. However, there was no correlation between these three parameters and pulmonary function tests (PFTs). The authors explained this lack of correlation by a too low number of slices acquired per patient, and by the absence of patients with mild and moderate emphysema in their study (RODARTE et al. 1989; HOFFMAN and MCLENNAN 1997).

More recently, MISHIMA et al. (1999a) attempted to detect early emphysema on the basis of fractal analysis. They quantified the size distribution of low attenuation area (i.e., lower than −960 HU) clusters on 2-mm thick HRCT slices obtained at full inspiration in healthy subjects (n=30) and in patients with chronic obstructive lung diseases (COPD) (n=73). All normal subjects had low attenuation areas smaller than 30% of total lung area. In patients with COPD, these areas varied from 2.6 to 67.6%. The authors observed that the cumulative size distribution of the low attenuation area clusters followed a power-law distribution characterized by an exponent D. Although the COPD group of patients with low attenuation areas smaller than 30% of total lung area and the normal subjects had similar low attenuation areas, the corresponding D values were significantly smaller in the COPD patients. On the basis of an elastic spring network model, these authors attributed this smaller value to the coalescence of smaller low attenuation area clusters into larger low attenuation area clusters in COPD patients. There was no correlation between the value of D and PFTs except for diffusion capacity of CO (D_LCO). Assuming that the exponent D is related to the fractal dimension of the alveolar surface (d_f), a measure of terminal airspace geometry complexity, a smaller D in a two-dimensional CT image is the consequence of larger low attenuation area clusters and represents the reduced d_f in COPD patients. These authors concluded that 30% could be the critical value of low attenuation area to discriminate normal and mild from severe COPD patients but that low attenuation area is not sufficiently accurate to distinguish early emphysematous patients from normal subjects. The value of D could be a sensitive parameter in order to detect terminal airspace enlargement that occurs in early emphysema.

5.1.3.5
Factors Influencing CT Densitometry

5.1.3.5.1
Age

Morphometric data from THURLBECK (1967a) and from GILLOOLY and LAMB (1993a) showed a significant correlation between airspace size and age. The increase of airspace size associated with advanced age could thus influence the CT density parameter and should be taken into account in longitudinal studies. In order to investigate the possible influence of age on density measurements, we measured the RA_{950} in 42 healthy subjects aged from 23 to 71 years old and we found a weak but significant correlation between age and the RA_{950} (r=0.328; p=0.034). These results are in accordance with those from SOEIJIMA et al. (2000) who investigated 36 symptom-free non-

smoking subjects with normal lung function during a 5-year follow-up period and showed that the percentage of RA_{960} increased with age, at least in the middle and lower lung zones.

5.1.3.5.2
CT Parameters

Since CT scanning parameters can influence the attenuation values and their distribution curve, MISHIMA et al. compared the low attenuation area (<−960 HU) obtained with various numbers of slices ranging from 3 to 10, various slice thicknesses from 2 to 5 mm, and various electric tube currents ranging from 50 to 250 mA. On the basis of image quality, exposure dose, and correlations with lung function tests, they suggested that three 2-mm thick CT sections, acquired with 200 mA tube current, are the most appropriate parameter to assess pulmonary emphysema (MISHIMA et al. 1999b).

5.1.3.5.3
Number of CT Sections

Pulmonary emphysema is heterogeneously distributed throughout the lung. Studies based on point-counting confirmed that an adequate assessment cannot be obtained from one lung slice alone (TURNER and WHIMSTER 1981). On the other hand, radiation exposure considerations are likely to favor sampling techniques rather than whole lung measurements. Depending on the presence of emphysema and on its spatial distribution, the minimum number of scans providing accurate results could change from patient to patient, but no CT study has defined the minimum number of scans necessary to provide accurate results. In a study based on comparisons between HRCT and macroscopy, the authors have attempted to define a maximum interval distance between HRCT scans providing valid results. They recalculated the RA_{950} by successively considering one scan of two, one of three, one of four and so on and compared these results to the results obtained with 1 cm intervals. The individual variability of the RA_{950} was very heterogeneous from patient to patient and no bend in the relationship linking the coefficient of variation of RA_{950} and the interval distance was found. Consequently, no particular interval distance could be proposed as on optimal standard (GEVENOIS et al. 1995b). More recently, MISHIMA et al. (1999b) attempted to define the influence of the slice number on RA_{960} in 30 patients with chronic obstructive pulmonary disease. These authors calculated correlation coefficients between RA_{960} measured on 5, 3, and

2 CT sections and RA_{960} measured on 10 CT sections, respectively. The corresponding coefficients were 0.976, 0.953, and 0.908, thus indicating highly significant correlations. The authors concluded that three slices were sufficient to obtain the overall extent of emphysema but they neither reported the severity of the disease in terms of functional deficit nor the heterogeneity of its spatial distribution within the lungs.

5.1.3.5.4
Spiral and Multislice CT

Spiral and multislice CT scanning has the major advantage that the entire thorax can be imaged during one single breath-hold. This method involves simultaneous transport of the patient at a constant speed through the CT gantry while data are continuously acquired over multiple gantry rotations (KOHZ et al. 1995). From spiral CT data, three-dimensional (3D) reconstructions, lung volume measurements and quantification of lung disorders can be obtained. By using dedicated software packages that are currently available on almost all modern scanners, KAUCZOR et al. (1998) compared lung volumes measured by spiral CT and by plethysmography. They found statistically significant correlations between both measurements and an underestimation of TLC by 12%, measured by spiral CT, likely due to the supine posture of the subject in the CT scanner compared to the seated posture in the plethysmography. Spiral CT data could be of great interest in the quantification of heterogeneously distributed lung disorders such as pulmonary emphysema, but to date no study has validated spiral CT parameters as referenced to histopathology. Dedicated software packages reconstructing a 3D model of the lungs, calculating their volume and providing the frequency distribution curve of attenuation values within this lung volume could be applied to spiral CT data sets. Such a program has been used by PARK et al. (1999) who compared the percentage volume of the lung occupied by attenuation values lower than three thresholds (–900 HU, –910 HU, –950 HU) with the percentage area of the lung occupied by attenuation values lower than these thresholds. The authors found highly significant correlations ($r=\sim0.98$) between lung attenuation measurements obtained with the 3D model and those from 2D images (SANDERS et al. 1988).

5.1.3.5.5
Lung Volume and Size

A potential role of expiratory CT in the assessment of emphysema was first suggested by KNUDSON et

al. (1991) because correlations between CT measurements and lung function tests were stronger in expiration than in inspiration. In 64 patients most of whom suffered from airflow obstruction, these authors obtained 8 -mm thick CT slices at two levels in upper lung zones at full inspiration and full expiration. Then they measured the percentage area of lung occupied by attenuation values lower than –900 HU. This percentage area was compared with various PFTs such as static lung compliance, D_LCO, and forced expiratory flow in 1 s (FEV_1). The strongest correlations between CT with physiologic variables consistent with emphysema were seen with CT measurements obtained at full expiration.

To investigate the possible role of quantitative CT during expiration, other authors measured relative areas of lung occupied by attenuation values lower than various thresholds ranging from –800 to –970 HU, at full inspiration and full expiration, in 89 patients who underwent surgical lung resection. These authors found two different thresholds, valid by comparisons with macroscopy (–910 HU) and microscopy (–820 HU), that were different from the threshold found valid for CT scans obtained at full inspiration (–950 HU) (GEVENOIS et al. 1995b, 1996b). In addition, multiple regression analysis showed that CT measurements obtained at full expiration did not yield any additional significant information compared to that obtained at full inspiration as to the prediction of the anatomic extent of emphysema. In a study based on visual scoring, NISHIMURA et al. (1998) showed that expiratory CT underestimates the degree of emphysema as compared with inspiratory CT. In summary, expiratory CT is not as accurate as inspiratory CT for measuring the extent of pulmonary emphysema. This conclusion suggests that possible errors secondary to variations of lung volume at which the CT scans are obtained could be avoided by using spirometric triggering (LAMERS et al. 1994; KOHZ et al. 1995; KAUCZOR et al. 1998; MISHIMA et al. 1999b).

In another study focused on expiratory CT, the authors studied correlations between PFTs and objective CT data obtained in deep expiration and in deep inspiration, respectively (GEVENOIS et al. 1996b). The authors observed higher correlations between PFTs reflecting the airflow obstruction and the CT data obtained in expiration than in inspiration but correlations were similar for diffusing capacity between inspiratory and expiratory data. The authors thus concluded that expiratory CT reflects more the expiratory airflow limitation and the subsequent air trapping than reduction of the alveolar wall surface. These findings were confirmed by EDA et al. (1997) and by LAMERS et al. (1994) who also found higher correlations between PFTs reflecting the airflow obstruction and CT data obtained in expiration than in inspiration, whereas the correlations between diffusing capacity and CT data were similar.

Independently from the lung volume at which the CT scan is obtained, the lung size could also influence CT parameters. Morphometric studies showed contradictory results, suggesting either that the number of alveoli in the human lung was (DUNNILL 1962a, 1964; WEIBEL 1963) or was not positively correlated with body length (ANGUS and THURLBECK 1972). In a cross-sectional study in 42 healthy subjects, the authors found a significant correlation between total lung capacity (TLC) and the mean lung attenuation (MLA) ($r = -0.42$; $p = 0.006$) as well as between TLC and the RA_{950} ($r = 0.39$; $p = 0.012$). The larger the TLC in absolute values, the lower the MLA and the higher the RA_{950}. These results suggest that the relative amount of lung tissue per unit of volume is lower in larger lungs than in smaller lungs. Accordingly, because the structure of the alveolar wall is unrelated to the lung size, the dimensions of the air spaces should be greater in larger lungs than in smaller lungs.

5.1.3.6
Comparison between CT Quantification and Pulmonary Function Tests

Although PFTs may be short-term and long-term reproducible tests, they represent global measurements resulting from more than 10 million airways that contribute unequally to airflow (GURNEY 1998). They are of limited value in the measurement of the obstruction of airways, notably small airways, that are predominantly affected in emphysema (GURNEY 1998). Autopsy studies have indeed shown that up to one-third of the lung can be destroyed by emphysema before respiratory function becomes impaired (UPPALURI et al. 1997). The poor sensitivity of PFTs to diagnose mild emphysema was confirmed by SANDERS et al. (1988) who found that emphysematous features of emphysema were visually detected on CT scans in 69% of smokers with normal $D_L CO$ with or without associated obstructive deficits. In this study, CT showed evidence of emphysema in 96% of patients selected on the basis of functional criteria of emphysema as suggested by the American Thoracic Society such as decreased $D_L CO$ (<80% predicted value) plus evidence of obstructive lung disease (decreased FEV_1 <80% and/or increased residual volume RV>120% predicted values). The authors concluded that CT may be more sensitive than PFTs in detecting mild emphysema.

The lack of sensitivity of PFTs to detect pulmonary emphysema is probably related to pulmonary zones in which ventilatory disorders are not assessed by conventional PFTs. First, the total airflow resistance of all respiratory bronchioles contributes little to the total airflow resistance of the lung (GURNEY 1998). Despite the high airflow resistance through one single respiratory bronchiole, the parallel connection of a high number of bronchioles leads to a wide total cross-sectional area and drastically reduced the airflow resistance (RIENMÜLLER et al. 1991). Second, the upper lung zone has a relatively high ventilation/perfusion (V/Q) ratio compared to lower lung zone. Thus, in the relatively hypoventilated upper lung zone, emphysema produces smaller measurable pulmonary dysfunction than in the lower zone. Consistently, GURNEY et al. (1992) and HARAGUSHI et al. (1998) showed that the extent of emphysema had higher correlations with $D_L CO$ in the lower lung zone than in the upper lung zone, even though the upper lung zone was more severely affected by emphysema. On the basis of the lobar distribution of emphysema as determined by CT, SAITOH et al. (2000) reported that the airflow limitation, the residual volume and the total lung capacity were higher in the predominantly lower-lobe emphysematous group than in the predominantly upper-lobe emphysematous group. However, NAKANO et al. reported that the correlations between the RA_{960}, and FEV_1 or RV/TLC were higher in the lower lobes but the correlation between RA_{960} and $D_L CO$ was higher in the upper lobes. The authors attributed this discrepancy between their results and the other studies, to the high incidence of severe emphysema in the upper lobe, which affects the $D_L CO$ (NAKANO et al. 1999). The central versus peripheral predominant location of the emphysematous area determines also the importance of functional impairment. HARAGUSHI et al. found higher correlations between PFTs, including FEV_1 and $D_L CO$, and attenuation values of the central region than the peripheral region of lungs (HARAGUSHI et al. 1998). This is in agreement with NAKANO et al. (1999) who reported a higher incidence of emphysema in central regions compared to peripheral regions (HARTLEY et al. 1994). These authors explained the results by the move of particle deposition from the outer to the inner lung, by a greater stratified distribution of pulmonary blood perfusion in the outer than in the inner lung, and by the lymphatic drainage of particles from the outer to the inner lung, which is favored by the ventilatory movements (NAKANO et al. 1999).

The correlation between CT indexes and $D_L CO$ or $D_L CO$/VA were extensively documented in numerous studies and ranged from -0.5 to -0.75 (SANDERS et al. 1988; GOULD et al. 1988; KINSELLA et al. 1990; MORRISON et al. 1989; WATANUKI et al. 1994). In a study comparing CT with microscopic morphometry, we obtained correlation coefficients between RA_{950} and microscopic indexes of the same degree of magnitude as between FEV_1/VC or D_LCO/VA and these microscopic indexes ($r=\sim0.70$). Considering the microscopic measurements of the reference method, we entered the RA_{950} and independent PFTs in stepwise procedures. The resulting relationships revealed that D_LCO and the RA_{950} were sufficient to predict the results of microscopic measurements (GEVENOIS et al. 1996a).

More recently, using the 3D model described above, PARK et al. (1999) investigated the relationships between PFTs and percentage of lung volume occupied by attenuation values lower than three thresholds (–900 HU, –910 HU, –950 HU). He found moderate to high correlations between these percentage volumes and TLC ($r=0.62-0.71$), FEV_1 ($r=-0.57$ to -0.60) and FEV_1/FVC ($r=-0.75$ to -0.82), and D_LCO ($r=-0.57$ to -0.64). The percentage volume lower than –950 HU was more closely correlated with D_LCO and FEV_1 than either the volumes lower than –910 HU or –900 HU. The authors concluded that lung densitometry derived from 3D lung models is an available alternative method compared to 2D models for quantifying emphysema.

5.1.3.7
Surgical Treatment of Emphysema

Lung volume reduction surgery (LVRS) is a therapeutic option for severe debilitating emphysema. The surgical intervention consists of bilateral wedge resection of emphysematous lung tissue by means of sternotomy, bilateral thoracotomy, or video-assisted thoracoscopic surgery (KAZEROONI 1999). This technique induces a one-time benefit improvement, peaking at 3–6 months after surgery, in terms of lung function, exercise tolerance, and quality of life (COOPER and LEFRAK 1996; GELB et al. 1998; RUSSI et al. 1999; GEDDES et al. 2000). The retrospective nature of patient selection and the inability to accurately quantify the amount of resected emphysematous lung tissue are two major obstacles to define criteria for candidate selection for LVRS (RUSSI et al. 1999). Patients are usually selected on the basis of clinical, physiological and radio-anatomical assessments. Intolerable dyspnea and exercise intolerance not palliated by medical therapy, and severe airflow obstruction are the main clinical criteria. On the other hand, anatomic features, consisting of lobar severity of emphysema, with a heterogeneous distribution of emphysema, are important for a better clinical outcome. Upper lobe predominance, higher amounts

of regional heterogeneity, and a larger percentage of normal or mildly emphysematous lung tissue showed the highest association with improvement of the quality of life and exercise tolerance (KAZEROONI 1999).

In a study based on visual CT analysis of emphysema, WEDER et al. showed that the mean increase in FEV_1 after LVRS was $\sim80\%$, 40%, and 35% for heterogeneous, intermediate heterogeneous and homogeneous emphysema, respectively (WEDER et al. 1997). However, this study revealed that preoperative characteristics of pulmonary function or chest CT morphology could not explain the postoperative improvement in terms of FEV_1.

In order to determine whether quantitative CT provides relevant information for guiding patient selection, GIERADA et al. (1997) compared CT quantification of pulmonary emphysema with FEV_1, arterial oxygen tension (P_aO_2), and a 6 -min walking distance before and after LVRS. They showed that the values of quantitative CT indexes of global and regional emphysema severity were related to outcome measures after LVRS. Indexes of global emphysema severity include the mean lung attenuation, the percentage of whole lung with attenuation below –900 HU (emphysema index), and the percentage of whole lung with attenuation below –960 HU (severe emphysema index). Indexes of regional emphysema severity include emphysema indexes in the upper and lower halves of the lungs and the ratio of the emphysema index of the upper lung to that of lower lung. Postoperative improvement was better with a mean lung attenuation greater than –900 HU, an emphysema index of 75% or greater, a severe emphysema index greater than 25%, a ratio of upper-lung and lower-lung emphysema indexes of 1.5 or greater and an upper lung emphysema index greater than 75%. Considering other simple characteristics of the attenuation distribution curve such as the standard deviation and the full width at half maximum, the authors did not identify a CT index of emphysema heterogeneity that could predict patient outcome.

As recently reviewed by KAZEROONI (1999), a high upper-lung and lower-lung emphysema ratio has been the best predictor of improvement in term of FEV_1 and a 6 -min walking distance until 2 years after bilateral apical LVRS. This ratio demonstrated a higher correlation with outcome than the percentage of emphysema of the whole lung and than the functional parameters reflecting hyperinflation such as RV, TLC, RV/TLC ratio and FEV_1, or D_LCO (KAZEROONI 1999). So, quantitative CT could play an important role in identifying potentially suitable candidates and standardizing the preoperative imaging evaluation of these candidates.

New Perspectives in Pharmacotherapy

For many years, it was hypothesized that in patients with a_1-antitrypsin deficiency, replacement therapy could prevent the progression of pulmonary emphysema. In a double-blind controlled study performed over a 3-year period and involving 56 patients with a_1-antitrypsin deficiency of PI *ZZ phenotype and moderate to severe emphysema, DIRKSEN et al. (1999) estimated the loss of lung tissue by calculating the 15th lowest percentile of the attenuation distribution curve. DIRKSEN et al. found that this loss tended to be higher in the group of untreated patients compared with the group of patients treated by a_1-antitrypsin infusion ($p=0.07$). In addition, power analysis showed that this protective effect would reach statistical significance is a similar trial with 130 patients in contrast to calculations based on annual decline of FEV_1 showing that 550 patients would be needed to show a 50% reduction on annual decline. These authors concluded that lung density measurements could be more sensitive than PFTs for detecting the progression of emphysema and that CT may facilitate future randomized clinical trials.

5.1.4
Conclusion

CT scanning is of particular interest for the in vivo diagnosis and quantification of pulmonary emphysema because this imaging technique offers measurements of both morphological and functional information, with the possibility to interrelate structure and function. The presence and extent of pulmonary emphysema can be roughly estimated by visual assessment of CT sections but objective quantification is more accurate and more reproducible. This review documents the role of CT in the diagnosis and quantification of pulmonary emphysema. Nevertheless, this technique is not yet standardized and important issues are still unsolved. Further studies are needed to address the following questions: 1) to establish normal values of reference, 2) to investigate the reproducibility of CT measurements, 3) to investigate the influence of image acquisition parameters on the CT measurements, 4) to compare the accuracy of various CT indexes based on a percentile of the frequency distribution curve such as the 5th or the 15th lowest one, on absolute thresholding such as RA_{950} or RA_{910}, on regional lung inflation, or on a fractal dimension, 6) to evaluate the accuracy of CT to quantify as well as to diagnose pulmonary emphysema and to recognize subtypes of emphysema, 7) to validate new CT techniques such as multidetector systems, 8) to investigate the ability of CT to discriminate emphysema from other chronic obstructive pulmonary diseases. Since inspiratory CT allows reasonably accurate objective quantification of the extent of pulmonary emphysema, the appropriate CT and post-processing techniques should be standardized.

5.2
Magnetic Resonance Imaging

5.2.1
Introduction

As compared to other organs of the human body, the lung parenchyma has a low tissue density and therefore yields a low magnetic resonance (MR) signal. Moreover, the anatomically complex structure of the lung leads to susceptibility artifacts that cause further MR signal degradation. In patients with emphysema, this situation is aggravated by the fact that emphysematous lung will have an even lower density than normal lung. At first sight, the emphysematous lung is thus no very appealing target for MRI. Despite all drawbacks, however, the lung parenchyma of patients suffering from emphysema has been a focus of interest for MRI almost since the very introduction of this diagnostic modality into clinical routine. Interest in MR imaging of the emphysematous lung parenchyma and of the entire emphysematous thorax in general was further vitalized by constant technical improvements and new developments such as gradient-echo techniques, intravenous and aerosolized gadolinium chelates, inhaled oxygen enhancement, and hyperpolarized noble gases (KAUCZOR et al. 1999, 2001, 2002). Because the individual MR techniques used in the imaging of the lung parenchyma have been described in detail previously, this chapter will concentrate on the application of these techniques to the imaging of pulmonary emphysema. Hereby, the preceding chapter about CT of pulmonary emphysema anticipates much of the fundamental and clinical research efforts that still remain to be done in MRI.

5.2.2
Proton (^{1}H) Magnetic Resonance Imaging

Given the hardly favorable MR imaging circumstances of the lung parenchyma itself, clinical studies performed in patients with emphysema rather concentrated on the chest wall and the diaphragm. The aim of these studies was to quantify changes in shape and rib cage dimensions between inspiration and expiration. This was mainly achieved by generating 3D reconstructions of the acquired MRI data and by correlating lung volumes and morphometric changes of the diaphragm. In a basic study, CLUZEL et al. (2000) performed MRI of the thorax in five volunteers at three predefined respiratory levels (TLC, FRC, and RV). After manual segmentation and 3D reconstruction, the diaphragm was divided into the central (dome) and the peripheral (apposition) zones. Thereby, the overall size of the diaphragm, and surface areas of the central and peripheral zones of the diaphragm could be measured in all three orientations. The comparison of MR data with respiratory volumes revealed that MRI tended to overestimate RV and to underestimate TLC. Between RV and TLC, the mean volume under the diaphragm decreased by 66%, whereas the mean total volume of the rib cage increased by 23%. The contribution of the diaphragm to the inspiratory capacity was 60%. Based on this finding, the authors termed the diaphragmatic contribution to inspiration as "inspiratory pump".

A complementary perspective on the "inspiratory pump" was given by SUGA et al. (1999). In patients assessed with MRI before and after lung volume reduction surgery, changes in chest wall dimensions were in good accordance with postoperative improvements in respiratory mechanics. In that study, dynamic MRI was applied to evaluate diaphragmatic and chest wall motion over several respiratory cycles in a group of 6 volunteers and 28 patients with emphysema, 9 of which were studied before and after lung volume reduction surgery. After baseline normalization in all subjects, the maximum excursion and the ratio between diaphragmatic and chest wall motion were compared with %FEV$_1$, %VC, %TLC, and %RV. All six volunteers showed synchronous excursion of both the diaphragm and the chest wall without substantial differences between the right and the left hemithorax. Chest wall motion was significantly higher at the lung apices than at the lung bases. In patients with emphysema, chest wall motion was markedly reduced and irregular. Moreover, the motion patterns at the apices and the bases of the lungs were dissociated. Differences between inspiration and expiration were significantly smaller in patients with emphysema than in healthy volunteers. The differences in chest wall motion between the apices and the bases of the lungs seen in the healthy volunteers were no longer apparent in the patients with emphysema. After lung volume reduction surgery, an obvious increase in diaphragmatic and chest wall excursion was found as compared to the preoperative status. The

overall correlations between functional data and MRI derived data were good. The authors concluded that MRI was helpful in quantifying impaired respiratory mechanics in patients with emphysema and suggested that MRI could play a role in assessing the impact of lung volume reduction surgery.

5.2.3
Contrast Agents: Intravenous Administration and Inhalation

Gadolinium chelates are the most widely applied intravenous contrast agents used in MRI (Fig. 5.3). Gadolinium chelates can also be aerosolized and inhaled, but experience with this technique in humans is limited. Besides gadolinium chelates, so-called blood-pool contrast agents, that consist of ultra-small super paramagnetic iron oxide particles, of coated gadolinium-complex compounds, or of albumin-binding agents, have been used in clinical pilot studies investigating patients with suspected pulmonary embolism. For both gadolinium chelates and blood-pool agents, however, there currently is no substantial scientific experience as to potential benefits of their use in patients with suspected or proven emphysema (KAUCZOR and KREITNER 1999; KAUCZOR et al. 2001, 2002). Results from animal studies and episodic reports from patients with non-emphysematous diseases nevertheless suggest that MRI has the potential to elucidate medically relevant issues in patients with emphysema (Fig. 5.3). These issues include perfusion analyses, notably in patients with early emphysema, and studies related to perfusion-ventilation inhomogeneities. In this context, recent experience with the combined use of perfusion and ventilation imaging appears promising. In animal studies, VIALLON et al. (2000) have developed an overlay technique resulting in "fusion" of MR perfusion images and MR ventilation images using hyperpolarized gases, generating "ventilation/perfusion" data. Similar data were acquired by CHEN et al. (1999) using oxygen enhancement as a surrogate for ventilation (Fig. 5.3). In this context, it can be expected that many of the MR techniques already available for lung imaging in general will contribute to solve the specific issues in patients with pulmonary emphysema.

5.2.4
Inhaled Oxygen

In proton MRI, oxygen can be used as a contrast agent because its paramagnetic properties affect

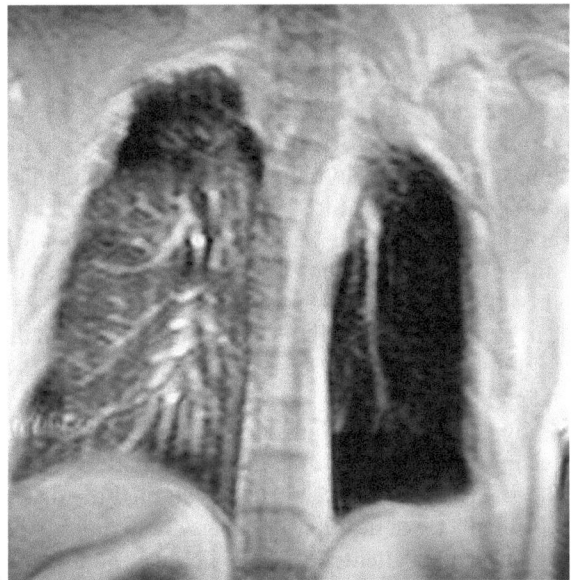
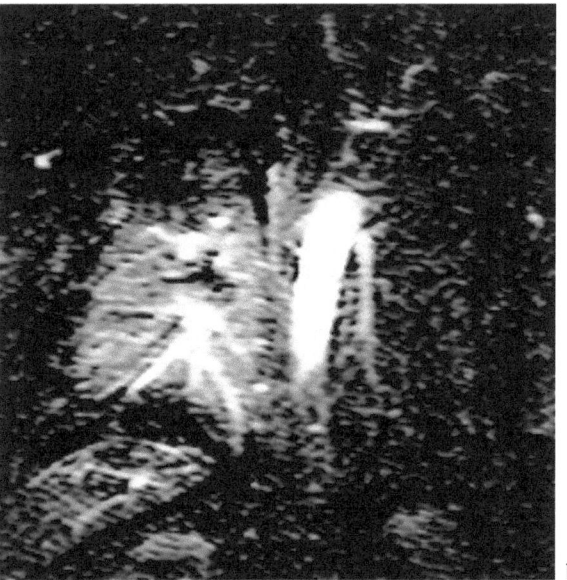

a

b

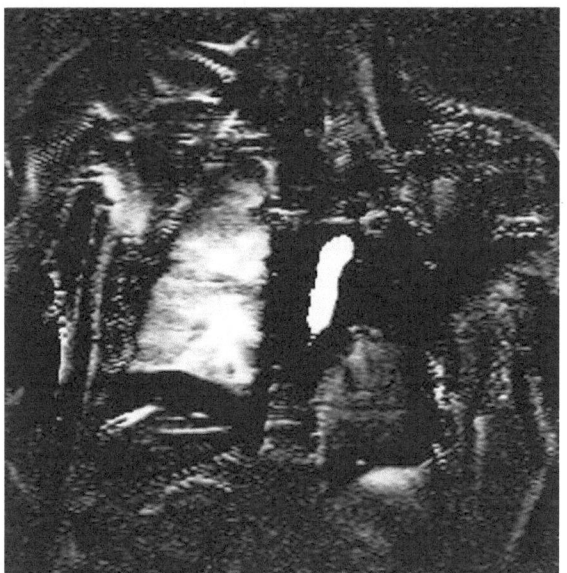

c

Fig. 5.3. MR imaging of a patient with Swyer-James syndrome. Gadolinium-enhanced T1-weighted image (**a**), FAIRER (**b**), and oxygen-enhanced image (**c**) show virtually normal right lung with some defects in the upper lobe, whereas the hypoplastic left lung is both underperfused and underventilated. A complete match of perfusion and ventilation can be appreciated

both lung parenchyma and blood by direct contact (Edelman et al. 1996). The dissolution of oxygen causes a T1-shortening in blood and tissue which results in a signal increase on T1-weighted images. By subtracting an image of a patient breathing 100% oxygen from an image of the same patient breathing room air, a surrogate of ventilation can be evaluated (Fig. 5.3) (Chen et al. 1999; Mai et al. 2000). Modifications of the MR acquisition sequence allow the image for either ventilation or perfusion to be weighted. Overall, oxygen imaging is an appealing alternative technique for imaging surrogates of ventilation and is more easy to handle that hyperpolarized noble gases. However, representative experience with this technique in patients with emphysema is still lacking. Potentially, this technique could be used to study regional oxygen uptake and diffusion in patients with known or suspected emphysema (Fig. 5.3).

5.2.5
Non-Proton Magnetic Resonance Imaging

Other than in proton-based techniques, non-proton-based techniques use hyperpolarized noble gases, mostly [3]He and [129]Xe, to generate their images. Instead of relying on the indirect enhancement effect on pulmonary parenchyma by inhaled or injected substances, the gas itself is directly imaged (BLACK et al. 1996; KAUCZOR et al. 1997). In normal individuals, ventilation is represented by an almost complete and homogeneous MR signal in the lung (KAUCZOR et al. 1997). In the posterior parts of the lung, however, transient ventilation defects, most likely corresponding to individual secondary pulmonary lobules, can sometimes be detected (ALTES et al. 2001). More widespread ventilation defects always represent impaired regional ventilation (KAUCZOR et al. 1997). The exact threshold between normal and abnormal patterns of regional ventilation, however, has not yet been determined.

DE LANGE et al. (1999) studied 13 healthy individuals and 3 individuals with a history of smoking and known COPD. Both COPD and emphysema were associated with multiple ventilation defects seen on MRI. The defects were round or wedge-shaped, patchy or widespread, and could involve distinct lung segments or entire lobes. Ventilation defects were characterized by either partial signal loss or a complete lack of signal. The authors speculated that certain patterns of ventilatory defects could be associated with either central or peripheral location of bronchial and bronchiolar obstruction. It also remained unclear how the varying patterns of ventilation defects are associated with functional indicators of bronchial obstruction or hyperinflation.

A key advantage of [3]He is that it can be used to assess MR diffusion. This capacity is based on both the properties of the gas and dedicated MRI acquisition techniques (KAUCZOR et al. 2001, 2002). In a rat model of emphysema, this technique has revealed significant differences in apparent diffusion coefficients (ADC) and airway sizes between healthy and diseased animals (CHEN et al. 2000). In patients with emphysema, however, ADC mapping is difficult given the limitations by concomitant ventilation defects. In the normal lung, diffusive gas movement is isotropic, whereas diffusion is anisotropic in fibrosis and emphysema. This could suggest a non-spherical alteration in alveolar geometry and a preferential direction for diffusive gas movement (KAUCZOR et al. 2002). The higher ADC values seen in patients with severe emphysema are consistent with increased alveolar size as a result of alveolar wall destruction (Fig. 5.4). This is congruent with results of a study by SALERNO et al. (2002). The authors compared [3]He diffusion MRI with lung function tests in normal individuals and in patients with emphysema. Their results confirmed that ADC images in healthy individuals were homogeneous, whereas they displayed large variability in patients

with emphysema. Moreover, the mean ADCs were larger in patients with emphysema as compared to normal individuals. There was a good correlation between ADC and predicted FEV_1 and the FEV_1/FVC ratio, respectively. MR diffusion measurements of the lungs may prove helpful in the assessment of emphysema and other diseases that alter alveolar size and alveolar elastic recoil. Moreover, the potential of MR diffusion measurements opens new perspectives on the follow-up of medical treatment in patients with early emphysema. Both the therapeutic effect and the long-term influence of these drugs could be assessed using MRI.

By combining [3]He inhalation with dynamic MRI techniques, real-time imaging of ventilation can be performed. GIERADA et al. (2000) studied the distribution of [3]He in healthy volunteers and patients with emphysema. Patients with centrilobular emphysema related to cigarette smoking had severe ventilation defects, predominantly in the upper lobes. The gas distribution during the first inhalation of [3]He was characterized by sequential filling of non-segmental lung regions interspersed with unfilled regions. The distribution of the MR signal became more homogeneous during rebreathing of the gas from a bag. This observation was attributed to collateral ventilation and thought to represent different intrapulmonary time constants. The authors showed that lung regions severely diseased on CT correlated with delayed or absent ventilation at [3]He MRI. Washout was substantially prolonged in emphysematous lung regions, most likely due to the retention of gas and the low local oxygen concentration.

Overall, the available evidence indicates that nonproton MRI holds substantial promise for the elucidation of pathomechanisms and pathomorphology in patients with emphysema. As in other fields of

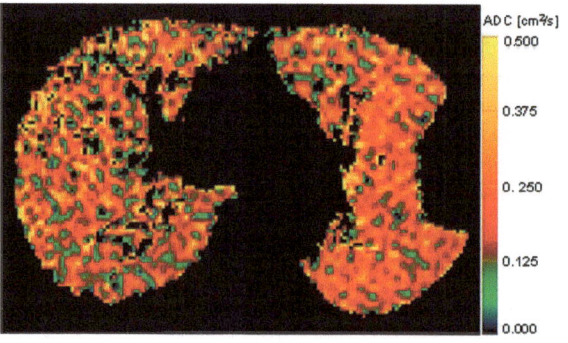

Fig. 5.4. Apparent diffusion coefficient (ADC) map based on [3]He MRI in a patient with emphysema. The map shows subtle areas of parenchymal destruction that correspond to increased values for the ADC (orange and yellow areas)

pulmonary MRI, however, there still is a flagrant lack of clinical studies and of uniform examination protocols. Once these issues are resolved, there should be little doubt that non-proton MRI will hold a key place in the preclinical and clinical evaluation of patients with emphysema.

5.2.6
Conclusion

MRI is of particular interest for the functional in vivo evaluation of patients with pulmonary emphysema. As for CT, MRI can provide both functional and morphological information. MRI, however, has the advantage of non-irradiation. Despite the promising results of first clinical studies documented in this report, the following issues still remain to be solved: 1) to establish normal reference values for non-proton MRI, notably as to ADC and MR diffusion, 2) to establish reference values related to age and gender of potential study collectives, 3) to establish standardized examination protocols for proton and non-proton MRI of the lungs, 4) to validate preliminary findings with the morphological and functional standards of reference, 5) to determine the clinical relevance of MRI in the assessment of early emphysema and in the follow-up of patients under medical therapy and 6) to establish standardized protocols for the post-processing of acquired MRI data.

References

Altes TA, Powers PL, Knight-Scott J et al. (2001) Hyperpolarized ³He MR lung ventilation imaging in asthmatics: preliminary findings. J Magn Reson Imaging 13:378–384

Angus GE, Thurlbeck WM (1972) Number of alveoli in the human lung. J Appl Physiol 32:483–485

Bae KT, Slone RM, Gierada DS, Yusen RD, Cooper JD (1997) Patients with emphysema: quantitative CT analysis before and after lung volume reduction surgery. Radiology 203: 705–714

Bankier AA, Maertelaer V de, Keyzer C, Gevenois PA (1999) CT of pulmonary emphysema: subjective assessment and objective quantification by densitometry and macroscopic morphometry. Radiology 211:851–858

Bergin CJ, Muller NL, Miller RR (1986a) CT in the quantitative assessment of emphysema. J Thorac Imaging 1:94–103

Bergin C, Müller N, Nichols DM et al. (1986b) The diagnosis of emphysema: a computed tomographic-pathologic correlation. Am Rev Respir Dis 133:541–546

Black RD, Middleton HL, Cates GD et al. (1996) In vivo He-3 MR images of guinea pig lungs. Radiology 199:867–870

Boren HG (1962) Alveolar fenestrae: relationship to the pathology and pathogenesis of pulmonary emphysema. Am Rev Respir Dis 85:328–344

Chen Q, Levin DL, Kim D et al. (1999) Pulmonary disorders: ventilation-perfusion MR imaging with animal models. Radiology 213:871–879

Chen XJ, Hedlund LW, Moller HE, Chawla MS, Maronpot RR, Johnson GA (2000) Detection of emphysema in rat lungs by using magnetic resonance measurements of ³He diffusion. Proc Natl Acad Sci USA 97:11478–11481

Cluzel P, Similowski T, Chartrand-Lefebvre C, Zelter M, Derenne JP, Grenier PA (2000) Diaphragm and chest wall: assessment of the inspiratory pump with MR imaging – preliminary observations. Radiology 215:574–583

Cooper JD, Lefrak SS (1996) Is volume reduction surgery appropriate in the treatment of emphysema? Yes. Am J Respir Crit Care Med 153:1201–1204

Cosio MG, Shiner RJ, Saetta M, Wang N, King M, Ghezzo H, Angus E (1986) Alveolar fenestrae in smokers. Relationships with light microscopic and functional abnormalities. Am Rev Respir Dis 133:126–131

Coxson HO, Rogers RM, Whittall KP, D'Yachkova Y, Paré PD, Sciurba FC, Hogg JC (1999) A quantification of the lung surface area in emphysema using computed tomography. Am J Respir Crit Care Med 159:851–856

De Lange EE, Mugler JP III, Brookeman JP et al. (1999) Lung air spaces: MR imaging evaluation with hyperpolarized ³He gas. Radiology 210:851–857

Dirksen A, Dijkman JH, Madsen F et al. (1999) A randomized clinical trial of a1-antitrypsin augmented therapy. Am J Respir Crit Care Med 160:1468–1472

Dunnill MS (1962a) Postnatal growth of the lung. Thorax 17: 329–333

Dunnill MS (1962b) Quantitative methods in the study of pulmonary pathology. Thorax 17:320–328

Dunnill MS (1964) Evaluation of a simple method of sampling the lung for quantitative histological analysis. Thorax 19: 443–448

Eda S, Kubo K, Fujimoto K, Matsuzawa Y, Sekiuchi M, Sakai F (1997) The relations between expiratory chest CT using helical scanning and pulmonary function tests in emphysema. Am J Respir Crit Care Med 155:1290–1294

Edelman RR, Hatabu H, Tadamura E, Li W, Prasad PV (1996) Noninvasive assessment of regional ventilation in the human lung using oxygen-enhanced magnetic resonance imaging. Nat Med 2:1236–1239

Geddes D, Davies M, Koyama H, Hansell D, Pastorino U, Pepper J, Agent P, Cullinan P, McNeill SJ, Glodstraw P (2000) Effect of lung-volume reduction surgery in patients with severe emphysema. N Engl J Med 343:239–245

Gelb AF, Hogg JC, Müller NL et al. (1996) Contribution of emphysema and small airways in COPD. Chest 109:353–359

Gelb AF, Brenner M, Mc Kenna RJ Jr, Fischel R, Zamel N, Schein MJ (1998) Serial lung function and elastic recoil 2 years after lung volume reduction surgery for emphysema. Chest 113:1497–1506

Gevenois PA, Yernault JC (1995) Can computed tomography quantify pulmonary emphysema? Eur Respir J 5:843–848

Gevenois PA, Zanen J, Maertelaer V de, Vuyst P de, Dumortier P, Yernault JC (1995a) Macroscopic assessment of pulmonary emphysema by image analysis. J Clin Pathol 48: 318–322

Gevenois PA, Maertelaer V de, Vuyst P de, Zanen J, Yernault JC (1995b) Comparison of computed density and macroscopic morphometry in pulmonary emphysema. Am J Respir Crit Care Med 152:653–657

Gevenois PA, Vuyst P de, Maertelaer V de, Zanen J, Jacobovitz D, Cosio MG, Yernault J-C (1996a) Comparison of computed density and microscopic morphometry in pulmonary emphysema. Am J Respir Crit Care Med 154:187–192

Gevenois PA, Vuyst P de, Sy M, Scillia P, Chaminade L, Maertelaer V de, Zanen J, Yernault J-C (1996b) Pulmonary emphysema: quantitative CT during expiration. Radiology 199:825–829

Gierada DS, Slone RM, Bae KT, Yusen RD, Lefrak SS, Cooper JD (1997) Pulmonary emphysema: comparison of preoperative quantitative CT and physiologic index values with clinical outcome after lung volume reduction surgery. Radiology 205:235–242

Gierada DS, Saam B, Yablonskiy D, Cooper JD, Lefrak SS, Conradi MS (2000) Dynamic echo planar MR imaging of lung ventilation with hyperpolarized (3)He in normal subjects and patients with severe emphysema. NMR Biomed 13:176–181

Gillooly M, Lamb D (1993a) Airspace size in lungs of lifelong non-smokers: effect of age and sex. Thorax 48:39–43

Gillooly M, Lamb D (1993b) Microscopic emphysema in relation to age and smoking habit. Thorax 48:491–495

Gould GA, Macnee W, McLean A, Warren M, Redpath A, Best JJK, Lamb D, Flenley DC (1988) CT measurements of lung density in life can quantitate distal airspace enlargement: an essential defining feature of human emphysema. Am Rev Respir Dis 137:380–392

Gurney JW (1998) Pathophysiology of obstructive airways disease. Radiol Clin North Am 36:15–27

Gurney JW, Jones KK, Robbins RA, Gossman GL, Nelson KJ, Daughton D, Spurzem JR, Rennard SI (1992) Regional distribution of emphysema: correlation of high-resolution CT with pulmonary function tests in unselected smokers. Radiology 183:457–463

Haragushi M, Shimura S, Hida W, Shirato K (1998) Pulmonary function and regional distribution of emphysema as determined by high-resolution computed tomography. Respiration 65:125–129

Hartley PG, Galvin JR, Hunninghake GW, Merchant JA, Yagla SJ, Speakman SB, Schwartz DA (1994) High-resolution CT-derived measures of lung density are valid indexes of interstitial lung diseases. J Appl Physiol 76:271–277

Hayhurst MD, Flenley DC, McLean A, Wightman AJA, MacNee W, Wright D, Lamb D, Best J (1984) Diagnosis of pulmonary emphysema by computerised tomography. Lancet 2: 320–322

Hoffman EA, Mc Lennan G (1997) Assessment of the pulmonary structure-function relationship and clinical outcomes measures. Acad Radiol 4:758–776

Hruban RH, Meziane MA, Zerhouni EA, Khouri NF, Fishman EK, Wheeler PS, Dumler JS, Hutchins GM (1987) High resolution computed tomography of inflation-fixed lungs: pathologic-radiologic correlation of centrolobular emphysema. Am Rev Respir Dis 136:935–940

Kalender WA, Fitche H, Bautz W, Skalej M (1991) Semiautomatic evaluation procedure for quantitative CT of the lung. J Comput Assist Tomogr 15:248–255

Kauczor HU, Kreitner KF (1999) MRI of the pulmonary parenchyma. Eur Radiol 9:1755–1764

Kauczor HU, Ebert M, Kreitner KF et al. (1997) Imaging of the lungs using ³He MRI: preliminary clinical experience

in 18 patients with and without lung disease. J Magn Reson Imaging 7:538–543

Kauczor HU, Heussel CP, Fisher B, Klamm R, Mildenberger P, Thelen M (1998) Assessment of lung volumes using helical CT at inspiration and expiration: comparison with pulmonary function tests. AJR Am J Roentgenol 171:1091–1095

Kauczor HU, Heussel CP, Thelen M (1999) Update on diagnostic strategies of pulmonary embolism. Eur Radiol 9: 262–275

Kauczor HU, Chen XJ, Beek EJ van, Schreiber WG (2001) Pulmonary ventilation imaged by magnetic resonance: at the doorstep of clinical application. Eur Respir J 17:1008–1023

Kauczor HU, Hanke A, Beek EJ van (2002) Assessment of lung ventilation by MR imaging: current status and future perspectives. Eur Radiol 12:1962–1970

Kazerooni EA (1999) Radiologic evaluation of emphysema for lung volume reduction surgery. Clin Chest Med 20:45–861

Kinsella M, Müller NL, Abboud RT, Morrison NJ, DyBuncio A (1990) Quantitation of emphysema by computed tomography using a "Density Mask" program and correlation with pulmonary function tests. Chest 97:315–321

Knudson RJ, Standen JR, Kaltenborn WT, Knudson DE, Rehm K, Habib MP, Newell JD (1991) Expiratory computed tomography for assessment of suspected pulmonary emphysema. Chest 99:1357–1366

Kohz P, Stäbler A, Beinert T, Behr J, Egge T, Heuck A, Reiser MF (1995) Reproductibility of quantitative, spirometrically controlled CT. Radiology 197:539–542

Kuwano K, Matsuba K, Ikeda T, Murakami J, Araki A, Nishitani H, Teruyoshi I, Yasumoto K, Shigematsu N (1990) The diagnosis of mild emphysema: correlation of computed tomography and pathology scores. Am Rev Respir Dis 141:169–178

Lamers RJ, Thelissen GR, Kessels AG, Wouters EF, Engelshoven JM (1994) Chronic obstructive pulmonary disease: evaluation with spirometrically controlled CT lung densitometry. Radiology 193:109–113

Mai VM, Chen Q, Bankier AA, Edelman RR (2000) Multiple inversion recovery MR subtraction imaging of human ventilation from inhalation of room air and pure oxygen. Magn Reson Med 43:913–916

McLean A, Warren PM, Gillooly M, MacNee W (1992) Microscopic and macroscopic measurements of emphysema: relation to carbon monoxide gas transfer. Thorax 47: 144–149

McNamee JE (1991) Fractal perspectives in pulmonary physiology. J Appl Physiol 71:1–8

Miller RR, Muller NL, Vedal S, Morrison NJ, Stapels CA (1989) Limitations of computed tomography in the assessment of emphysema. Am Rev Respir Dis 139:980–983

Mishima M, Hirai T, Itoh H et al. (1999a) Complexity of terminal airspace geometry assessed by lung computed tomography in normal subjects and patients with chronic obstructive pulmonary disease. Proc Natl Acad Sci USA 96: 8829–8834

Mishima M, Itoh H, Sakai H et al. (1999b) Optimized scanning conditions of high resolution CT in the follow-up of pulmonary emphysema. J Comput Assist Tomogr 23:380–384

Morgan MDL (1992) Detection and quantification of pulmonary emphysema by computed tomography: a window of opportunity. Thorax 47:1001–1004

Morrison NJ, Abboud RT, Ramadan F, Miller RR, Gibson NN, Evans KG, Nelems B, Müller NL (1989) Comparison of single breath carbon monoxide diffusing capacity and pressure-volume curves in detecting emphysema. Am Rev Respir Dis 139:1179–1187

Müller NL, Thurlbeck WM (1996) Thin-section CT, emphysema, air-trapping, and airway obstruction. Radiology 199: 621–622

Müller NL, Stapels CA, Miller RR, Abboud RJ (1988) "Density Mask": an objective method to quantitate emphysema using computed tomography. Chest 94:782–787

Nagai A, Inano H, Matsuba K, Thurlbeck WM (1994) Scanning electron microscopy morphometry of emphysema in humans. Am J Respir Crit Care Med 150:1411–1415

Nakano Y, Sakai H, Hirai T, Oku Y, Nishimura K, Mishima M (1999) Comparison of low attenuation areas on computed tomographic scans between inner and outer segments of the lung in patients with chronic obstructive pulmonary disease: incidence and contribution to lung function. Thorax 54:384–389

Nishimura K, Murata K, Yamagishi M et al. (1998) Comparison of different computed tomography scanning methods for quantifying emphysema. J Thorac Imaging 13:193–198

Park KJ, Bergin CJ, Clausen JL (1999) Quantification of emphysema with three-dimensional CT densitometry: comparison with two-dimensional analysis, visual emphysema scores, and pulmonary function test results. Radiology 211:541–547

Rienmüller RK, Behr J, Kalender A, Schätzl M, Altmann I, Merin M, Beinert T (1991) Standardized quantitative high resolution CT in lung diseases. J Comput Assist Tomogr 15:742–749

Rodarte JR, Chaniotakis M, Wilson TA (1989) Variability of parenchymal expansion measured by computed tomography. J Appl Physiol 67:226–231

Russi EW, Bloch KE, Weder W (1999) Functional and morphological heterogeneity of emphysema and its implication for selection of patients for lung volume reduction surgery. Eur Respir J 14:230–236

Saetta M, Shiner RJ, Angus GE, Kim WD, Wang NS, King M, Ghezzo H, Cosio MG (1985) Destructive index: a measurement of lung parenchyma destruction in smokers. Am Rev Respir Dis 131:764–769

Saitoh T, Koba H, Shijubo N, Tanaka H, Sugaya F (2000) Lobar distribution of emphysema in computed tomographic densitometric analysis. Invest Radiol 35:235–243

Salerno M, Lange EE de, Altes TA, Truwit JD, Brookeman JR, Mugler JP III (2002) Emphysema: hyperpolarized helium 3 diffusion MR imaging of the lungs compared with spirometric indexes – initial experience. Radiology 222: 252–260

Sanders C, Nath PH, Bailey WC (1988) Detection of emphysema with computed tomography: correlation with pulmonary function tests and chest radiography. Invest Radiol 23:262–266

Snider GL, Kleinerman JL, Thurlbeck WM, Bengali ZH (1985) The definition of emphysema. Report of a national Heart, Lung, and Blood Institute, Division of Lung Disease Workshop. Am Rev Respir Dis 132:182–185

Soeijima K, Yamaguchi K, Kohda E et al. (2000) Longitudinal follow-up study of smoking-induced lung density changes by high resolution computed tomography. Am J Respir Crit Care Med 161:1264–1273

Stern EJ, Franck MS (1994) CT of the lung in patients with pulmonary emphysema: diagnosis, quantification, and correlation with pathologic and physiologic findings. AJR Am J Roentgenol 162:791–798

Stockley RA (2000) Alpha-1-antitrypsin deficiency: what next? Thorax 55:614–618

Stoel BC, Vrooman HA, Stolk J, Reiber JHC (1999) Sources of error in lung densitometry with CT. Invest Radiol 34:303–309

Suga K, Tsukuda T, Awaya H et al. (1999) Impaired respiratory mechanics in pulmonary emphysema: evaluation with dynamic breathing MRI. J Magn Reson Imaging 10:510–520

Tepper J, Pfeiffer J, Aldrich M, Tumas D, Kern J, Hoffman E, McLennan G, Hyde D (2000) Can retinoic acid ameliorate the physiologic and morphologic effects of elastase instillation in the rat? Chest 117:242S–244S

Thurlbeck WM (1967a) The internal surface area of nonemphysematous lungs. Am Rev Respir Dis 95:765–773

Thurlbeck WM (1967b) Internal surface area and other measurements in emphysema. Thorax 22:483–496

Thurlbeck WM, Müller NL (1994) Emphysema: definition, imaging, and quantification. AJR Am J Roentgenol 163: 1017–1025

Thurlbeck WM, Dunnill MS, Hartung W, Heard BE, Heppleston AG, Ryder RC (1970) A comparison of three methods measuring emphysema. Hum Pathol 1:215–226

Turner P, Whimster WF (1981) Volume of emphysema. Thorax 36:932–937

Uppaluri R, Mitsa T, Sonka M, Hoffman EA, McLennan G (1997) Quantification of pulmonary emphysema from lung computed tomography images. Am J Respir Crit Care Med 156:248–254

Viallon M, Berthezene Y, Decorps M et al. (2000) Laser-polarized ^{3}He as a probe for dynamic regional measurements of lung perfusion and ventilation using magnetic resonance imaging. Magn Reson Med 44:1–4

Watanuki Y, Shunsuke S, Nishikawa M, Miyashita A, Okubo T (1994) Correlation of quantitative CT with selective alveolobronchogram and pulmonary function tests in emphysema. Chest 106:806–813

Weder W, Thurnheer R, Stammberger U, Bürge M, Russi EW, Bloch KE (1997) Radiological emphysema morphology is associated with outcome after surgical lung volume reduction. Ann Thorac Surg 64:313–320

Weibel ER (1963) Morphometry of the human lung. Springer, Berlin Heidelberg New York

6 Lung Fibrosis

Sujal R. Desai and Hans-Ulrich Kauczor

6.1 CT Imaging

Sujal R. Desai

CONTENTS

6.1.1 Introduction

Lung fibrosis occurs in diverse clinical settings (Colby and Carrington 1995). Interstitial fibrosis is a non-specific pathological end-point that can occur as a consequence of granulomatous diseases (most commonly, sarcoidosis and tuberculosis), certain drug therapies and repeated exposure to organic or inorganic antigens (Seal et al. 1968; Fitzgerald

S. R. Desai, MD, MRCP, FRCR
Consultant Radiologist, Department of Radiology, King's College Hospital, Denmark Hill, London SE5 9RS, UK

et al. 1973; Gillett and Ford 1978). When there is widespread fibrosis, pulmonary physiology is generally abnormal and the typical finding on lung function testing is a restrictive functional deficit with abnormalities of gas transfer (Gibson 1996).

It has been known for centuries that there is a strong relationship between form and function: normal physiology generally indicates that there is no gross structural abnormality. The corollary is that abnormalities of structure are frequently associated with alterations of physiology. Therefore, studying the structural and physiological alterations of disease, in concert, makes intuitive sense. In the present chapter the value of structure-function investigations and specifically those in which computed tomographic (CT) changes are correlated with physiological indices, is discussed; idiopathic pulmonary fibrosis (IPF) is used as the model fibrotic lung disease in this chapter. The utility of structure-function studies in increasing our understanding of the pathophysiology of other diseases that also cause pulmonary fibrosis (namely systemic sclerosis, sarcoidosis and extrinsic allergic alveolitis) is also presented.

6.1.2 Issues Relating to the Study of Structural-Functional Correlations in Patients with IPF

Studies of structure versus function generally evaluate the relationship between individual physiological indices [i.e. pulmonary function tests (PFTs)] and some index of altered anatomy (whether that be surgical biopsy, the appearances on plain chest radiography or the changes at CT). Before the relative merits and demerits of each method are considered, it is important to realise that, as has been stressed previously (Hansell 2001), the strength of correlation between indices is critically dependent on the *intrinsic* relationship between those two variables. Furthermore, the "noise" of measurement can significantly influence the magnitude of the relationship.

6.1.3
Pulmonary Function Tests versus Histo-pathological Estimates of Disease Severity

The natural tendency for those undertaking structure-function studies has always been to compare functional abnormalities with the changes seen at microscopic examination. Because histopathological analysis is widely regarded as the final arbiter in diagnosis, the usual logic is that it is appropriate to correlate the magnitude of pathological derangement with indices of functional impairment. However, there are flaws in this reasoning which should be emphasised. Firstly, pulmonary function tests provide a "global statement" of pulmonary dysfunction at many (but not necessarily similarly afflicted), lung units. In contrast, biopsy by its very nature is a measure of *local* disease severity. Thus, the presumption that biopsy will reflect the global picture is unjustified. This is particularly true in diseases like IPF where there is considerable regional heterogeneity.

The second problem is the assumption that a biopsy cohort is representative of the general population of patients with IPF. The majority of patients with suspected IPF currently do not undergo biopsy either because the degree of respiratory disability contraindicates surgery or that a confident diagnosis has already been made using non-invasive tests (GAENSLER and CARRINGTON 1980; TURNER-WARWICK et al. 1980; JOHNSTON et al. 1997). Thus, biopsy tends to be reserved for younger patients or for those with "atypical" disease (GILLETT and FORD 1978). Thirdly, the terminology of pathological scoring may also lead to discrepant observations: the term "disease severity", for example, has been applied loosely in histopathological studies to mean either the extent or the cellular activity of disease (FULMER et al. 1979; WRIGHT et al. 1981; WATTERS et al. 1986). Finally, there is the issue of observer disagreement between histopathologists, an important consideration for a test that is widely regarded as the "gold standard". Observer variation for semi-quantitative histopathological grading is of sufficient magnitude to introduce measurement "noise" (ASHCROFT et al. 1988; CHERNIACK et al. 1991).

For all the above reasons, it comes as little surprise that studies comparing histopathological scores with PFTs have produced widely disparate results. In one early study of 29 patients, there was good correlation between the severity of pulmonary fibrosis and some (but certainly not all) functional indices (CRYSTAL et al. 1976). In another, where both the degree of fibrosis and cellularity were quantified separately, there was a significant negative correlation between the fibrosis score and compliance but none with the more conventional tests of pulmonary function, including total lung capacity and the diffusing capacity for carbon monoxide (FULMER et al. 1979). Against these are the results of another study in which there were no observed relationships between histopathological features and functional parameters (WRIGHT et al. 1981).

6.1.4
Pulmonary Function Tests versus Chest Radiographic Quantification of Disease Severity

The advantages of conventional chest radiography are well known and should be listed: relatively low cost, technical simplicity and limited radiation dose are perhaps the most important and have ensured a lasting role for the technique in clinical practice. However, as a tool for undertaking quantitative studies, chest radiography has significant limitations. Because of the two-dimensional nature of the radiographic image, structural superimposition is a problem. Thus, accurate characterisation and quantification is difficult. Another disadvantage is the relative insensitivity of chest radiography to diffuse lung disease so that it is not always possible to exclude significant lung disease on conventional radiographs (HARGREAVE et al. 1972; EPLER et al. 1978; DAVISON et al. 1983). Furthermore, as with pathological quantification, there is the question of observer variability (not insignificant for chest radiography), compounded by the variable influences of observer experience and film quality (TUDDENHAM 1963; REGER et al. 1972).

There have been frequent attempts to extract pathophysiological information from chest radiography in a variety of clinical scenarios. In patients with pulmonary oedema, for example, the radiographic score of extravascular lung water correlates well with physiological estimates of lung fluid and negatively with arterial oxygenation (PISTOLESI et al. 1985, 1988). However, in patients with fibrosing lung disease, the correlations between radiographic severity and functional indices are generally weak (GAENSLER et al. 1972; CARRINGTON et al. 1976; TURNER-WARWICK et al. 1980). Thus, the general consensus is that the chest radiograph is only capable of providing limited functional information in patients with diffuse lung disease.

6.1.5
Pulmonary Function Tests versus Computed Tomographic Quantification of Disease Severity

Because the whole lungs are imaged, CT can (unlike histopathological estimation) quantify *global* disease extent. Furthermore, since there is no superimposition of anatomical structures (unlike at chest radiography), it is possible to accurately characterise *and* quantify the extent of parenchymal abnormality; CT features in IPF have been shown to reflect macroscopic features (MÜLLER et al. 1986). Thus, the relationships between *individual* CT patterns and functional indices may be evaluated.

A good maxim in quantitative studies, particularly where subjective estimation is used, is that the magnitude of measurement noise, due to observer disagreement, should be estimated (see below). In this regard, it is also worth understanding that the methods for scoring CT abnormalities can significantly influence the results. The danger with too coarse a scoring system is that potential relationships between CT patterns and functional indices may be masked. Conversely, an over complex method for quantification may obfuscate relationships because of the noise from observer disagreement (NG et al. 1999). This notwithstanding, COLLINS et al. (1994) have shown that observer agreement for scoring CT patterns is clinically acceptable in patients with lung fibrosis. Importantly, observer variation for the grading of abnormalities in patients with fibrosing alveolitis is lower on CT than plain chest radiography. More significantly, there was is effect of observer experience: intra-observer variability for grading the extent of lung involvement is lower for experienced radiologists.

An important and inevitable drawback of CT is that patients are exposed to ionising radiation. Valid concerns about the radiation burden to the population stemming from the widespread availability and use of CT are regularly voiced (MOLYNEUX 1991; DIMARCO and BRIONES 1993). In the United Kingdom, it has been estimated that CT contributes about one quarter of the annual collective dose from all radiological examinations (HUGHES and O'RIORDAN 1993). Plainly, methods for reducing dose are to be commended. One of the simplest ways is to restrict the duration of volume or individual scans. In this regard, because of interspaced sections, high resolution CT (HRCT) can have an intrinsic advantage over conventional CT, in which a continuous "volume" of lung is imaged (MAYO et al. 1993; VAN DER BRUGGEN-BOGAARTS et al. 1995). Volumetric high resolution scanning with Multidetector CT, however, offers new exciting possibility for diagnosis and categorisation of lung fibrosis on the basis of typical CT and distribution patterns. The clinical impact of this novel approach still has to be investigated. Reducing milliamperage and kilovoltage can also markedly limit patient exposure without significant detriment on image quality (NIADICH et al. 1990; ZWIREWICH et al. 1991) and optimal tube currents for so-called "low dose CT" protocols have been studied (LEE et al. 1994; MAYO et al. 1995). Dose modulation is another promising approach to reduce radiation exposure without a loss in image quality. Levels which require a higher level of radiation are the lung apices with the shoulders whereas the lower lung with a large volume of air requires significantly less radiation for the same image quality.

6.1.6
Methods for Quantifying CT Abnormalities in Patients with Pulmonary Fibrosis

In essence, there are two methods for quantifying the extent and severity and CT abnormalities in patients with pulmonary fibrosis: visual grading and automated (computerised) methods. In both, the aim is to estimate the extent and severity of individual CT patterns at a lobar level, in representative sections through the lungs or in the whole lung volume.

The advantages of visual estimation are obvious: once mastered the technique is relatively simple and fast. Moreover, compared to computer-based quantification, visual scoring is cheap. The extent of abnormal lung can be expressed as a percentage of the total on a continuous scale (to the nearest 5 or 10%) or, alternatively, using a coarser semi-quantitative grading system (Fig. 6.1.1). Furthermore, the proportions of abnormal lung comprising different CT patterns can also be estimated: for example, within a region of fibrotic lung the majority may be a reticular pattern but there may be areas of ground-glass opacification (Fig. 6.1.2), and foci of admixed emphysema. Ancillary features such as the coarseness of fibrosis and the presence/severity of associated traction bronchiectasis can also be graded (DESAI et al. 2003). The usual practice, in most studies, is to estimate the extent of disease at five representative sections, typically at the level of the great vessels, aortic arch, carina, pulmonary venous confluence and one centimetre above the dome of the right hemidiaphragm.

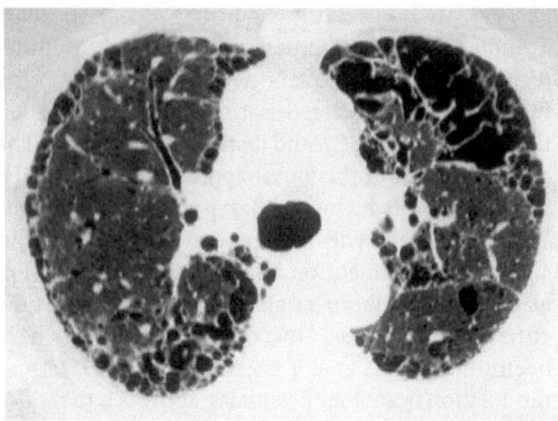

Fig. 6.1.1. CT at the level of the carina in idiopathic pulmonary fibrosis. There is a coarse subpleural reticular pattern; the visual extent of disease was estimated as 45% and 50% by two independent observers

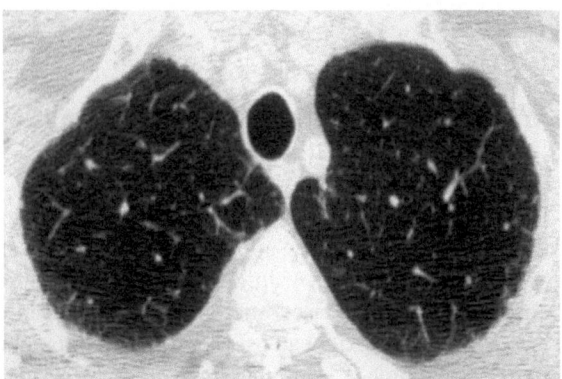

a

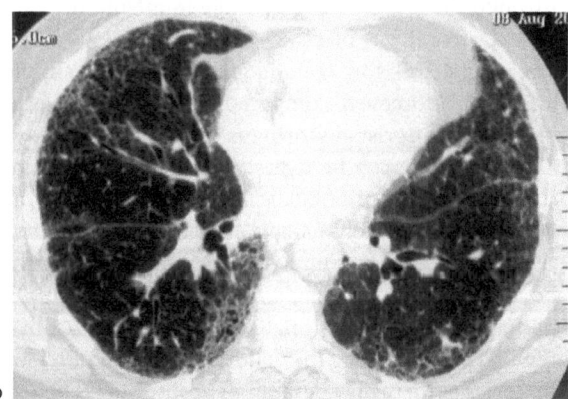

b

Fig. 6.1.2a, b. CT in a patient with idiopathic pulmonary fibrosis. a At the level of the great vessels, there is relatively limited disease (visual estimate of extent = 10%). b At the level of the pulmonary venous confluence there is more extensive involvement (visual estimate of extent = 25%). Within areas of abnormal lung, there is a predominant reticular pattern but also admixed ground-glass opacification. The proportions of these patterns making up the abnormal lung (up to a total of 100%) can also be quantified (for example at this level, visual estimate of the reticular pattern = 90% and ground-glass opacification = 10%)

The obvious disadvantage of visual scoring is subjectivity, again highlighting the importance of quantifying observer variation.

Techniques for more objective quantification of CT images have been devised but an important caveat with all such methods is the likely impact on time and resources. One of the simplest approaches involves quantifying the overall change in parenchymal attenuation. In patients with fibrotic sarcoidosis, for example, mean lung density is higher than in normal subjects (GILMAN et al. 1983). In these patients, the increase in attenuation is significantly and negatively correlated with spirometric and plethysmographic lung volumes. In addition to the straightforward measurement of overall lung density, CT density histograms can also be plotted. In this way, a variety of indices may be extracted, including the mean and median grey scale density, skewness and kurtosis (the latter a measure of the "peakedness" of the histogram). In one study of patients with IPF and asbestosis, there were significant correlations between these measures of grey scale density and physiological parameters (HARTLEY et al. 1994). It will be obvious that an important disadvantage of such an analysis is the known propensity for parenchymal density to vary during breathing (ROBINSON and KREEL 1979; KALENDER et al. 1990); semi-automatic techniques to address this problem have been devised but also have the drawback of extra expense and time (KALENDER et al. 1991).

Another objective method for quantification is to quantify the proportions of CT patterns underlying equally-spaced grid points (HIROSE et al. 1993). In an animal model, there were good correlations between the CT extent of fibrosis and morphometric estimates of lung abnormality. It is understandable that such grid systems, whilst promising objectivity, are time-consuming and likely to limit the number of patients that can be studied (SAKAI et al. 1987). The value of sophisticated image processing protocols, which enhance the demonstration of subtle CT abnormalities, have recently been propounded (NAPEL et al. 1993); in diffuse infiltrative diseases, thin-section maximum intensity projection images may aid the demonstration of subtle micronodules (REMY-JARDIN et al. 1996). Similarly, minimum intensity projection reconstruction has been shown to be superior to conventional HRCT images for the demonstration of low attenuation features (BHALLA et al. 1996). Despite their attractiveness, the utility of such techniques in diffuse diseases, in which different CT patterns often coexist, is unclear. An intelligent post-processing tool to differentiate vasculature, bronchi, emphysema and fibrosis is based on a combination of different techniques, including thresholding, texture-based pattern recognition and, most important, multivariate discrimination analysis. Such a system can reach a correct categorisation in up to 81% of scans (DELORME et al. 1997).

Multiple neural networks complemented by an expert rule are capable to automatically detect ground glass opacities and quantify their extent (HEITMANN et al. 1997).

Such a tool might be very helpful to grade disease activity in lung fibrosis and objectively quantify the success of therapy. Multiple neural networks perform significantly better than a traditional density mask. When compared with an expert radiologist the neural network reaches a sensitivity of 99% and a specificity of 83%. Typical errors are cardiac motion artefacts and partial volume effects (HEITMANN et al. 1997).

6.1.7
Statistical Analysis

The choice of statistical test with which to express the results of structure-function studies is crucial. In this regard, there are two important considerations: firstly, which test to use to describe the magnitude of observer disagreement and secondly the appropriate statistical methods to evaluate the relationship between individual CT patterns and physiological parameters. A detailed account of statistical tests is beyond the scope of the present chapter. However, a brief description of statistical techniques is appropriate.

6.1.7.1
Statistical Tests for Quantifying the Magnitude of Observer Variation

For data comprising a series of "continuous" variables, the most simple way to describe the degree of observer agreement is the *correlation coefficient*. However, there are problems with such an approach: the assumption that a high correlation coefficient equates with low observer variability is incorrect. This can be illustrated by the simple example of the scenario in which one observer consistently overscores a CT pattern: in such a case the *correlation* is high but observer agreement is not (BRENNAN and SILMAN 1992). An alternative test, for continuous data is the *coefficient of variation*, which is a measure of the standard deviation of observations as a percentage of the sample mean (KIRKWOOD 1988a). There are advantages and disadvantages to using

the coefficient of variation. One benefit is that the coefficient is independent of the units of measurement. However, a problem is the presumption that the magnitude of variation is related to the magnitude of the measurements; it will be clear to the reader that higher scores of disease extent do not necessarily lead to greater observer variation. Another approach (and perhaps the most appropriate) to expressing observer disagreement for continuous variables is the *single determination standard deviation*. With this statistical test, the *range* within which most disagreements occur is calculated (CHINN 1991). More specifically, single determination standard deviation it is the standard deviation of the mean difference between observers (BRENNAN and SILMAN 1992).

The determination of observer variation for non-continuous (categorical) data is more problematic. A statement about the *percentage agreement* between observers is generally meaningless, because there is no distinction between true and chance agreement (BRENNAN and SILMAN 1992; COCHRANE and GARLAND 1952). A more powerful tool is the *kappa coefficient* which, put simply, is the ratio of the proportion of times that observers agree (corrected for chance agreement) to the proportion of times that observers could agree (also corrected for chance agreement) (SIEGEL and CASTELLAN 1989). The theoretical range of the calculated kappa coefficient is from 1.0 (complete agreement) to 0.0 (no agreement) and indicates how far observers have moved away from chance agreement. Although there are no strict definitions as to what values of the coefficient can be considered good or clinically unacceptable variation, a kappa coefficient of greater than 0.4 can be said to reflect reasonable observer agreement (BRENNAN and SILMAN 1992). For categorical data that is ordinal (i.e. ranked) rather than nominal (i.e. unranked), there is a modification to the standard kappa coefficient: the *weighted kappa coefficient* is sensitive to differences between a set of readings in which one observer gives a grade of 1 and another observer gives a grade of 2 and another set of readings in which, for example, observers separately assign grades 1 and 4; the greater observer discrepancy in the latter case would not be manifest in the unweighted kappa statistic.

6.1.7.2
Statistical Tests for Evaluating the Relationships Between CT Patterns and Physiological Indices

At its simplest, the relationships between individual CT patterns and pulmonary function indices can be tested using univariate techniques (Lowe 1993). A scattergram of data points can be plotted and will show, graphically, the relationship (if any exists), between two variables. The regression equation for the graph will describe the best-fitting straight line through the data (Fig. 6.1.3). The strength of the relationship between the two variables is expressed in the *correlation coefficient* (Pearson's product-moment correlation coefficient for data that are normally distributed or the Spearman's correlation coefficient, for non-normally distributed data). The correlation coefficient ranges from +1 (indicating full positive correlation) to –1 (full negative correlation) and is a measure of how much the data points are scattered around the regression line. Where there is considerable scatter, the correlation coefficient approaches zero.

A more useful measure of scatter around the regression line is the *square of the correlation coefficient* (R^2) (Lowe 1993). In the specific context of structure-function studies, the R^2 value describes the proportion of variation in a given lung function index that is accounted for by changes in the extent or severity of a morphological pattern on CT. Thus, a correlation coefficient of 0.70, for example, between two variables (y and x) gives an R^2 value of 0.49 and indicates that just under one half of the variation in y can be predicted by knowing the values of x.

The relationship between a single dependent but multiple independent variables is best evaluated with *multiple regression analysis* (Kirkwood 1988b; Lowe 1993). In this technique, separate multivariate models are constructed for each dependent variable. In the context of the computed tomographic-physiological studies, individual pulmonary function indices are the *dependent* variables, whereas CT patterns are the *independent* variables. The R^2 value in multivariate analysis has the same significance as in univariate techniques (see above) and highlights the proportion of the variation in individual lung function indices that can be accounted for by CT patterns.

6.1.8
Relationships between CT Features and Functional Indices in Patients with IPF

Idiopathic pulmonary fibrosis (IPF; also called cryptogenic fibrosing alveolitis) is the paradigm of a fibrotic lung disease. Patients with IPF are usually aged over 50 (Turner-Warwick et al. 1980; Johnston et al. 1997), and have progressive

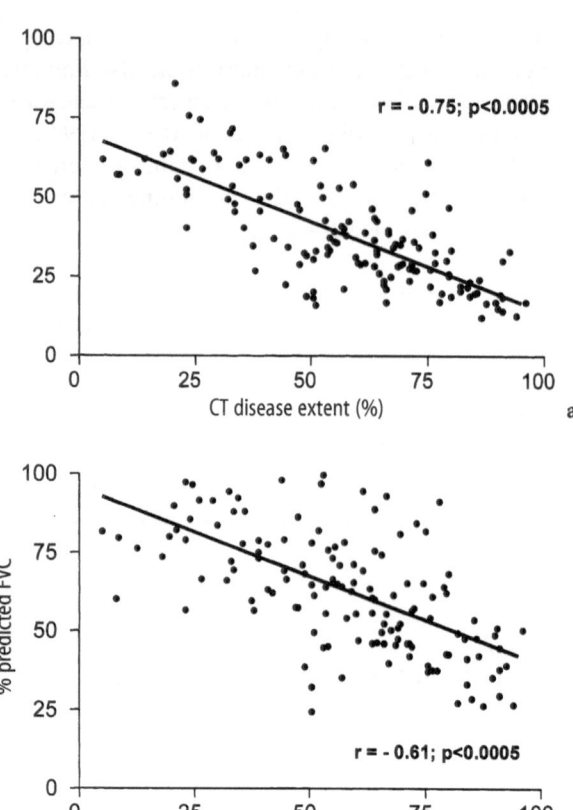

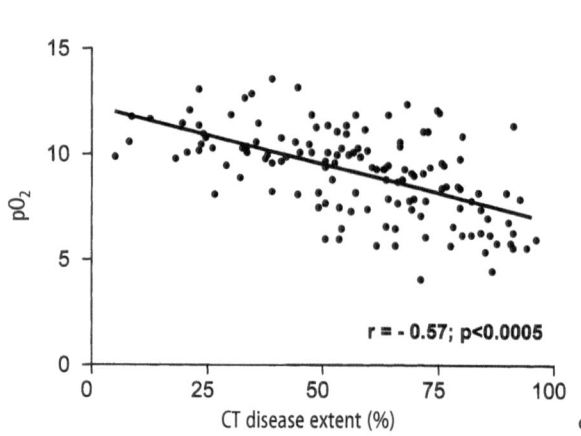

Fig. 6.1.3a–c. Univariate correlations demonstrating the relationship between the extent of disease on CT and (**a**) the percent predicted single-breath diffusing capacity for carbon monoxide (Dlco), (**b**) the percent predicted forced vital capacity (FVC) and (**c**) the partial pressure of oxygen (pO$_2$) in 212 patients with IPF (*unpublished data*)

symptoms of dyspnoea and cough (Travis et al. 2002). As with other causes of pulmonary fibrosis, the characteristic deficit on lung function testing is restrictive: in the typical patient spirometric and plethysmographic volumes are reduced and gas transfer is usually low (GIBSON 1996). Historically, physiological tests have been relied upon to "objectively" grade the disease severity. However, most physicians who care for patients with IPF will admit that the interpretation of lung function tests in patients with IPF is rarely straightforward. The problem of knowing which of a plethora of functional indices most reliably reflects disease extent (which is, incidentally, an important prognostic factor; GAY et al. 1998) is a vexing one.

The relationships between CT patterns and PFTs have been explored in previous series. In one of the earliest studies, SIDER et al. (1987) compared the appearances on CT with the results of pulmonary function tests in a variety of lung diseases. The authors showed that in the 15 patients with restrictive indices, there was evidence of interstitial lung disease on CT; the use of relatively thick sections (1 cm) and slow scanning times (3 seconds) probably precluded more sophisticated analyses in this study.

In subsequent HRCT series, it has been possible to evaluate the relationships between CT patterns and functional indices more thoroughly (STAPLES et al. 1987; WELLS et al. 1997b). In one, the images of 68 patients with IPF were scored for the extent of disease and, where present, emphysema (WELLS et al. 1997b). The authors showed that the percent predicted DLco (but not other physiological indices) was the best *single* marker of disease extent. Moreover, the extent of disease on CT was the most important determinant of functional impairment, a relationship borne out on multivariate analysis. There is some clinical utility in this result because (perhaps for the first time) it has added some precision to the interpretation of lung function tests in IPF (Figs. 6.1.3, 6.1.4).

More recently it has been surmised that a *combination* of functional indices may be a better index of disease extent in patients with IPF than individual physiological parameters. By using CT estimates of disease extent, a composite physiological index (CPI; comprising the *weighted* forced expiratory volume in one second, forced vital capacity and Dlco) has been developed (WELLS et al. 2003). It has been shown that, compared to individual indices, the CPI is a more powerful predictor of disease extent and more importantly, prognosis in IPF (DESAI et al. 2000; WELLS et al. 2003) (Fig. 6.1.4). The benefits of the CPI appear to be threefold: firstly, it is derived from routinely available lung function tests. Secondly, there is no reliance exercise data (unlike a previously devised physiological index; WATTERS et al. 1986); this is an advantage because exercise tests are not undertaken, or indeed feasible, in all patients. Finally, the CPI also appears to take account of coexistent emphysema which would normally perturb the relationship between Dlco and disease extent (WELLS et al. 1997b).

6.1.9
Morphological-Functional Relationships in Other Fibrosing Lung Diseases

The ability of CT to accurately depict anatomical changes can be exploited in other diseases in which fibrosis is known to occur. Structural-functional relationships have been studied in patients with pulmonary fibrosis secondary to systemic sclerosis, sarcoidosis and extrinsic allergic alveolitis. Some of these studies are briefly discussed below.

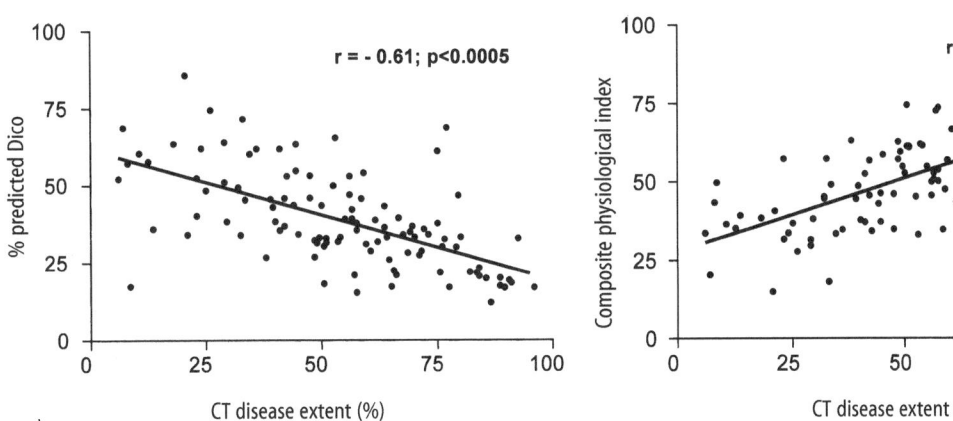

a   b

Fig. 6.1.4a, b. Univariate correlations between CT disease extent and (a) % predicted single-breath diffusing capacity for carbon monoxide (Dlco) and (b) the composite physiological index in 109 patients with IPF. The data plots show that the composite physiological index, made up of a weighted combination of the forced expiratory volume in one second, the forced vital capacity and the Dlco, is a better indicator of the extent of lung involvement than the Dlco in isolation (*unpublished data*)

6.1.9.1
Morphological-Functional Relationships in Systemic Sclerosis

Until recently, it has been assumed that the fibrosing alveolitis of systemic sclerosis (FASSc) is histopathologically identical to that in IPF (N. K. HARRISON et al. 1991). However, it is known that patients with FASSc survive longer than those with IPF; five year survival (from the onset of dyspnoea) in patients with IPF and FASSc is 50% and 86% respectively (WELLS et al. 1994). It has also been shown recently that the histopathological pattern in the majority of patients with systemic sclerosis is different to that in IPF (BOUROS et al. 2002). Non-specific interstitial pneumonia (NSIP), as opposed to usual interstitial pneumonia (UIP, the expected pattern in patients with IPF), is the commonest finding in systemic sclerosis. However, whether this alone accounts for the prognostic advantage of systemic sclerosis patients has not been resolved (BOUROS et al. 2002).

Important differences between patients with IPF and FASSc have been unravelled in CT-functional studies. WELLS et al. (1997a) have shown that for the same disease extent on CT, patients with FASSc have less depression of arterial oxygenation, a smaller arterial-alveolar oxygen gradient and less oxygen desaturation on exercise, than in IPF. The results suggest that there is greater perfusion of regions of under- and non-ventilated lung (either due to loss of the normal hypoxic vasoconstrictor response or new vessel formation) in IPF, thus leading to increased shunting. Thus, it was not unreasonable for the authors to postulate that differences in prognosis between IPF and FASSc may be attributed to the effects of more profound shunt-related hypoxia in the former (WELLS et al. 1997a).

6.1.9.2
Morphological-Functional Relationships in Sarcoidosis

Abnormalities of pulmonary function are common in patients with sarcoidosis. When there is "end-stage" fibrotic lung disease, the expected finding on lung function testing is a restrictive deficit. However, it may come as a surprise to some readers that there is frequently physiological evidence of airflow limitation, even in the early stages of sarcoidosis (MILLER et al. 1974; KANEKO and SHARMA 1977; SCANO et al. 1986; B. D. W. HARRISON et al. 1991). The pathophysiological basis of obstruction in sarcoidosis is poorly understood, but theoretically could be at the level of large or small airways and either because of surrounding granulomata or peribronchiolar fibrosis (LONGCOPE and FREIMAN 1952; MILLER et al. 1974; UDWADIA et al. 1990).

With CT it has been possible to evaluate the relationship between individual CT patterns and functional indices. In a series of 45 patients with parenchymal sarcoidosis of varying severity, CT patterns were quantified and correlated with the results of lung function tests (HANSELL et al. 1998). A reticular pattern, regions of decreased attenuation on expiratory CT images (denoting obstruction of small airways) and a nodular pattern were the most frequent CT features. The lung function abnormalities in this cohort ranged from marked restriction to predominant airflow obstruction. The authors showed that the extent of a reticular pattern was the main determinant of physiological impairment. However, an intriguing finding of the study was that the extent of a reticular pattern was independently correlated with indices of airflow obstruction. Although the extent of air trapping on expiratory scans was correlated with indices of obstruction, the relationship was less impressive. These results are surprising because they suggest that the obstructive component of sarcoidosis is related *primarily* to a CT feature (i.e. the reticular pattern) that is generally taken to indicate pulmonary fibrosis (HANSELL et al. 1998). Whether obstruction is due to fibrosis and obliteration of small airways is not known. However, such CT-functional studies do suggest further avenues for investigation.

6.1.9.3
Morphological-Functional Relationships in Extrinsic Allergic Alveolitis

A lymphocytic and plasma cell interstitial infiltrate is the typical finding on histopathological examination of the lungs in patients with extrinsic allergic alveolitis (EAA) (HAPKE et al. 1968; SEAL et al. 1968). The intensity of pathological changes varies form case to case but in general there are poorly-formed, non-necrotising granulomas and evidence of a bronchiolitis (COLBY and CARRINGTON 1995). Not surprisingly, the functional impairment in patients with EAA is variable and, as with sarcoidosis, airflow obstruction occurs. The mechanisms of airflow limitation in EAA have been the subject of debate and, over time, have been ascribed to a variety of causes including chronic bronchitis/ emphysema, asthma and bronchiolitis (HAPKE et al. 1968; SUTINEN et al. 1983; SELMAN-LAMA and PEREZ-PADILLA 1993).

In the subacute phase, CT demonstrates variable proportions of poorly-defined centrilobular opacities (a reflection of the propensity for the disease to be bronchiolocentric), ground-glass opacification (presumable due to the lymphocytic/plasma cell infiltrate) and regions of decreased attenuation (a testament to the bronchiolitis) (SILVER et al. 1989; HANSELL and MOSKOVIC 1991; REMY-JARDIN et al. 1993) (Fig. 6.1.5). The lobular regions of decreased attenuation (incidentally, also evident in patients with chronic EAA where there is supervening fibrosis (REMY-JARDIN et al. 1993) are functionally significant; there is correlation between the extent of regions of decreased attenuation on CT and indices of airflow obstruction (HANSELL et al. 1996). More importantly, such morphological-functional data have challenged the notion that airflow limitation in EAA is due to emphysema. Indeed, there is no independent relationship between the extent of decreased attenuation on CT and indices of gas transfer (which would be expected if the areas of low density were due to emphysema) (HANSELL et al. 1996).

6.1.10
Summary

The study of structural changes in fibrotic lung disease together with the functional abnormalities is a valuable exercise. Anatomical changes are frequently evident

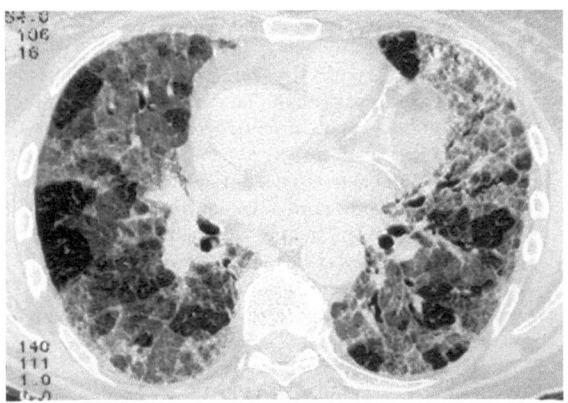

Fig. 6.1.5. CT in chronic extrinsic allergic alveolitis. There is diffuse abnormality in both lungs comprising widespread ground-glass opacification and a superimposed reticular pattern; dilated segmental and subsegmental airways are demonstrated within areas abnormal lung parenchyma. Many foci of lobular decreased attenuation are also noted, indicating associated small airways disease (*courtesy of Dr Simon PG Padley, London*)

on chest radiography in patients with diffuse fibrotic lung diseases. However, accurate characterisation and quantification is difficult because of the inherent problems of limited contrast resolution (compared to CT) and superimposition. Although open biopsy is considered the gold standard for diagnosis, for reasons discussed in this chapter, quantification and correlation with functional indices is fraught with difficulty. In contrast, CT provides a more robust *global* measure of the extent and severity of parenchymal disease. The study of morphological changes on CT together with functional alterations, has not only provided intriguing (and sometimes unexpected) insights but also challenged some strongly held dogmas in patients with fibrosing lung diseases.

References

Ashcroft T, Simpson JM, Timbrell V (1988) Simple method of estimating severity of pulmonary fibrosis on a numerical scale. J Clin Pathol 41:467–470

Bhalla M, Naidich DP, McGuinness G et al (1996) Diffuse lung disease: assessment with helical CT-preliminary observations of the role of maximum and minimum intensity projection images. Radiology 200:341–347

Bouros D, Wells AU, Nicholson AG et al (2002) Histopathologic subsets of fibrosing alveolitis in patients with systemic sclerosis and their relationship to outcome. Am J Respir Crit Care Med 165:1581–1586

Brennan P, Silman A (1992) Statistical methods for assessing observer variability in clinical measures. Br Med J 304: 1491–1494

Bruggen-Bogaarts BAHA van der, Broerse JJ, Lammers J-WJ et al (1995) Radiation exposure in standard and high-resolution chest CT scans. Chest 107:113–115

Carrington CB, Gaensler EA, Mikus JP et al (1976) Structure and function in sarcoidosis. The BAL cooperative group steering committee. Ann NY Acad Sci 278:265–283

Cherniack RM, Colby TV, Flint A et al (1991) Quantitative assessment of lung pathology in idiopathic pulmonary fibrosis. Am Rev Respir Dis 144:892–900

Chinn S (1991) Repeatability and method comparison. Thorax 46:454–456

Cochrane AL, Garland LH (1952) Observer error in the interpretation of chest films: an international investigation. Lancet 2:505–509

Colby TV, Carrington CB (1995) Interstitial lung disease. In: Thurlbeck WM, Churg AM (eds) Pathology of the lung. Thieme, New York, pp 589–737

Collins CD, Wells AU, Hansell DM et al (1994) Observer variation in pattern type and extent of disease in fibrosing alveolitis on thin section computed tomography and chest radiography. Clin Radiol 49:236–240

Crystal RG, Fulmer JD, Roberts WC et al (1976) Idiopathic pulmonary fibrosis: clinical, histologic, radiographic, physiologic, scintigraphic, cytologic, and biochemical aspects. Ann Intern Med 85:769–788

Davison AG, Haslam PL, Corrin B et al (1983) Interstitial lung disease and asthma in hard-metal workers: bronchoalveolar lavage, ultrastructural, and analytical findings and results of bronchial provocation tests. Thorax 38:119–128

Delorme S, Keller-Reichenbecher MA, Zuna I, Schlegel W, Van Kaick G (1997) Usual interstitial pneumonia. Quantitative assessment of high-resolution computed tomography findings by computer-assisted texture-based image analysis. Invest Radiol 32:566–574

Desai SR, Hansell DM, Rubens MB et al (2000) A CT-derived composite physiological score as an indicator of prognosis in usual interstitial pneumonitis. Eur Radiol 10 (Suppl 1): 186(A)

Desai SR, Wells AU, Rubens MB et al (2003) Traction bronchiectasis in cryptogenic fibrosing alveolitis: associated computed tomographic features and physiological significance. Eur Radiol 13:1801–1808

DiMarco AF, Briones B (1993) Is chest CT performed too often? Chest 103:985–986

Epler GR, McLoud TC, Gaensler EA et al (1978) Normal chest roentgenograms in chronic diffuse infiltrative lung disease. N Engl J Med 298:934–939

Fitzgerald MX, Carrington CB, Gaensler EA (1973) Environmental lung disease. Med Clin North Am 57:593–622

Fulmer JD, Roberts WC, von Gal ER (1979) Morphologic-physiologic correlates of the severity of fibrosis and degree of cellularity in idiopathic pulmonary fibrosis. J Clin Invest 63:665–676

Gaensler EA, Carrington CB (1980) Open biopsy for chronic diffuse infiltrative lung disease: clinical, roentgenographic and physiological correlations in 502 patients. Ann Thorac Surg 30:411–426

Gaensler EA, Carrington CB, Coutu RE et al (1972) Pathological, physiological, and radiological correlations in the pneumoconioses. Ann NY Acad Sci 200:574–607

Gay SE, Kazerooni EA, Toews GB et al (1998) Idiopathic pulmonary fibrosis: predicting response to therapy and survival. Am J Respir Crit Care Med 157:1063–1072

Gibson GJ (1996) Alveolar diseases. In: Gibson GJ (ed) Clinical tests of respiratory function. Chapman and Hall, London, pp 223–247

Gillett DG, Ford GT (1978) Drug-induced lung disease. Monogr Pathol 19:21–42

Gilman MJ, Laurens RG Jr, Somogyi JW et al (1983) CT attenuation values of lung density in sarcoidosis. J Comput Assist Tomogr 7:407–410

Hansell DM (2001) Computed tomography of diffuse lung disease: functional correlates. Eur Radiol 11:1666–1680

Hansell DM, Moskovic E (1991) High-resolution computed tomography in extrinsic allergic alveolitis. Clin Radiol 43:8–12

Hansell DM, Wells AU, Padley SPG et al (1996) Hypersensitivity pneumonitis: correlation of individual CT patterns with functional abnormalities. Radiology 199:123–128

Hansell DM, Milne DG, Wilsher ML et al (1998) Pulmonary sarcoidosis: morphologic associations of airflow obstruction at thin-section CT. Radiology 209:697–704

Hapke EJ, Seal RME, Thomas GO et al (1968) Farmer's lung: a clinical, radiographic, functional and serological correlation of acute and chronic stages. Thorax 23:451–468

Hargreave F, Hinson KF, Reid L et al (1972) The radiological appearances of allergic alveolitis due to bird sensitivity (bird's fancier's lung). Clin Radiol 23:1–10

Harrison BDW, Shaylor JM, Stokes TC et al (1991) Airflow limitation in sarcoidosis: a study of pulmonary function in 107 patients with newly diagnosed disease. Respir Med 85:59–64

Harrison NK, Myers AR, Corrin B et al (1991) Structural features of interstitial lung disease in systemic sclerosis. Am Rev Respir Dis 144:706–713

Hartley PG, Galvin JR, Hunninghake GW et al (1994) High-resolution CT-derived measures of lung density are valid indexes of interstitial lung disease. J Appl Physiol 76: 271–277

Heitmann KR, Kauczor H-U, Mildenberger P, Uthmann T, Perl J, Thelen M (1997) Automatic detection of ground glass opacities on lung HRCT using multiple neural networks. Eur Radiol 7:1463–1472

Hirose N, Lynch DA, Cherniack RM et al (1993) Correlation between high resolution computed tomography and tissue morphometry of the lung in bleomycin-induced pulmonary fibrosis in the rabbit. Am Rev Respir Dis 147: 730–738

Hughes JC, O'Riordan MC (1993) Radiation exposure of the UK population – 1993 review, 5th edn. National Radiation Protection Board. HMSO, Oxon

Johnston IDA, Prescott RJ, Chalmers JC et al (1997) British Thoracic Society study of cryptogenic fibrosing alveolitis: current presentation and initial management. Thorax 52: 38–44

Kalender WA, Reinmuller R, Seissler W et al (1990) Measurement of parenchymal attenuation: use of spirometric gating with quantitative CT. Radiology 175:265–268

Kalender WA, Fichte H, Bautz W et al (1991) Semiautomatic evaluation procedures for quantitative CT of the lung. J Comput Assist Tomogr 15:248–255

Kaneko K, Sharma OP (1977) Airway obstruction in pulmonary sarcoidosis. Bull Eur Physiopath Resp 13:231–240

Kirkwood BR (1988a) Means, standard deviations and standard errors. In: Kirkwood BR (ed) Essential of medical statistics. Blackwell, Oxford, pp 12–20

Kirkwood BR (1988b) Multiple regression. In: Kirkwood BR (ed) Essential of medical statistics. Blackwell, Oxford, pp 65–72

Lee KS, Primack SL, Staples CA et al (1994) Chronic infiltrative lung disease: comparison of diagnostic accuracies of radiography and low- and conventional-dose thin-section CT. Radiology 191:669–673

Longcope WT, Freiman DG (1952) A study of sarcoidosis based on a combined investigation of 160 cases including 30 autopsies from the Johns Hopkins Hospital and the Massachusetts General Hospital. Medicine (Baltimore) 31: 1–132

Lowe D (1993) Data summary. In: Lowe D (ed) Planning for medical research: a practical guide to research methods. Astraglobe, Middlesborough, pp 94–127

Mayo JR, Jackson SA, Müller NL (1993) High-resolution CT of the chest: radiation dose. AJR 160:479–481

Mayo JR, Hartman TE, Lee KS et al (1995) CT of the chest: minimal tube current required for good image quality with the least radiation dose. AJR 164:603–607

Miller A, Teirstein AS, Jackler I et al (1974) Airway function in chronic pulmonary sarcoidosis with fibrosis. Am Rev Respir Dis 109:179–189

Molyneux AJ (1991) Computed tomography and radiation doses. Lancet 337:1164

Müller NL, Miller RR, Webb WR et al (1986) Fibrosing alveolitis: CT-pathologic correlation. Radiology 160:585–588

Naidich DP, Marshall CH, Gribbin C et al (1990) Low-dose CT of the lungs: preliminary observations. Radiology 175: 729–731

Napel S, Rubin GD, Jeffrey RB Jr (1993) STS-MIP: a reconstruction technique for CT of the chest. J Comput Assist Tomogr 17:832–838

Ng CS, Desai SR, Rubens MB et al (1999) Visual quantification and observer variation of signs of small airways disease at inspiratory and expiratory CT. J Thorac Imag 14:279–285

Pistolesi M, Miniati M, Milne ENC et al (1985) The chest roentgenogram in pulmonary edema. Clin Chest Med 6:315–344

Pistolesi M, Miniati M, Giuntini C (1988) A radiographic score for clinical use in the adult respiratory distress syndrome. Intensive Crit Care Dig 7:2–4

Reger RB, Smith CA, Kibelstis JA et al (1972) The effect of film quality and other factors on the roentgenographic categorization of coal workers' pneumoconiosis. AJR 115: 462–472

Remy-Jardin M, Remy J, Wallaert B et al (1993) Subacute and chronic bird breeder hypersensitivity pneumonitis: sequential evaluation with CT and correlation with lung function tests and bronchoalveolar lavage. Radiology 189:111–118

Remy-Jardin M, Remy J, Artaud D et al (1996) Diffuse infiltrative lung disease: clinical value of sliding-thin-slab maximum intensity projection CT scans in the detection of mild micronodular patterns. Radiology 200:333–339

Robinson PJ, Kreel L (1979) Pulmonary tissue attenuation with computed tomography: comparison of inspiration and expiration scans. J Comput Assist Tomogr 3:740–748

Sakai F, Gamsu G, Im J-G et al (1987) Pulmonary function abnormalities in patients with CT-determined emphysema. J Comput Assist Tomogr 11:963–968

Scano G, Monechi GC, Stendardi L et al (1986) Functional evaluation in stage I pulmonary sarcoidosis. Respiration 49:195–203

Seal RME, Hapke EJ, Thomas GO et al (1968) The pathology of the acute and chronic stages of farmer's lung. Thorax 23:469–489

Selman-Lama M, Perez-Padilla R (1993) Airflow obstruction and airway lesions in hypersensitivity pneumonitis. Clin Chest Med 14:699–714

Sider L, Dennis L, Smith LJ et al (1987) CT of the lung parenchyma and the pulmonary function test. Chest 92:406–410

Siegel S, Castellan NJ Jr (1989) Measures of association and their tests of significance. In: Siegel S, Castellan NJ Jr (eds) Nonparametric statistics for the behavioural sciences. McGraw-Hill, New York, pp 284–291

Silver SF, Müller NL, Miller RR et al (1989) Hypersensitivity pneumonitis: evaluation with CT. Radiology 173:441–445

Staples CA, Müller NL, Vedal S et al (1987) Usual interstitial pneumonia: correlation of CT with clinical, functional, and radiologic findings. Radiology 162:377–381

Sutinen S, Reijula K, Huhti E et al (1983) Extrinsic allergic bronchiolo-alveolitis: serology and biopsy findings. Eur J Respir Dis 64:271–282

Travis WD, King TE Jr, and the Multidisciplinary Core Panel (2002) American Thoracic Society/European Respiratory Society international multidisciplinary consensus classification of idiopathic interstitial pneumonias. Am J Respir Crit Care Med 165:277–304

Tuddenham WJ (1963) Problems in perception in chest roentgenology: facts and fallacies. Radiol Clin North Am 1:277–289

Turner-Warwick M, Burrows B, Johnson A (1980) Cryptogenic fibrosing alveolitis: clinical features and their influence on survival. Thorax 35:171–180

Udwadia ZF, Pilling JR, Jenkins PF et al (1990) Bronchoscopic and bronchographic findings in 12 patients with sarcoidosis and severe or progressive airways obstruction. Thorax 45:272–275

Watters LC, King TE, Schwarz MI et al (1986) A clinical, radiographic, and physiologic scoring system for the longitudinal assessment of patients with idiopathic pulmonary fibrosis. Am Rev Respir Dis 133:97–103

Wells AU, Cullinan P, Hansell DM et al (1994) Fibrosing alveolitis associated with systemic sclerosis has a better prognosis than lone cryptogenic fibrosing alveolitis. Am J Respir Crit Care Med 149:1583–1590

Wells AU, Hansell DM, Rubens MB et al (1997a) Functional impairment in lone cryptogenic fibrosing alveolitis and fibrosing alveolitis associated with systemic sclerosis: a comparison. Am J Respir Crit Care Med 155:1657–1664

Wells AU, King AD, Rubens MB et al (1997b) Lone cryptogenic fibrosing alveolitis: a functional-morphologic correlation based on the extent of disease on thin-section computed tomography. Am J Respir Crit Care Med 155:1367–1375

Wells AU, Desai SR, Rubens MB et al (2003) Idiopathic pulmonary fibrosis: a composite physiologic index devived from disease extent observed by computed tomography. Am J Respir Crit Care Med 167:962–969

Wright PH, Heard BE, Steel SJ et al (1981) Cryptogenic fibrosing alveolitis: assessment by graded trephine lung biopsy histology compared with clinical, radiographic, and physiological features. Br J Dis Chest 75:61–70

Zwirewich CV, Mayo JR, Müller NL (1991) Low-dose high-resolution CT of lung parenchyma. Radiology 180:413–417

6.2 MR Imaging

Hans-Ulrich Kauczor

CONTENTS

6.2.1
Introduction

Although CT is the imaging gold standard of pulmonary structure its potential to assess pulmonary function is limited. In lung fibrosis, estimation of disease activity, i.e. active inflammation, and severity of restrictive compromise of lung function are important parameters to make treatment decisions. MRI as an alternative radiological modality to CT has shown great potential for functional investigations of different organ systems, especially brain and heart. Recently, studies of different facets of lung function have received increased attention (Kauczor et al. 1998, 2001; Leutner and Schild 2001). For example, MR-based strategies provide adequate surrogates to estimate the extent of active inflammation or impaired ventilation distribution in fibrotic lung disease. Improved understanding of the underlying disease process and assessment of the clinical impact of such surrogate measurements are among the next steps before MRI will be widely accepted as a helpful tool in the clinical arena of lung fibrosis.

H.-U. Kauczor, MD
Professor of Radiology, Innovative Krebsdiagnostik und Therapie, Deutsches Krebsforschungszentrum (DKFZ), Im Neuenheimer Feld 280, 69120 Heidelberg, Germany

6.2.2
MRI in Lung Disease

In general, MRI of the lung is extremely challenging due to low spin density, significant susceptibility artifacts and motion artifacts caused by breathing, cardiac pulsation and blood flow (Kauczor and Kreitner 2000). In fibrotic lung disease, the disease itself is associated with an increase in spin density which provides some disease-related contrast. At the same time susceptibility artifacts which are caused by multiple air-tissue interfaces are reduced by the disease itself. Both relate to improved visualization of fibrotic changes when compared to normal lung parenchyma (Fig. 6.2.1). MRI was evaluated in the differentiation of various causes of airspace disease (Moore et al. 1986), however the increase in signal intensity by itself is highly unspecific. Fibrotic lung disease is represented by parenchymal bands, reticulation, nodules, interlobular septal thickening as well as ground-glass opacities. All of them are well visualized on proton density and T1-weighted images (Fig. 6.2.1). These findings at MRI correlate well with gross morphologic features at pathology (Primack et al. 1994). Parenchymal opacification represented an active inflammatory process, such as alveolitis, pneumonia and granulomatous inflammation in 12 out of 14 cases. In 2 out of 14 cases it represented fibrosis. The underlying diseases of these false positive findings were talcosis in one patient and silicosis with a consolidation due to conglomerate fibrosis in another patient. Reticular patterns represented fibrosis in 5 out of 5 patients (Primack et al. 1994). The administration of a contrast agent is capable of significantly improving the detection of honeycombing but not the detection of parenchymal ground-glass opacities (King et al. 1996). The differentiation of various causes of airspace disease is not successful, because there is considerable overlap in the measured T1 and T2 values between different underlying diseases. Relaxometry alone is not able to establish a specific diagnosis. As is well-known from CT, assessment of the extent and distribution of lung

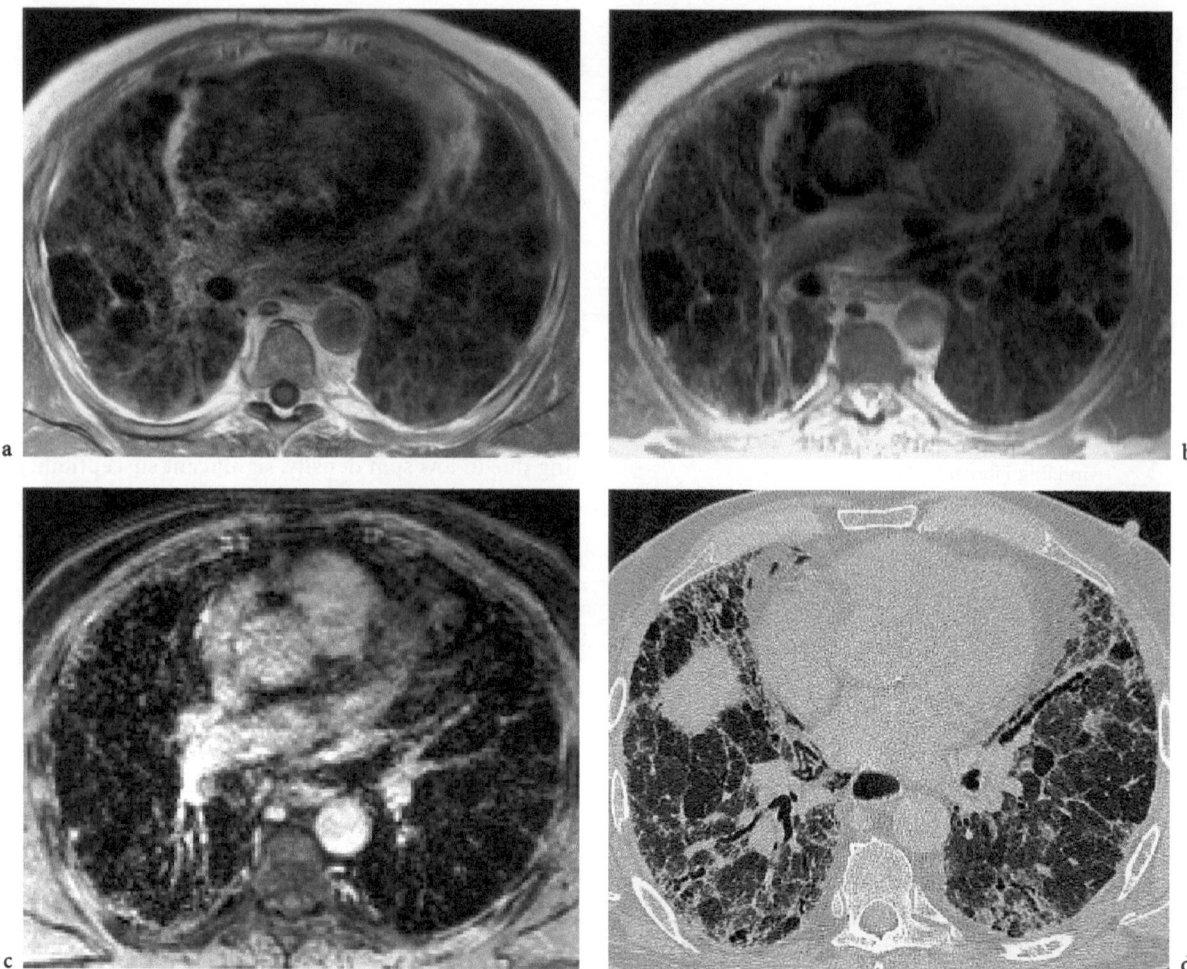

Fig. 6.2.1a–d. Patient with usual interstitial pneumonitis. **a** T1-weighted turbo spin echo (TSE) shows parenchymal opacities and reticular abnormalities. **b** T2-weighted HASTE shows areas with active inflammation with higher and honeycombing with low signal intensity. **c** T1-weighted gradient echo sequence post-contrast shows areas of active inflammation with slight contrast enhancement. **d** High-resolution CT demonstrates signs of severe fibrosis with intralobular septa, ground-glass opacities and traction bronchiectasis

fibrosis is more helpful in indicating towards the final diagnosis of interstitial lung disease.

In lung fibrosis, the assessment of disease activity is of great importance to decide whether and which anti-inflammatory treatment should be initiated. For this purpose active inflammation, such as alveolitis, must be differentiated from fibrosis. Alveolitis is represented by parenchymal opacification and ground-glass opacities (PRIMACK et al. 1994). It exhibits even higher signal intensity on T1-weighted and T2-weighted images than fibrosis (Fig. 6.2.1). It has been demonstrated that signal intensity is related to clinical severity of disease. Signal intensity also serves as a potential indicator for the response to anti-inflammatory therapy (McFADDEN et al. 1987). Areas affected by active inflammation, such as

alveolitis, are highly perfused. Thus, they also show marked enhancement after contrast administration. Successful anti-inflammatory treatment reduces the activity of alveolitis as well as the development of fibrosis. It is associated with a marked decrease in signal intensity. Bleomycin-induced lung damage is used as a model for interstitial lung disease in experimental studies. After intratracheal instillation of bleomycin in rabbits acute inflammation ("alveolitis") is induced. After a phase of active alveolitis fibrotic changes occur and become more and more prominent. Repeated MRI showed that signal intensities on T1-weighted images post-contrast as well as on T2-weighted images were markedly higher in the alveolitic phase than in later fibrotic stages (KERSJES et al. 1999). When comparing both sequences, changes were even more pronounced on T2-weighted images than after contrast administration. A similar time course of signal intensities was observed in radiation pneumonitis. Looking at patients receiving radiotherapy for bronchogenic carcinoma, the pulmonary parenchyma surrounding the tumor showed a steady increase in signal intensity on T1-weighted and T2-weighted images over several months. During the further course after irradiation, signal intensities slowly decreased (YANKELEVITZ et al. 1994). These observations have been confirmed in animal experiments. Again measurements of T2 relaxation times were the most sensitive (SHIOYA et al. 1997). Macromolecular contrast agents, which are still not available for patient use, may improve the differentiation between alveolitis and fibrosis even further (BERTHEZENE et al. 1992). In the active alveolitic phase, leakage of macromolecules from the intravascular space into the extravascular compartment was observed. In the fibrotic phase, enhancement was markedly diminished which indicates a decrease in plasma volume within the fibrotic lung (BERTHEZENE et al. 1992).

Silicosis and silicotuberculosis can result in progressive massive fibrosis, which appeared hypointense on T2-weighted images as compared to skeletal muscle with some high signal areas centrally. These areas most likely corresponded to necrosis (MATSUMOTO et al. 1998b). Since the behavior of signal intensities is different in fibrotic and neoplastic tissue, lung cancer can be differentiated as a lesion with high signal on T2-weighted images (MATSUMOTO et al. 1998a). Pre-contrast, the consolidations were isointense to skeletal muscle on T1-weighted images in 70% of cases. Post-contrast, about 50% of the consolidations showed rim enhancement whereas the central areas, which most likely corresponded to necrosis, will not be enhanced (MATSUMOTO et al. 1998b). In a different report progressive massive fibrosis showed high signal intensity on T1-weighted images in 14 out of 18 lesions whereas only 4 out of 18 exhibited low signal intensity (JUNG et al. 2000). After contrast administration, there was a gradual increase in signal intensity up to 3 min and a later plateau up to 15 min (JUNG et al. 2000). The findings encountered in Wegener's granulomatosis, which is characterized by solid and cavitated intrapulmonary nodules, are similar to progressive massive fibrosis. The walls of larger nodules showed marked enhancement after contrast administration whereas the central, necrotic areas did not enhance (KALAITZOGLOU et al. 1998). In lipoid pneumonia a pulmonary consolidation with high signal intensity on T1-weighted images corresponding to the lipid content has been described (LAURENT et al. 1999).

6.2.3
Comparison of CT and MRI in Lung Fibrosis

CT is the gold standard for the morphological assessment of lung disease. Although the extent and distribution of airspace disease can also be clearly depicted at MRI, it is inferior to CT regarding anatomic assessment and demonstration of fibrosis, i.e. parenchymal bands and reticular abnormalities. But there is a high rate of agreement in the assessment of parenchymal opacification and ground glass opacities (Fig. 6.2.1) (MÜLLER et al. 1992). Thus, MRI is able to demonstrate the areas with active alveolitis, i.e. treatable disease, with a sufficiently high sensitivity and as well as CT does (PRIMACK et al. 1994). In a recent retrospective study with patients suffering from a wide array of chronic infiltrative lung diseases, MRI has confirmed its potential to visualize disease activity by contrast enhancement on T1-weighted images. Of the 17 patients with active disease mainly diagnosed by bronchoalveolar lavage, 14 had enhanced lesions at MRI, and 3 had non-enhanced lesions. Pulmonary lesions were not enhanced in any patients with inactive disease (GAETA et al. 2000). Although there are no prospective clinical studies available so far, it seems highly promising to use T2-weighted and contrast-enhanced MRI for the assessment of active, i.e. treatable, disease, especially interstitial pneumonia and sarcoidosis.

6.2.4
³He-MRI in Lung Fibrosis

MRI using inhaled contrast agents has not been investigated systematically in patients with lung fibrosis (KAUCZOR et al. 1997; GAST et al. 2002). ³He-MRI can be used to visualize ventilation distribution and detect ventilation defects. In such studies fibrosis has been more closely associated with a heterogeneous distribution of signal intensities rather than with distinct defects (Figs. 6.2.2, 6.2.3) (KAUCZOR et al. 1997). The most likely cause is the lack of ventilation defects caused from bronchial obstruction. The heterogeneity indicates regional differences in ventilatory states, time constants, compliance and oxygen concentration. Functional and clinical significance of these findings are still to be investigated. The use of

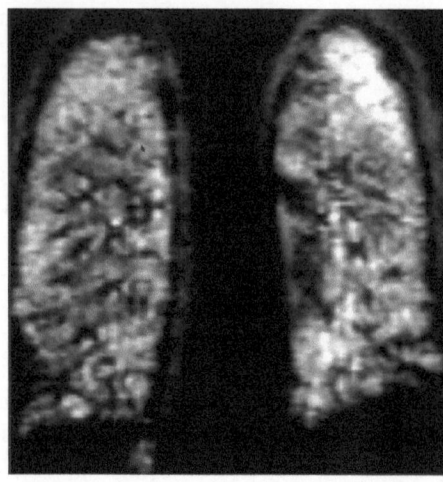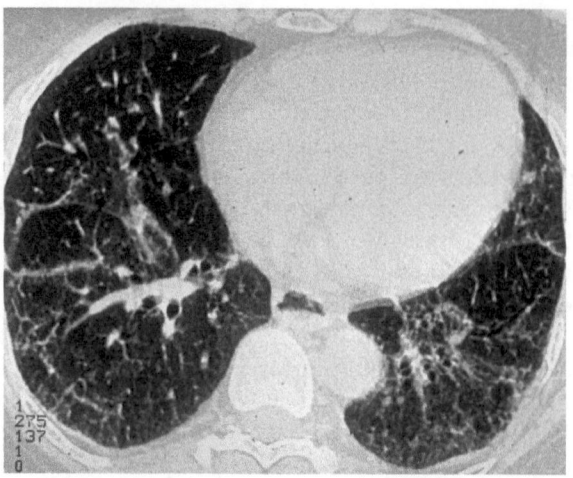

a
b

Fig. 6.2.2a, b. Patient with lung fibrosis due to systemic sclerosis. **a** ³He-MRI shows heterogeneous distribution of signal intensity and some small linear ventilation defects due to fibrotic bands. **b** High-resolution CT demonstrates signs of moderate fibrosis with intralobular septa and curvilinear bands with a typical predominance in the lower and posterior lung regions

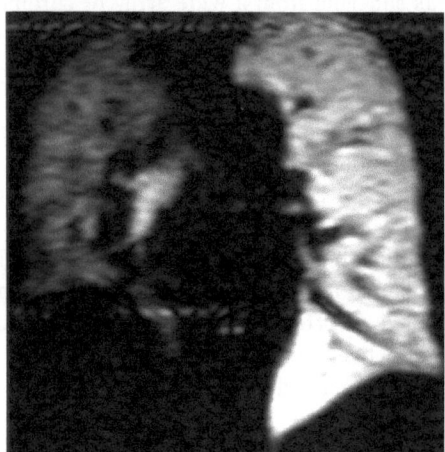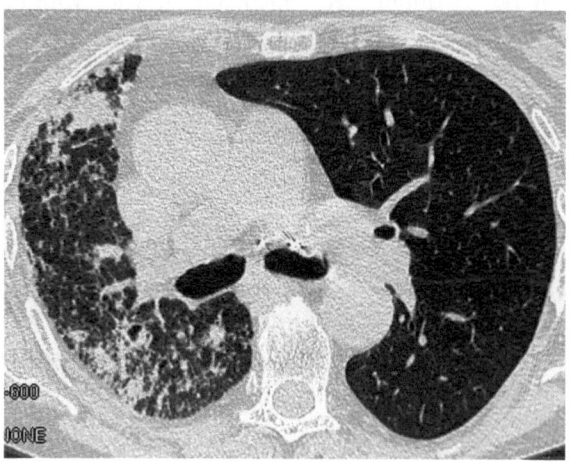

a
b

Fig. 6.2.3a, b. Patient with idiopathic lung fibrosis and a lung transplant on the left. **a** ³He-MRI shows homogeneous ventilation of the lung transplant with high signal intensity, whereas the smaller fibrotic lung shows heterogeneous ventilation with a much lower signal intensity. **b** High-resolution CT demonstrates normal parenchyma on the left and fibrotic destruction on the right

diffusion-weighted ³He-MRI allows for assessment of pulmonary microstructure. The results of these measurements are given as the apparent diffusion coefficient (ADC) which is a measure of the size of the peripheral airspaces within the lung. With normal values being below 0.24 cm²/s they can be significantly increased (0.35 cm²/s) in lung fibrosis. The increase in size of peripheral airspaces in fibrosis corresponded to the presence and extent of honeycomb cysts (HANISCH et al. 2000).

References

Berthezene Y, Vexler V, Kuwatsuru R et al. (1992) Differentiation of alveolitis and pulmonary fibrosis with a macromolecular MR imaging contrast agent. Radiology 185:97–103

Gaeta M, Blandino A, Scribano E et al. (2000) Chronic infiltrative lung diseases: value of gadolinium-enhanced MRI in the evaluation of disease activity – early report. Chest 117: 1173–1178

Gast K, Viallon M, Eberle B et al. (2002) MR imaging in lung transplant recipients using hyperpolarized 3He: comparison with CT. J Magn Reson Imaging 15:268–274

Hanisch G, Schreiber W, Diergarten T et al. (2000) Investigation of intrapulmonary diffusion by 3He MRI. Eur Radiol 10:S345

Jung J, Park S, Lee J et al. (2000) MR characteristics of progressive massive fibrosis. J Thorac Imaging 15:144–150

Kalaitzoglou I, Drevelengas A, Palladas P et al. (1998) MRI appearance of pulmonary Wegener's granulomatosis with concomitant splenic infarction. Eur Radiol 8:367–370

Kauczor H-U, Kreitner K-F (2000) Contrast-enhanced MRI of the lung. Eur J Radiol 34:196–207

Kauczor H-U, Ebert M, Kreitner K-F et al. (1997) Imaging of the lungs using 3He MRI: preliminary clinical experience. J Magn Reson Imaging 7:538–543

Kauczor H-U, Surkau R, Roberts T (1998) MRI using hyperpolarized noble gases. Eur Radiol 8:829–827

Kauczor H-U, Chen X, Beek E van et al. (2001) Pulmonary ventilation imaged by magnetic resonance: at the doorstep of clinical application. Eur Respir J 17:1–16

Kersjes W, Hildebrandt G, Cagil H et al. (1999) Differentiation of alveolitis and pulmonary fibrosis in rabbits with magnetic resonance imaging after intrabronchial administration of bleomycin. Invest Radiol 34:13–21

King M, Bergin C, Ghadishah E et al. (1996) Detecting pulmonary abnormalities on magnetic resonance images in patients with usual interstitial pneumonitis: effects of varying window settings and gadopentetate dimeglumine. Acad Radiol 3:300–307

Laurent F, Philippe J, Vergier B et al. (1999) Exogenous lipoid pneumonia: HRCT, MR, and pathologic findings. Eur Radiol 9:1190–1196

Leutner C, Schild H (2001) MRI of the lung parenchyma. Fortschr Roentgenstr 173:168–175

Matsumoto S, Miyake H, Oga M et al. (1998a) Diagnosis of lung cancer in a patient with pneumoconiosis and progressive massive fibrosis using MRI. Eur Radiol 8:615–617

Matsumoto S, Mori H, Miyake H et al. (1998b) MRI signal characteristics of progressive massive fibrosis in silicosis. Clin Radiol 53:510–514

McFadden R, Carr T, Wood T (1987) Proton magnetic resonance imaging to stage activity of interstitial lung disease. Chest 92:31–39

Moore H, Webb W, Müller N et al. (1986) MRI of pulmonary airspace disease: experimental model and preliminary clinical results. AJR Am J Roentgenol 146:1123–1128

Müller N, Mayo J, Zwirewich C (1992) Value of MR imaging in the evaluation of chronic infiltrative lung disease: comparison with CT. AJR Am J Roentgenol 158:1205–1209

Primack S, Mayo J, Hartman T et al. (1994) MRI of infiltrative lung disease: comparison with pathologic findings. J Comput Assist Tomogr 18:233–238

Shioya S, Tsuji C, Kurita D et al. (1997) Early damage to lung tissue after irradiation detected by the magnetic resonance T2 relaxation time. Radiat Res 148:359–364

Yankelevitz D, Henschke C, Batata M et al. (1994) Lung cancer: evaluation with MR imaging during and after irradiation. J Thorac Imaging 9:41–46

7   Analysis of Distribution of Ventilation

Edwin J. R. van Beek, Andrew Swift, Jim M. Wild

7.1   Helium MR Imaging

Edwin J. R. van Beek, Andrew Swift, Jim M. Wild

CONTENTS

7.1.1
Introduction

Conventional proton MR images of the lungs are hampered by the low proton density in the lungs and by artifacts caused by inhomogeneous static

E. J. R. van Beek, MD, PhD, FRCR
Clinical Senior Lecturer/Honorary Consultant Radiology, Academic Unit of Radiology, Floor C, Royal Hallamshire Hospital, Glossop Road, Sheffield S10 2JF, UK
A. Swift, MD
Academic Unit of Radiology, Floor C, Royal Hallamshire Hospital, Glossop Road, Sheffield S10 2JF, UK
J. M. Wild, MD
Academic Unit of Radiology, Floor C, Royal Hallamshire Hospital, Glossop Road, Sheffield S10 2JF, UK

fields within the thorax. Therefore, the lung has traditionally been considered a black hole in terms of MR imaging. In recent years, the introduction of hyperpolarized noble gases, such as helium-3 (3-He) and xenon-129 (129-Xe), has shown promise for functional imaging of the pulmonary air spaces. These isotopes have a nuclear spin of 1/2 and are thus sensitive to nuclear magnetic resonance techniques. In the presence of a strong magnetic field (B_0) these spins in a given atom can exist in one of either two ground states aligned either parallel to or anti-parallel to B_0. Conventional proton MRI utilizes the fact that in thermal equilibrium a slightly larger proportion of the spins adopt the lower energy ground state (parallel to B_0) and this population can be subsequently excited into the higher state by radio frequency excitation. The size of this fractional population imbalance is governed by Boltzmann thermal statistics and is directly proportional to the field strength and the gyromagnetic ratio of the nucleus and inversely proportional to the temperature. The subsequent return of spins to the lower ground state is what produces the coherent radio frequency discharge, which we measure in MRI as the free induction decay (FID). The low spin density of the gaseous state and the lower gyromagnetic ratio of these noble gas nuclei, when compared to the protons, limit the sensitivity of the thermal equilibrium noble gas NMR. However, in recent years laser optical pumping techniques (Kastler 1950; Colegrove et al. 1963; Miron et al. 1984) have been used to temporarily enhance the spin population beyond that attainable in thermal equilibrium. The production of a large ensemble equilibrium spin population is called hyperpolarization. Under the right experimental conditions, it is possible to achieve hyperpolarization of 70% and above, which is in the order of 100,000 times that which can be attained with noble gases at thermal equilibrium. Thus, although the overall spin density introduced into the lungs is low, the hyperpolarization compensates to give ample signal for MR imaging when the gas is inhaled in vivo.

Although 129-Xe was first used to produce ventilation MRI images in animals and humans (ALBERT et al. 1994; SAKAI et al. 1996; WAGSHUL et al. 1996; MUGLER et al. 1997), the use of 3-helium soon followed and has made more impact since in the field of lung MRI. Hyperpolarized 3-He MR imaging was successfully performed ex-vivo, in animals and in humans in rapid succession between 1994 and 1996 (MIDDLETON et al. 1995; BLACK et al. 1996; EBERT et al. 1996; KAUCZOR et al. 1996; MACFALL et al. 1996). The main advantage of 129-xenon is its wide availability (natural abundance in the atmosphere), while 3-He is a limited resource (produced as a by-product of tritium decay in the nuclear arms industry) and should involve an active recycling program. Pure ^{129}xenon is also slightly easier to produce, albeit at considerable costs. However, the main disadvantages of ^{129}xenon are its anesthetic properties stemming from its lipophilic nature, its rapid uptake into the blood through the lung membrane-capillary barrier and its relatively inferior polarization levels. Furthermore, the polarization of ^{129}xenon is more technically complex, the polarization levels that can be achieved are lower as a result of the smaller spin-exchange cross sections and the T1-relaxation times are shorter than for ^{3}helium, meaning the polarization has a shorter half-life. Finally the gyromagnetic ratio of ^{129}xenon is smaller than ^{3}helium so its inherent sensitivity to NMR is lower. Thus, the production and application of hyperpolarized 3-helium gas has gained momentum more quickly, and this is currently the preferred method for lung imaging. Although the use of hyperpolarized 129-Xe has huge potential, it is likely that its applications will be outside the area of lung ventilation. Therefore, this chapter will focus on the hyperpolarized 3-helium MR imaging applications.

7.1.1.1
Processes to Produce Hyperpolarization

An excellent review of gas polarization by optical pumping techniques is given by GOODSON (2002). Optical pumping involves the transfer of angular momentum from circularly polarized light to the electron and nuclear spins of atoms, and was first demonstrated by the physicist KASTLER (1950). Initially, most of the work using optical pumping for nuclear spin polarization revolved around the requirements for nuclear physics experiments (BECKER et al. 1998). However, in recent years the potential applications in biomedical sciences were recognized and have been rapidly developed.

There are currently two main methods to obtain a hyperpolarized state of 3-He gas. The first uses a method called "spin exchange optical pumping", which uses an alkali metal vapor as an intermediary for spin polarization transfer (MIRON et al. 1984). This technique was recently reviewed by some of the pioneers in this work, WALKER and HAPPER (1997). Initially, prototypes of these systems were developed in a research setting, the technology developed has subsequently led to a commercial set-up, which is currently undergoing clinical tests (Fig. 7.1.1). This system can be adapted for production of either hyperpolarized 3-He or 129-Xe, and work is in progress to make this commercially viable (DRIEHUYS et al. 1996). We outline in brief the physics involved in such a spin-exchange polarizer. Firstly; irradiation of rubidium vapor with circularly polarized photons of wave-length 794.8 nm promotes the valence electron of the metal to an excited state. This process takes place at a pressure of 3 bars of 3-He gas mixed with a small amount of nitrogen gas, within a homogeneous magnetic field. Collisions between alkali atoms (in a polarized state) and 3-He atoms lead to the polarization transfer from rubidium electrons to 3-He nuclei. The costs of these individual systems are in the order of 75–150 k US dollars. Such devices require skilled operation and are capable of producing 1 l of gas polarized to around 35–40% at 1 atm pressure (sufficient gas to image a single patient with 3 separate imaging protocols) in a period of 24 h.

The second method is known as "metastability exchange optical pumping", which takes place entirely between 3-He atoms without the need for an intermediary alkali metal atom (COLEGROVE et al. 1963). This set-up takes place at low pressure, in the presence of a weak radio frequency discharge which causes some of

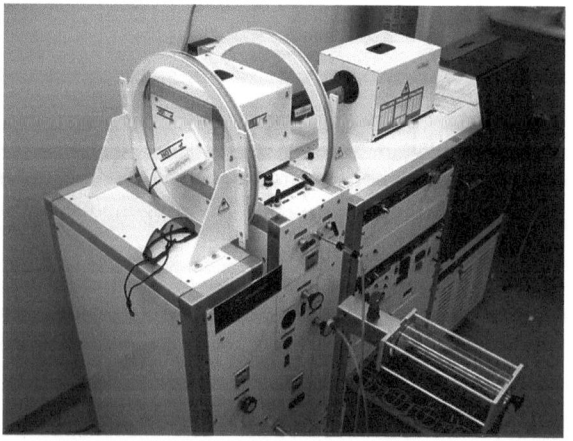

Fig. 7.1.1. Commercial polarizer using rubidium exchange technique (Amersham Health, Princeton, N.J.). This system is capable of producing 1 l of hyperpolarized 3-helium per 24 h

the valence electrons to be excited from ground state to a metastable state. These metastable electrons are then suitable for direct optical pumping with lasers at a frequency of 1083 nm. This produces an increased population or polarization of the excited electron energy levels. This polarization subsequently gets transferred to the nuclear spin of the 3-He atom by the hyperfine coupling interaction, causing polarization of the 3-He nucleus itself. Collisions between polarized metastable and non-polarized atoms in the ground state result in the transfer of polarization throughout the ensemble. Once polarization is achieved, the 3-He gas needs to be compressed to reach atmospheric pressure. This process of metastable optical pumping and compression of polarized gas has been refined at the University of Mainz, Germany (Fig. 7.1.2), where a large system has lent itself to up-scaling for production of large quantities (in excess of 1 bar-liter of gas per hour) of highly polarized (>60%) 3-He (BECKER et al. 1994). Smaller scale systems have been developed, using metastable optical pumping with dedicated peristaltic (NACHER et al. 1999) or diaphragmatic (GENTILE et al. 2000) pumps for compression of smaller quantities of gas.

There are advantages and disadvantages of the different technologies. Although the spin-exchange system and the small sized metastable systems lend themselves to on-site production of hyperpolarized 3-He, the disadvantage is that this requires space, expertise, manpower and the compliance with pharmaceutical production requirements. This may be more difficult to realize in smaller thoracic imaging centers. The scaled-up metastability system offers a central production facility, which has to be complemented by a distribution network. This would allow a purchase-to-need arrangement, similar to that used in the isotope industry. If gas is to be transported effectively in a polarized state, then care needs to be taken to preserve the polarization. This can be depleted as a result of several physical interactions. The first is any dipole-dipole interaction with proximal paramagnetic substances such as oxygen gas inside the glass cells. This puts huge demands on the cleanliness and vacuum preparation of production and transport systems. The second reason for loss of polarization is the introduction of stray magnetic fields, which will induce transitions and reduce the polarized population. Finally, there is a gradual polarization loss due to interactions between the atoms themselves and the surface of the storage cells. This can be minimized by reducing the presence of ferromagnetic and paramagnetic impurities in the glass. With careful choice of cell material this surface interaction contribution is minimal when compared to the first two factors (HEIL et al. 1995).

Naturally, these factors necessitate a transport system with shielding from external magnetic fields and glass cells with long relaxation times of the hyperpolarized 3-He gas. Such a system was recently tested, where hyperpolarized 3-He gas was produced at the University of Mainz, Germany, and transported by air and road to the University of Sheffield, United Kingdom (WILD et al. 2002b). Polarization levels at departure were approximately 40%, and imaging took place with an approximate 25% loss of polarization over the 12 h shipping time (Fig. 7.1.3).

Fig. 7.1.2. Large-scale polarizer for use as a central production facility (Institut für Physik, University of Mainz, Germany). This system is capable of producing many liters of hyperpolarized 3-helium per 24 h

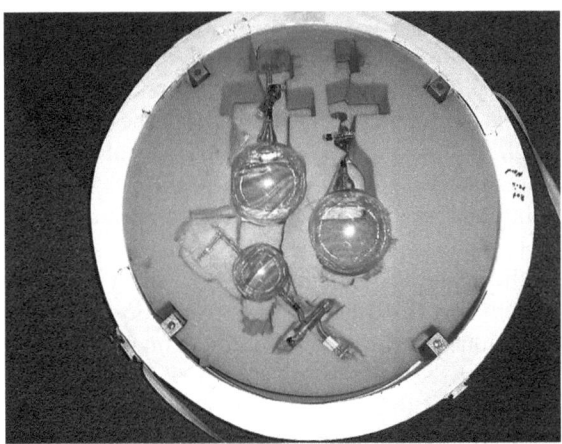

Fig. 7.1.3. Self-contained transport box with glass cells for long-range transportation of centrally produced hyperpolarized 3-He gas (Institut für Physik, University of Mainz, Germany)

7.1.1.2
Technical Requirements for MR Imaging of Hyperpolarized 3-He

The Larmor frequency of hyperpolarized 3-He is approximately 3/4 that of 1-H, i.e., 48 MHz at 1.5 T. Thus, a standard MR system needs to be adjusted to transmit and receive at the new frequency, requiring broadband receive/transmit amplifiers and RF coils. Volume coils of birdcage design have been employed most successfully in small animal studies (JOHNSON et al. 1997; BLACK et al. 1996), while simple surface coil designs have been used in human studies (KAUCZOR et al. 1996; MACFALL et al. 1996; DARASSE et al. 1997; DE LANGE et al. 1999; SAAM et al. 1999). More recently, quadrature coils have been designed, which have improved signal-to-noise ratios (DRIEHUYS et al. 1996; WALKER and HAPPER 1997; WILD et al. 2002a).

The introduction of an external source of polarization with optical pumping (alternative to that produced by the Boltzmann equilibrium of the Zeeman effect in the B_0 field) renders the signal-to-noise ratio less dependent upon field strength. Thus, hyperpolarized 3-He imaging lends itself to low field strength MR applications (TSENG et al. 1998; WONG et al. 1999; DURAND et al. 2002). Although this would have the great advantage of reduced susceptibility artifacts during imaging, most of the developments of hyperpolarized 3-He MR imaging have made use of 1.5 T systems. This is largely because this is the clinical standard field strength and the scanner manufacturers have to date focused their MR technology developments, especially broadband capabilities, at this field strength.

There are two key constraints in the choice of an optimum imaging strategy with inhaled hyperpolarized 3-He gas. The first is the non-recoverable hyperpolarization of the spin population once magnetization has been tipped into the transverse plane by the RF excitation pulse. This limits both the size of flip angle and the number of applicable RF pulses (phase encode views n) in the conventional Cartesian sampling of k-space. The fact that thermal recovery of polarization is negligible when compared to the size of the hyperpolarization allows ultra short repetition times (TR). Furthermore, the introduction of oxygen will reduce the T1-relaxation time from several hours to a few seconds (SAAM et al. 1995), which when combined with the finite time of a patients breath-hold, requires fast imaging sequences. The second constraint results from the high diffusion coefficient of ^{3}He at atmospheric pressure ($D \approx 1.8 \times 10^{-4}\,cm^2\,s^{-1}$ for pure ^{3}He at standard temperature and pressure).

This could potentially lead to significant signal dephasing as a function of the view-dependent MR imaging gradients. When combined, these two effects impose a complicated filter of the k-space data in all three dimensions (WILD et al. 2002a) and the imaging sequence requires careful choice. Thus, basic spin warp proton MR sequences with a succession of 90°RF pulses are not feasible, and fast low flip angle gradient echo sequences (FLASH) have been used most widely to date for in vivo imaging of 3-He gas (FRAHM et al. 1986). 2D imaging methodologies have been found to be more effective in that they depolarize the spins from a selected slice at any one time whereas a 3D sequence excites the whole lung thus depolarizing the whole gas reservoir. The choice of flip angle is important as is the phase encoding strategy (WILD et al. 2002a). Centric encoding results in a higher signal-to-noise ratio than sequential encoding with the expense of some blurring as the flip angle is increased. Several groups have utilized an increasing flip angle throughout the image acquisition to try and ensure a constant magnetization in order to circumvent these k-space filtering effects (ALBERT et al. 1994; MUGLER et al. 1997), however this requires very accurate calibration of the flip angle in vivo which in turn requires the subject to inhale a separate calibration dose of gas. In a recent report MUGLER et al. demonstrated promising results with a steady-state free precession (FISP) gradient echo sequence (MUGLER et al. 2002). The FISP sequence refocuses any residual transverse magnetization and recycles it with the stored longitudinal magnetization potentially offering a more efficient use of polarization and enhanced signal-to-noise when compared to FLASH. Problems arise with field inhomogeneity in FISP imaging, indeed the FISP images show a substantial artifact at the base of the lung close to the diaphragm where background susceptibility gradients are large. Alternative single shot sequences have also been implemented whereby all of the polarization is converted to transverse magnetization by a 90° pulse and k-space is rapidly encoded with either oscillating readout gradients in the case of EPI (SAAM et al. 1999) or a series of 180° refocusing pulses in the case of single shot fast spin echo or RARE sequences (DURAND et al. 2002). EPI techniques are potentially very attractive in that they are rapidly executed enabling multiple images to be acquired in the space of one breath-hold however these sequences are inherently sensitive to off-resonance effects induced by field inhomogeneities in the chest and have only been demonstrated to be free of artifacts in an axial plane (SAAM et al. 1999).

Furthermore the rapidly oscillating readout gradient causes substantial diffusion dephasing and limits a lower limit of the spatial resolution attainable. Similarly with RARE images diffusion of gas in the period between the 180° refocusing pulses also imposes a limit on the spatial resolution achievable. RARE offers efficient usage of the magnetization with the ability to refocus dephasing due to field inhomogeneities. However, care needs to be taken to ensure the pulses are efficiently calibrated and of homogeneous flip angle as non-ideal 180° pulses will soon depolarize the gas polarization. To our knowledge no known human in vivo applications of spin echo imaging such as RARE with 3-He gas have been reported at 1.5 T. Groups imaging at lower field strengths have presented in vivo spin echo images (DURAND et al. 2002), where the B_1 power dependence with B_0^2 means that at lower B_0 the power demands on the RF amplifiers for a homogeneous body transmit helium coil are less than those at 1.5 T.

Developments of these techniques and the interpretation of the physical properties of the interaction between hyperpolarized 3-He with RF pulses, lung tissue and oxygen have led to several, highly interesting sequences which have the potential to reveal exciting information pertaining to lung function and will be discussed in more detail below.

7.1.1.3
Delivery of Hyperpolarized Gas into MR Environment

Once the 3-He gas has been hyperpolarized, it needs to be transported into the MR system. Several methods may be employed, ranging from a simple bag (Fig. 7.1.4) to a more elaborate delivery system (Fig. 7.1.5).

The advantages of a simple bag are that it can be introduced quickly, for individual sequences and for single use. However, disadvantages are that it requires a completely anoxic breath (which may be more difficult in more severely affected lung patients) and it does not lend itself to easy recycling of the 3-He gas.

The advantages of a gas delivery system are that it can be accurately dosed, does not require an anoxic breath (rather the dose can be followed by air chaser) and the 3-He gas can be delivered within a chosen point of the respiratory cycle (EBERLE et al. 1999). This allows for standardization and individual quantification of gas delivery. This may become more important when the technique expands into clinical situations where patients have a limited cardiopulmonary reserve or in those with limited lung func-

tion and/or cooperation, such as children. Finally, the use of this system makes it relatively easy to recycle 3-He gas, as a valve system can be used to allow exhaled breaths to be collected.

7.1.2
Hyperpolarized 3-Helium MR Imaging: the Four Step Protocol

The development of hyperpolarized 3-helium MR imaging has led to four different techniques, which all yield different information on lung function. The

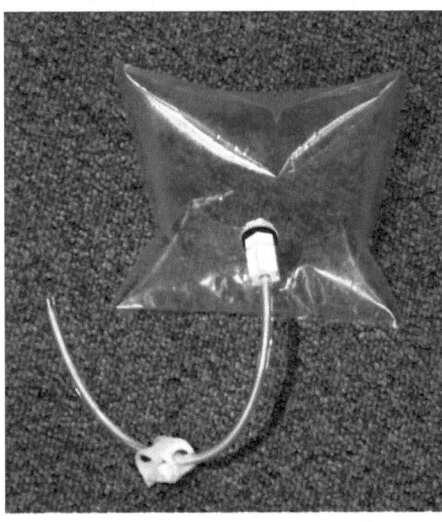

Fig. 7.1.4. Simple Tedlar bag with mouth piece for administration of hyperpolarized 3-He gas mixed with nitrogen (Amersham Health, Princeton, N.J.)

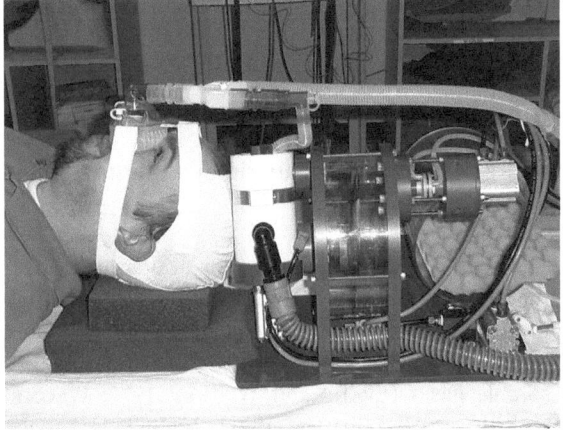

Fig. 7.1.5. Respirator controlled system for administration of hyperpolarized 3-He gas within a normal air-containing breath (University of Mainz, Germany)

techniques developed are: static ventilation distribution ("spin density"), apparent diffusion coefficient ("small airway size"), dynamic ventilation imaging ("gas inflow and outflow") and oxygen partial pressure ("oxygen uptake"). To understand how these techniques can improve our understanding of the pathophysiology of lung function, one has to be aware of the available techniques for clinicians at present and a brief summary is warranted at this point.

7.1.2.1
Pulmonary Function Tests

Pulmonary function tests (pulmonary spirometry) are routinely used for determination of lung volumes in a clinical setting (QUANJER et al. 1993). These consist of static inspiratory volumes, static expiratory volumes and dynamic expiratory volumes. Total lung capacity (TLC) is the volume of gas at the end of maximal inspiration, and is achieved when the forces opposing expansion balance maximal force generated by the inspiratory muscles. Residual volume (RV) is the volume remaining after a maximal expiration, and is governed by the balance between the maximal force generated by expiratory muscles and the elastic forces opposing reduction in lung volume. Functional residual capacity (FRC) is the volume of air in the lungs at the end of a normal expiration, and is determined by a balance between the inward elastic forces of the lung and the outward forces of the respiratory cage. Vital capacity (VC) is the volume from full inspiration to full expiration. Forced expiratory volume (FEV_1) consists of the volume that can be exhaled in one second. The main problems of standard spirometry are the lack of determination of which parts of the lungs contribute, while the results depend on patient cooperation.

7.1.2.2
Multiple Inert Gas Elimination Technique (MIGET)

The standard technique used for evaluation of ventilation-perfusion (V_A/Q) inequalities is the multiple inert gas elimination technique (MIGET), which is based on the gas exchange behavior of six inert soluble gases (EVANS and WAGNER 1977; WAGNER and RODRIGUEZ-ROISIN 1991). MIGET is capable of estimating alveolar ventilation and pulmonary blood flow patterns without interfering with ongoing physiological processes, while it facilitates the

determination of extrapulmonary influencing factors of abnormal gas exchange (RODRIGUEZ-ROISIN 1994). MIGET calculations are quite complicated, and are based upon assumptions such as a homogeneous compartmental lung model and homogeneous perfusion in lung areas that are similarly ventilated (ROCA and WAGNER 1994). Using these premises, MIGET can yield information on V_A/Q distribution, shunting, dead space and cardiac output. Some of the disadvantages of the technique are that it is firstly a whole lung measurement having no anatomical specificity, requires gas chromatography and carries the need for intravenous and intra-arterial access lines, albeit that later modifications suggest that intra-arterial blood analysis may be omitted (WAGNER et al. 1985). These factors explain why the technique has not been widely introduced in respiratory medicine. The four-step 3-He MRI protocol, which was proposed by the Mainz group (KAUCZOR et al. 2001) and has been adopted by leading world sites for clinical trials, attempts to address the parameters described above. Furthermore, additional information will need to be shown of interest to physiologists, pathologists and clinicians if the technology is to make an impact on patient management. This chapter will deal with static and dynamic ventilation distribution. We will briefly describe the other two parts of the four-step protocol, but for more detailed information the reader is referred to Chaps. 5.2 and 8 in this book.

7.1.3
Static Ventilation Imaging (Spin Density)

As discussed above, the use of standard spin warp sequences which utilize large flip angles is limited due to the non-renewable signal of hyperpolarized 3-He. Thus, fast gradient echo-type sequences with small flip angles have generally been employed. For example 15 slices could be acquired with 128 views per image with a sequence TR of 7 ms in approximately 14 s allowing a spin density study to be completed within a single breath-hold.

The spatial resolution of these images is in the order of $2.5 \times 2.5 \times 10$ mm, which is considerably better than what can be achieved using lung scintigraphy methods. The speed with which the images are obtained and the lack of ionizing radiation are further advantages of 3-He imaging. It is anticipated that the spatial resolution will increase further, thus approaching the spatial resolution of high resolution CT ($1 \times 1 \times 1$ mm).

Recently a lot of interest has been expressed in oxygen-enhanced proton MRI (see also Chap. 5.2, 2.4 and 8.4) as a means of measuring ventilation (MAI et al. 2002). The technique uses single shot fast spin echo sequences with an inversion recovery pre-pulse to introduce some T1 contrast. As oxygen is paramagnetic the T1 of water in the lungs when breathing pure oxygen should be lower than when breathing air. By comparing the signal intensities measured whilst breathing oxygen and air any statistical differences in contrast can be inferred in the way that parametric maps in functional MRI of brain activation are produced. The technique is less involved in terms of hardware than hyperpolarized gas MRI in that it can be performed with a standard proton MRI scanner. However, there is some uncertainty as to whether the oxygen enhancement truly represents ventilation as the images are built up as a statistical average from multiple images. Furthermore it is not clear whether dissolution of oxygen in extracellular lung water and the vessels causing a T1 reduction, truly represents regional ventilation. Hyperpolarized 3-He gas on the other hand gives us an instantaneous snapshot of gas ventilation in the lungs at a higher spatial resolution.

CHEN et al. performed high-resolution images in a living guinea pig, using a 2 T experimental MR system and a non-slice-selective coronal oriented radial sequence (CHEN et al. 1998). They demonstrated the feasibility to obtain in-plane resolution of 188 188 μm. Further developments by the same group have resulted in true lung MR microscopy, with a pixel size of 117 117 μm and a slice thickness of 468 μm (JOHNSON et al. 2001).

Within spin density studies, it is common to see areas that are better ventilated than others. Generally, this distribution is gravity-dependent, leading to increased signal intensity due to higher gas concentrations in the posterior and inferior lung zones with the patient in the supine position (KAUCZOR et al. 1997; GUENTHER et al. 2000; MATA et al. 2001). Volume measurements can be performed, which have been shown to correlate well with pulmonary function tests, but the reproducibility is still problematic (KAUCZOR et al. 1997; RIZI et al. 2001). Indeed care should be taken when inferring ventilation volume measurements from these thick slice FLASH images as they are particularly susceptible to signal dephasing due to background field gradients which could be misdiagnosed as ventilation defects. These background gradients are particularly strong around the vessels, close to the diaphragm and in the apexes. WILD et al. (2003a) recently proposed an effective method of restoring some of this signal loss by combining three images acquired as interleaves in the same breath-hold, with varied slice refocusing gradients. The resulting susceptibility compensated images, highlight the sensitivity of FLASH images to signal loss in these areas stressing that care should be taken when inferring volumetric measurements of lung ventilation from segmented 3-He FLASH images (Fig. 7.1.6) (WILD et al. 2003a).

In a preliminary study of 18 subjects more defects were demonstrated in patients with obstructive lung diseases, although some patients with normal lung function also showed either a large defect or fine linear defects (these latter patients suffered from bronchogenic carcinoma or pulmonary fibrosis, respectively) (KAUCZOR et al. 1997). This demonstrated, to some extent, the potential sensitivity of this technique for ventilation abnormalities when compared to lung function tests. A study of 32 imaging studies in 12 non-smoking healthy volunteers, 1 mild asthmatic and 3 smokers showed excellent homogeneous filling of the air spaces in the normal volunteers, small peripheral defects which varied over time in the asthmatic patient and more extensive defects which increased with severity of emphysema (DE LANGE et al. 1999).

A normal spin density series will demonstrate homogeneous signal in the lungs, with filling of all the air spaces (KAUCZOR et al. 1997; GUENTHER et al. 2000). Small (less than 2 cm) filling defects may be detected, which are usually transient and located in the posterior lung fields. It has been postulated that these represent positional subsegmental atelectasis due to gravitational effects (MATA et al. 2001). However, sometimes small peripheral defects may occur in patients with mild asthma (DE LANGE et al. 1999). These defects should be distinguished from larger and/or more widespread ventilation defects or inhomogeneous distribution of the 3-He gas, which are considered pathological (KAUCZOR et al. 1997). However, the exact definition of what is within the normal range or what is pathological is sometimes difficult, and is still under discussion (DE LANGE et al. 1999).

One study investigated the repeatability of hyperpolarized ^{3}He MR in 6 asthmatic subjects and 4 normal volunteers, by repeated imaging 30–60 min apart (ALTES et al. 2001a). The findings were concordant in 9 subjects, while 1 subject had a substandard second investigation resulting in diminished sensitivity for ventilation defects.

Several disease processes and the effects of smoking on the airways have been studied in small numbers of patients. Some example images are shown here (Figs. 7.1.7, 7.1.8). The findings of 3-He MR imaging in these settings will be discussed in more detail below.

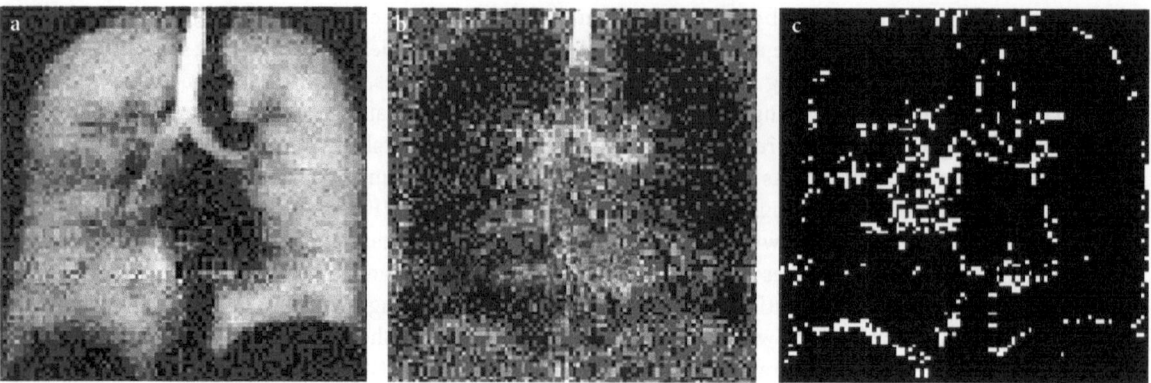

Fig. 7.1.6a, b. The effects of susceptibility artifacts in gradient echo imaging of HP gas can be reduced by careful sequence design – see WILD et al. (2003a). The images shown are based on a FLASH acquisition from a 10 mm coronal slice. **a** The standard FLASH ventilation image acquired with a balanced slice refocus from a central coronal slice. Low signal intensity is evident proximal to the blood vessels and heart with a fading of signal close to the diaphragm. **b** The difference between the compensated image and the standard image of **a**, with signal recovery in the trachea, around the blood vessels and at the diaphragm. The image of **c** is the difference between the segmented compensated image and the segmented standard image (threshold of 4 times the noise standard deviation) indicating the degree of potential underestimation of lung ventilation volume with the uncompensated images

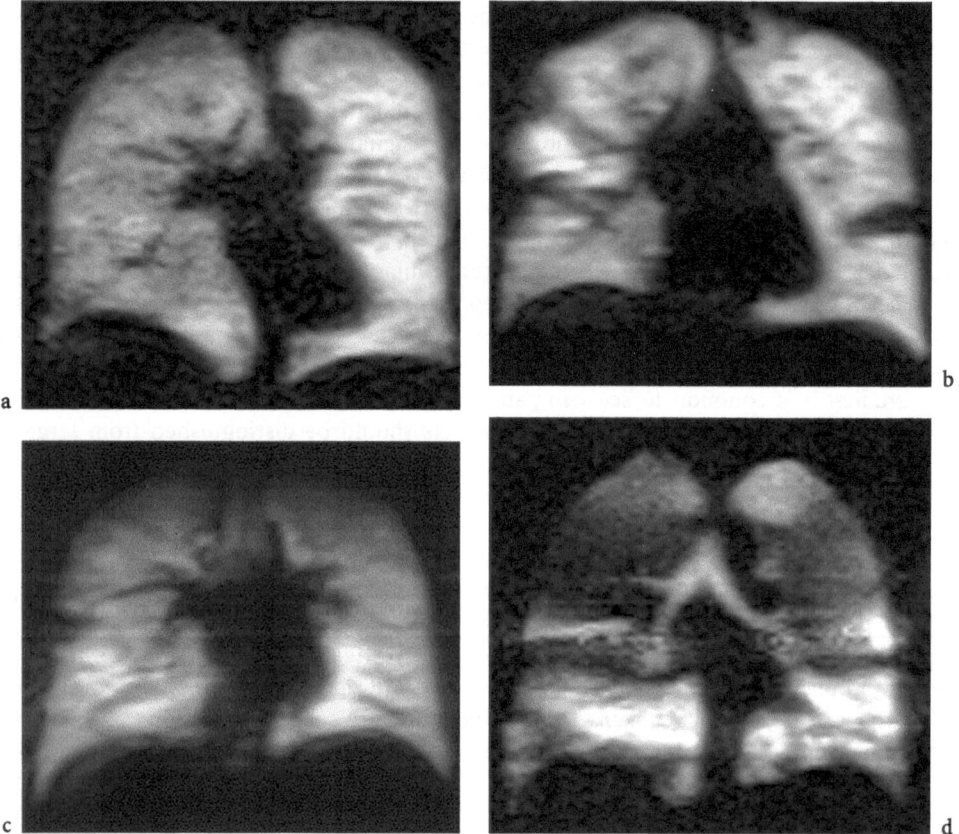

Fig. 7.1.7. HP ^{3}He images of ventilation in a healthy normal (**a**), an asthmatic (**b**), a smoker (**c**) and a patient with early chronic obstructive pulmonary disease (**d**). The ability of the technique to diagnose obstruction in the peripheral airways is immediately apparent with small wedge-shaped ventilation defects visible in the asthmatic and smoker (**b, c**) and large regions of impaired ventilation in the COPD patient (**d**). The images were acquired during breath-hold of 200 ml of HP ^{3}He gas using a FLASH sequence with a flip angle of 9°, 128 views, and a 10 mm slice thickness in a coronal plane with the subject lying supine using the 1.5 T clinical MRI system at the University of Sheffield

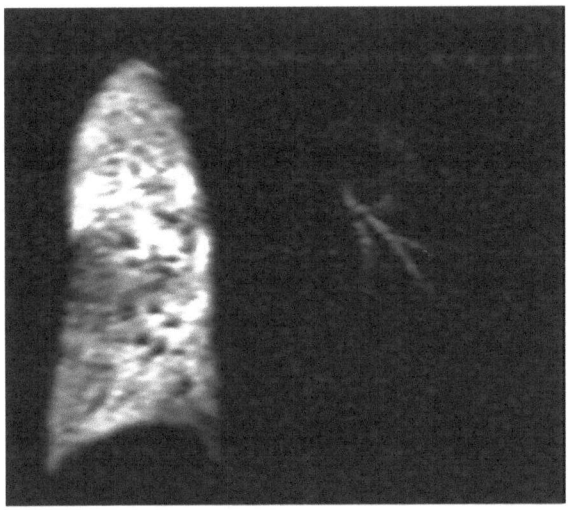
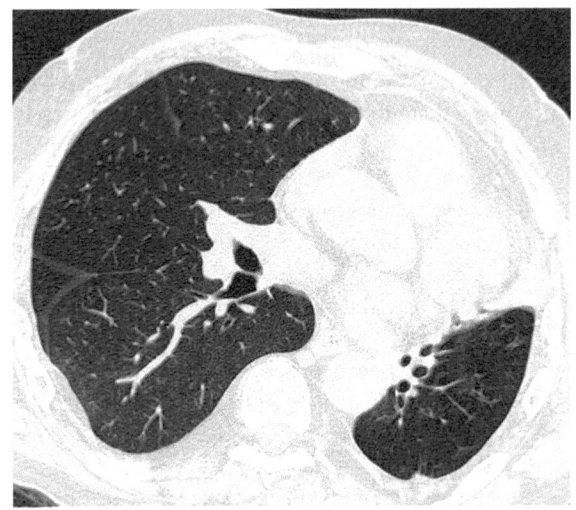

a   b

Fig. 7.1.8a, b. Patient with emphysema following single lung transplant on the right and lobectomy on the left. HP ^{3}He image (a) shows normal ventilation of the grafted lung without signs of rejection and filling of the central bronchi of the residual left lung but no ventilation of the alveolar space. CT for comparison (b) shows normal pulmonary parenchyma on the right and emphysematous destruction of the residual left lung

7.1.4
Dynamic Ventilation Imaging

Dynamic ventilation imaging using 3-He allows the visualization of the respiratory cycle. Using suitable imaging pulse sequences, the comparison of normal against abnormal ventilation can be shown. There are some parameters that can be varied to acquire different types of image, these include flip angle, length of acquisition window and positioning of the acquisition window (CHEN et al. 1998). Several pulse sequences have been used so far including gradient-echo FLASH, gradient echo-planar imaging (EPI), radial and spiral techniques.

MacFall et al. were the first to demonstrate dynamic images in humans using a fast gradient echo technique, thick slices in coronal orientation at a temporal resolution of 1800 ms (MACFALL et al. 1996). The minimum TR of a FLASH sequence was subsequently improved by SCHREIBER et al. (2000) by using an asymmetric echo with a non-selective RF pulse, giving a frame time of 130 ms for a 65 view FLASH image. Thus, it became feasible to observe in and outflow of gas, and trace the gas flow through the different airways (SCHREIBER et al. 2000). The drawback is that each line in k-space requires a RF pulse and each time a RF pulse is applied, the polarization of the 3-He is reduced. Furthermore the rectilinear Cartesian sampling used means that the center of k-space is only sampled once per frame, meaning that without keyhole type undersampling strategies the dynamic update rate of contrast in the image is limited by the overall frame rate (VAN VAALS et al. 1993).

For efficient use of the non-equilibrium magnetization EPI has the advantage over gradient echo because only one RF pulse is required per image meaning that the TR is the limiting factor on the temporal resolution. Also changes in image appearance can be directly equated to freshly inhaled polarized gas without having to consider the deconvolution of the temporal variation of signal which results in a multiple shot sequence from both depolarization and fresh influx (GIERADA et al. 2000). The drawback is the spatial resolution is limited by the high diffusion coefficient of 3-He and it's short T2* in the spatially inhomogeneous static field of the lungs. GIERADA et al. (2000) used EPI with a temporal resolution of 40 ms in healthy and emphysematous lungs, however the technique was limited by its spatial resolution. The pulse sequence requires an echo time of approx. T2*, as EPI has a longer echo time, marked signal loss around vessels was noticed due to field inhomogeneity effects and SAAM et al. (1999) observed a limited success with the pulse sequence in planes other than axial. Using an interleaved EPI sequence seems to lessen the diffusion and susceptibility limitations of EPI (MUGLER et al. 1998), however a similar study suffered greatly from motion artifacts (RUPPERT et al. 1998). It showed homogeneous and steady influx of gas into the peripheral air spaces, using axial slices, and confirmed a substantial gravity-dependent inflow into the posterior lung zones in a supine position.

An alternative means of sampling k-space in multiple views is offered by non-Cartesian strategies of radial projection and spiral imaging. These have the advantage that both central and outer parts of k-space are sampled per RF excitation (view) meaning that a fast refresh rate in the reconstructed image can be achieved by employing a "sliding window" reconstruction and sampling k-space in a continuously revolving manner (RIEDERER et al. 1998). VIALLON et al. (1999) presented both radial projection and interleaved spiral cine sequences (VIALLON et al. 2000) of inhaled 3-He gas in animal experiments. When combined with sliding window reconstruction these experiments gave high quality images with a fast pseudo-temporal image refresh rate.

Several studies have been performed in humans using interleaved spiral pulse sequences, the design of which offers a compromise between spin-warp gradient echo and EPI in terms of number of RF pulses, effectively a trade-off between consumption of hyperpolarization and spatial resolution (SALERNO et al. 2001). The short T2* for 3-He in the lung is accounted for by the shorter echo time that can be achieved with interleaved spiral imaging when compared to EPI. RUPPERT et al. (1998) performed an interleaved spiral 2D sequence with a temporal resolution of 240 ms and some distortion from susceptibility effects was noted.

More recently, WILD et al. (2003b) demonstrated radial projection imaging in humans with sliding window reconstruction. The short echo times offered with a radial sequence means firstly that artifacts due to field inhomogeneity are kept to a minimum and secondly the TR can be minimized to as low as 2 ms without slice selection and 4 ms with slice selection leading to data refreshment rates of 4 ms/frame. By over-sampling in the radial direction combined with angular undersampling, the time taken to acquire a complete image data set was further reduced without much compromise to spatial resolution (Figs. 7.1.9, 7.1.10). The information that can be obtained from dynamic 3-He MR imaging is potentially similar to that obtained from lung function tests. However, one has to realize that lung function tests yield an overall lung result, while 3-He MR imaging has the capability to obtain information on regional gas flow. The high temporal resolution achieved by this new technique could provide new understanding. The sensitivity of the technique has been shown by the detection of ventilation defects in healthy volunteers, which were attributed to normal lung physiology (KAUCZOR et al. 1997; GUENTHER et al. 2000; MATA et al. 2001), and more global defects would suggest impaired ventila-

tion (DE LANGE et al. 1999; ALTES et al. 2001a). There are several potential clinical applications for 3-He dynamic ventilation imaging. In asthma, insight may be gained from imaging dynamic gas flows during interventions such as giving beta-2 agonists. Some evidence suggests that the technique could even prove useful in management of difficult pulmonary air leaks (ROBERTS et al. 2000). Quantitative evaluation of the dynamic images can be performed by the generation of signal time curves in defined lung regions. For this the images need to be post-processed to correct for lung motion in order for the defined lung region to contain the same anatomy throughout the respiratory cycle. It has been shown that in some lung regions there is delayed ventilation, opening up the possibility that the technique can detect delayed expiratory time constants in patients with COPD (GIERADA et al. 2000).

A recent paper by GAST et al. (2002a) evaluated the possibility to derive quantifiable parameters of airflow using hyperpolarized 3-He MRI. A FLASH sequence with an image refreshment rate of 128 ms was used, and signal characteristics were observed and plotted. A motion correction algorithm was applied to try and maintain the region of interest constant. The following parameters were proposed and reference ranges obtained: tracheal transit time (0.11–1.21 s), tracheo-alveolar interval (0.0–0.02 s), alveolar rise time (0.22–0.62 s) and alveolar amplitude (0–76.6 arbitrary units). This latter measure will be difficult to implement, as the signal will depend on many external factors, such as level of polarization, amount of gas inhaled and dilution volume. This paper should be regarded as an initial attempt, as the temporal resolution is improving and therefore the observation potential increases.

7.1.5
Apparent Diffusion Coefficient

A unique difference between conventional proton MRI and noble gas MRI, is the degree of diffusive motion. This property is characterized by the diffusion coefficient, from which the average distance a molecule travels per unit time, and the diffusion length can be deduced. When the 3-He gas molecules are in the lung, the diffusion length of gas particles in restricted regions of the lung, for instance the small airways or alveoli, will have a smaller apparent diffusion coefficient (ADC) than in the larger airways. This contrast is captured through the application of a

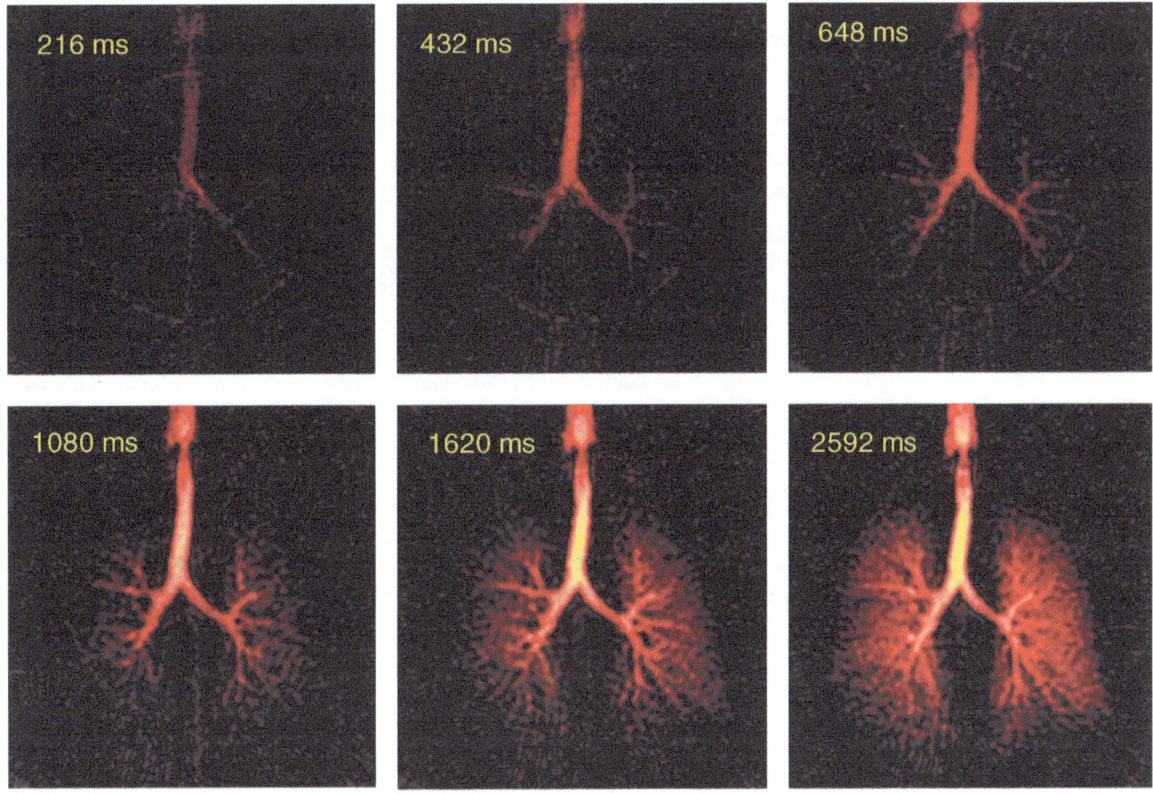

Fig. 7.1.9. Selected dynamic images from the time series acquired with a radial projection sequence (TE=2 ms, TR 5.4 ms with 128 views per frame) from the first part of an inhalation of 300 ml of ^{3}He polarized to 40% in a healthy subject. The images are reconstructed with a sliding window giving an effective temporal resolution in image contrast refreshment of the TR of the sequence (5.4 ms). The temporal passage of gas; down the trachea, into the bronchi and peripheral lung is clearly resolved in this selection of snapshots

coherent magnetic field gradient to disrupt the signal followed by another magnetic field gradient in the opposite direction. If the gas particles have moved between application of the two gradients the signal from that region will be attenuated, as particles in the large airways have a high diffusion length there will be more attenuation relative to the smaller more restricted airways. From this a diffusion map, called an ADC map, can be calculated.

The ADC of hyperpolarized 3-He has been measured in healthy and diseased lungs. It has been shown that ADC values are increased in emphysema, likely to be due to the disruption of the lung parenchyma resulting in alveolar destruction and enlargement. The normal ranges used in healthy volunteers can be used as a measure for the early diagnosis of emphysema in at risk groups and as new drugs become available it could be possible to test there efficacy in long term studies using ADC values. The interested reader is referred to Chap. 5.2.5.

7.1.6
Regional Oxygen Partial Tension MR Imaging

This is the final step in the four-step protocol, and aims to yield a regional map of hyperpolarized 3-helium signal over time. As the signal decay is linearly proportional to the concentration of oxygen in the airspaces, the map can be applied as a regional indicator of the uptake of oxygen from lung regions (DENINGER et al. 1999; EBERLE et al. 1999). Oxygen uptake is dependent on lung perfusion, and thus the map can be seen as a regional V_A/Q map of the lung. It is for the first time that direct measurements of the regional V_A/Q distribution can be obtained. Furthermore, this assessment is entirely non-invasive. Compared with MIGET, this technology further enhances our understanding of lung (patho-) physiology. Further description of the technology and its potential impact on understanding of lung function can be read in Chap. 8.3.

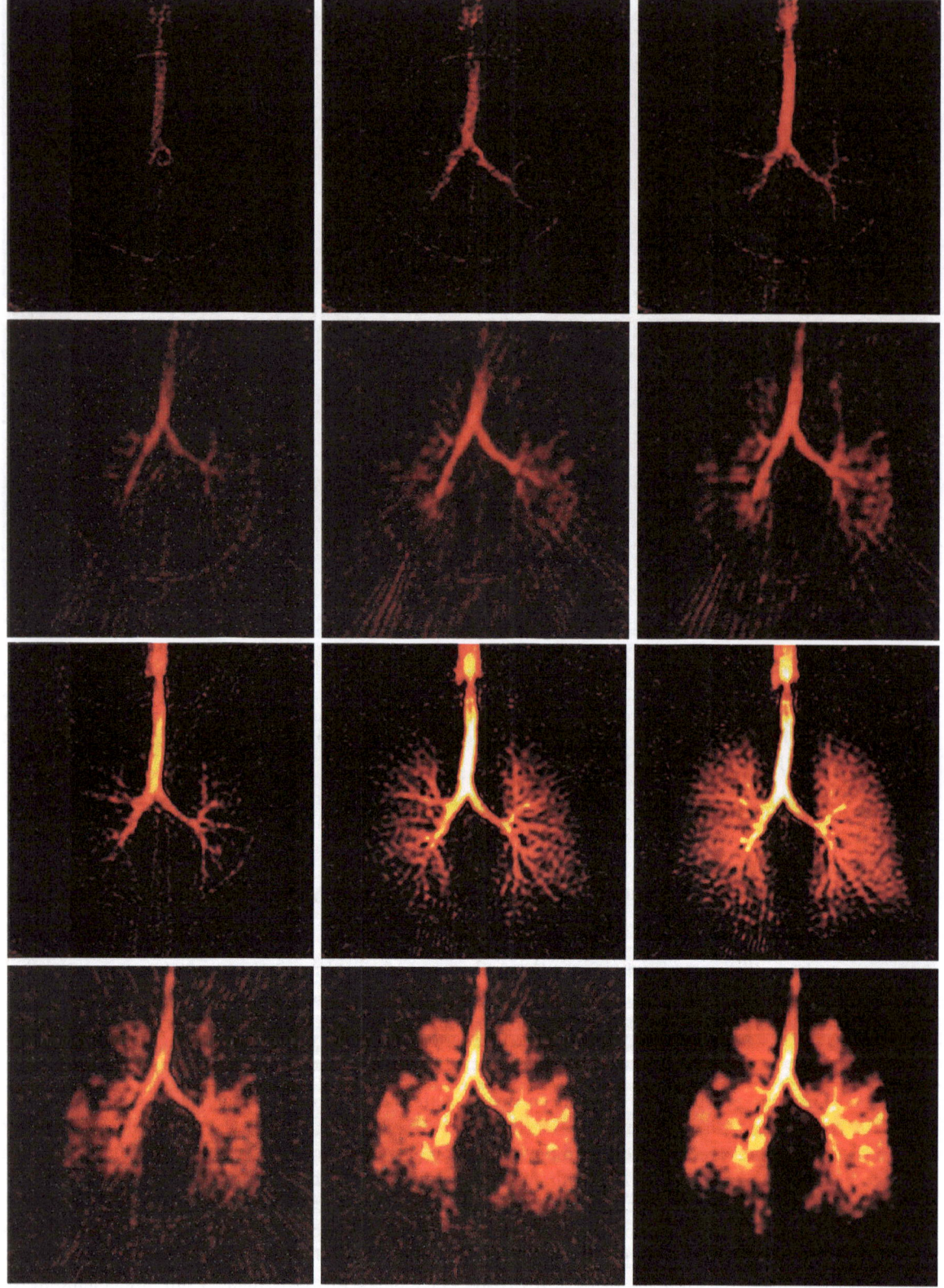

Fig. 7.1.10. Same series as in Fig. 7.1.9 in a patient with emphysema. Note delayed lung filling and multiple peripheral defects

7.1.7
Early Results of 3-He MR Imaging in Different Settings

Clinical information on the use of hyperpolarized 3-He MR imaging is still being received from ongoing studies. Thus, the information on clinical utility is still relatively limited. However, what seems of eminent importance is the enhanced information and integration of functional with morphological data required for treatment planning and assessment of treatment effectiveness. Some of the areas that have received some attention thus far are chronic obstructive pulmonary diseases (especially emphysema), asthma, cystic fibrosis and lung transplantation. Others are certain to follow, such as patients requiring lung resection (either for lung volume reduction surgery or for lung cancer).

7.1.7.1
Chronic Obstructive Pulmonary Diseases

Chronic obstructive pulmonary diseases are a mixture of lung diseases that are characterized by obstruction of air flow and air trapping. Emphysema is the most common disorder, and affects a large number of patients. Emphysema is defined as a lung condition which is characterized by permanent enlargement of air spaces distal to terminal bronchioles, accompanied by destruction of their walls (AMERICAN THORACIC SOCIETY 1962). Smoking is closely associated with emphysema (see below) as is ageing. There are other causes such as inherited alpha-1-antitrypsin deficiency, which causes emphysema at a young age, and industrial causes where inhaled toxic gases may cause enhanced protease activity leading to destruction of elastin and collagen fibers in the alveolar walls.

Several studies have shown that the extent of emphysema as shown by spirometry is well correlated with the extent and size of ventilation defects (KAUCZOR et al. 1996, 1997; MACFALL et al. 1996; DE LANGE et al. 1999; GIERADA et al. 2000). However, as most studies have been performed in relatively healthy patients (mild emphysema), the exact role of hyperpolarized 3-He MR imaging is still under discussion. There certainly seems evidence that 3-He MR imaging is capable of demonstrating early emphysematous changes better than what is feasible by routine lung function tests.

The possible role of hyperpolarized 3-He MRI for the preoperative and postoperative assessment of patients with emphysema who undergo lung volume reduction surgery is an interesting potential area of application. Lung volume reduction surgery is based on the principle that the respiratory pump becomes deficient as a result of overexpansion of the chest due to air trapping (COOPER et al. 1995). Thus, by reducing the overall lung volume, the hyperexpansion is reduced and the respiratory dynamics of the chest wall and diaphragm are restored. Although no studies have been performed to establish the role of 3-He MRI for this indication, the enhanced information that can be obtained by hyperpolarized 3-He MR imaging, such as improved quantification of emphysema and regional functional information, makes this an interesting area to pursue. Furthermore, both lung volumes and regional information on air space size and ventilation-perfusion mismatch would be available within the four-step protocol. In addition, proton MR imaging can yield information on movement of chest wall and diaphragm and can accurately define lung volumes in patients before and after lung volume reduction surgery (GIERADA et al. 1998).

A subtraction technique of calculated proton MR volumetry and hyperpolarized 3-He MR imaging, resulting in "effective ventilated lung volume", has been proposed (WOODHOUSE et al. 2002). One study compared hyperpolarized 3-He MRI with high resolution CT in six patients following single lung transplant (three patients had idiopathic pulmonary fibrosis and three suffered from emphysema) (ZAPOROZHAN et al. 2002). The techniques showed good correlation with pulmonary function tests, but significant differences were observed between the two groups and between the native and transplanted lung in terms of signal change following inhalation of hyperpolarized 3-He gas. This latter observation points to better functional assessment of individual lungs with the use of hyperpolarized 3-He MR imaging.

7.1.7.2
Smoker's Lungs

Smoking has several effects on the lungs, including increasing mucus production and enzymatic changes. Initially the small airway inflammation dominates, but soon the changes become chronic and involve the major airways. Chronic obstruction of the small airways leads to bronchiolitis, while enzymatic and toxic changes result in destruction of alveolar walls and emphysema. As described above, several studies have evaluated the potential role of hyperpolarized 3-He MR imaging in this setting.

A study on 12 volunteers counted the slice-by-slice ventilation defects and their size in 7 smokers and

5 non-smokers with normal mean forced expiratory volume in 1 s (GUENTHER et al. 2000). Smokers showed a greater number of defects, which were also more extensive than those seen in non-smokers.

A second study by the University of Virginia group also showed increasing extent of ventilation defects correlated with smoking history and extent of emphysema (DE LANGE et al. 1999). These studies suggest that early detection of lung damage due to smoking may be possible and that the extent of disease (and possible progression or treatment effects) may be quantifiable.

A recent study in 13 volunteers (7 smokers and 6 non-smokers) compared proton lung volumetry with hyperpolarized 3-He MR imaging (WOODHOUSE et al. 2002). The smokers exhibited a reduction in mean effective lung volume of 37%, which correlated well to the number of ventilation defects identified in the individual 3-He ventilation images. The non-smokers exhibited lower reductions in mean effective lung volume of 12%. The 3-He ventilation images showed fewer ventilation defects in this group.

7.1.7.3
Cystic Fibrosis and Bronchiectasis

Cystic fibrosis (CF) is an inherited disorder which traditionally led to early death of patients. The introduction of screening programs and supportive therapy such as antibiotic prophylaxis, mucolytics, physiotherapy and enzyme substitution therapy have improved the survival beyond the age of 30 years.

Image scoring systems have been used clinically and as objective measurement tools for assessment of therapeutic effects in research trials for CF (BHALLA et al. 1991; NATHANSON et al. 1991; DONNELLY et al. 1997). At present, evaluation of lung function in patients with CF is largely limited to chest radiography and (global) pulmonary function tests. More sophisticated tests, such as high resolution computed tomography (HRCT) and nuclear medicine techniques are considered less appropriate due to the amounts of ionizing radiation involved (especially if performed serially).

CF is a disease process that generally gives rise to non-uniform lung injury. Global assessment is therefore less helpful, as compensatory effects are not easily defined. Chest radiography is generally of lower sensitivity for imaging airway disease and changes due to injury. HRCT is much more sensitive to address lung fibrosis, cystic changes and bronchiectasis, but the accompanying ionizing radiation is of concern (BHALLA et al. 1991; NATHANSON et al. 1991; DONNELLY et al. 1997). Furthermore, HRCT is mor-

phological imaging rather than functional imaging. Isotope techniques are available that improve functional assessment, but availability, ionizing radiation and limited spatial and temporal resolution make it of limited value (PIEPSZ et al. 1980).

The introduction of screening programs has led to early intervention in CF. As a result, minimal changes will be present, which will be more difficult to assess in terms of therapeutic effects. The routine of chest radiography and global lung function tests will probably not yield sufficient discriminatory value to determine whether new treatments are effective. Thus, it seems of utmost importance to develop novel techniques, which have no ionizing radiation burden, and which can provide regional function lung data.

A small study in four subjects with CF demonstrated good acceptance of 3-He MRI (DONNELLY et al. 1999). Wide-spread ventilation defects were observed in all subjects with a predominance in the upper lung zones. There was good correlation between the extent of ventilation defects and pulmonary function tests.

A second study in three subjects with CF compared 3-He MR imaging with spirometry (ALTES et al. 2001b). There was good correlation between 3-He MR imaging findings and spirometry and two subjects were re-evaluated immediately after treatment for mucus clearance. Only slight improvement was seen in spirometry in one subject, but significant improvement of ventilation defects was noted by ^{3}He MR imaging in both subjects. Thus, this technology is capable of evaluating treatment response in terms of ventilation defect reversibility, whereas spirometry seems too insensitive to be applied as a tool.

7.1.7.4
Asthma

Asthma is a chronic disease that affects large numbers of predominantly young people. It is defined as a combination of reversible small airway obstruction, inflammation and increased airway responsiveness. Spirometry tends to be used for management, but its sensitivity for early detection of airway inflammation has been questioned. Furthermore, regional differences are not evaluable using standard techniques.

One study compared findings of 10 asthmatics and 10 healthy volunteers of hyperpolarized 3-He MR imaging and spirometry immediately prior to imaging (ALTES et al. 2001a). Of the asthmatics, seven had (generally small) ventilation defects distributed throughout the lungs, while no defects were present in the normal controls. Two patients were symptomatic, and their defects were more numerous and larger than in the remaining patients. The defects did not change over a 1-h period. One subject was studied a second time after 3 weeks, and showed that the defects had resolved. This subject also demonstrated that defects could resolve after bronchodilator therapy (ALTES et al. 2001a).

7.1.7.5
Lung Transplant Patients (Bronchiolitis Obliterans)

Bronchiolitis obliterans is one of the major complications of lung transplantation. The illness is the direct result of a chronic host versus graft reaction, and ultimately leads to failure and rejection of the grafted lung (PARADIS 1998). The early diagnosis is difficult, as the disease has a patchy distribution pattern and transbronchial biopsy is relatively insensitive until fibrosis, which is irreversible, has occurred. Hyperpolarized ^{3}He MR imaging may be able to assist in the early diagnosis, by revealing ventilation defects at a stage when the disease is not yet irreversible and may be treated by increasing immunosuppressive drugs. It may also guide transbronchial biopsy and improve its diagnostic yield in this patchy disease.

A study of six lung transplant patients correlated the findings of 3-He MR with a clinical grading system for bronchiolitis obliterans (MCADAMS et al. 1999). Coronal spin density MR was tolerated well and showed focal ventilation defects in all patients. Two patients also underwent 133-Xe scintigraphy, while HRCT was available in three patients. The 3-He MR imaging findings showed more extensive abnormalities in all circumstances and correlated well with the clinical grading system.

Another study in 14 patients compared the hyperpolarized 3-He MR findings with both 5 mm CT and 1 mm HRCT imaging (GAST et al. 2002b). Ventilation defects correlated with CT findings in 37 out of 59 defects, whereas the remaining 22 defects (37%) could not be explained by CT/HRCT. This study demonstrates that 3-He MRI is more sensitive for the detection of ventilation abnormalities than HRCT in the identification of early bronchiolitis obliterans and associated diseases.

7.1.7.6
Pulmonary Air Leaks

Pneumothorax is a frequent clinical condition, and is mainly due to trauma. However, a subset of patients will sustain spontaneous pneumothorax due to an air leak. If this leak persists in spite of conservative therapy, surgical intervention may be required and exact localization helpful. Conventional techniques are very sensitive for detection of pneumothorax, but unable to yield positional information on the air leak (CARR et al. 1992).

A study in two pigs with experimental air leaks used image sequences at 5 s intervals (ROBERTS et al. 2000). The images showed that hyperpolarized 3-He MR imaging is capable of detecting the leak and showing the localization. No human studies have been performed to date.

7.1.8
Conclusions

Hyperpolarized 3-He MR imaging of the lungs is capable of providing new insights into pulmonary pathophysiology. The technique offers a non-invasive method of assessment, which allows integration of lung morphology and lung function on a regional basis. As shown above, developments are taking place at great speed, and various disease states are being investigated using this technique. It is expected that it will find its way into routine clinical use within the next 5 years. Furthermore, it will keep offering a research tool to investigate pathophysiological parameters, such as lung function but also oxygen uptake. Finally, it will be capable of yielding quantitative information of disease states. This may in turn result in the opportunity to use hyperpolarized 3-He MRI as a measure for the assessment of novel therapies for both lung diseases and the delivery of inhaled medications for systemic diseases.

References

Albert MS, Cates GD, Driehuys B et al. (1994) Biological magnetic resonance imaging using laser-polarized 129Xe. Nature 370:199–201

Altes TA, Powers PL, Knight-Scott J et al. (2001a) Hyperpolarized 3-He MR lung ventilation imaging in asthmatics: preliminary findings. J Magn Reson Imaging 13: 378–384

Altes T, Froh DK, Salerno M et al. (2001b) Hyperpolarized Helium-3 MR imaging of lung ventilation changes following airway mucus clearance treatment in cystic fibrosis. Proc ISMRM 9:2003

American Thoracic Society (1962) Chronic bronchitis, asthma and pulmonary emphysems. A statement by the committee on diagnostic standards for non-tuberculous respiratory disease. Am Rev Respir Dis 85:762–768

Becker J, Heil W, Krug B et al. (1994) Study of mechanical compression of spin-polarized 3He gas. Nucl Instrum Methods A 346:45–51

Becker J, Bermuth J, Ebert M et al. (1998) Interdisciplinary experiments with polarized 3He. Nucl Instrum Methods A 402:327–336

Bhalla M, Turcios N, Aponte V et al. (1991) Cystic fibrosis: scoring system with thin section CT. Radiology 179:783–788

Black RD, Middleton HD, Cates GD et al. (1996) In vivo He-3 MR images of guinea pig lungs. Radiology 199:867–870

Carr JJ, Reed JC, Choplin RH, Pope TL Jr, Case LD (1992) Plain and computed radiography for detecting experimentally induced pneumothorax in cadavers: implications for detection in patients. Radiology 183:193–199

Chen XJ, Chawla MS, Hedlund LW, Möller HE, MacFall JR, Johnson GA (1998) MR microscopy of lung airways with hyperpolarized 3-He. Magn Reson Med 39:79–84

Colegrove FD, Schearer LD, Walters GK (1963) Polarisation of He3 gas by optical pumping. Phys Rev 132:2561–2572

Cooper JD, Trulock EP, Triantafillou AN et al. (1995) Bilateral pneumonectomy (volume reduction) for chronic obstructive pulmonary disease. J Thorac Cardiovasc Surg 109:106–119

Darasse L, Guillot G, Nacher PJ, Tastevin G (1997) Low-field 3He nuclear magnetic resonance in human lungs. CR Acad Sci II B 324:691–700

DeLange EE, Mugler JP, Brookeman JR et al. (1999) Lung air spaces: MR imaging evaluation with hyperpolarized 3He gas. Radiology 210:851–857

Deninger AJ, Eberle B, Ebert M et al. (1999) Quantification of regional intrapulmonary oxygen partial pressure evolution during apnea. J Magn Reson Imaging 141:207–216

Donnelly LF, Gelfand MJ, Brody AS, Wilmont RW (1997) Comparison between morphologic changes seen on high-resolution CT and regional pulmonary perfusion seen on SPECT in patients with cystic fibrosis. Pediatr Radiol 27:920–925

Donnelly LF, MacFall JR, McAdams HP et al. (1999) Cystic fibrosis: combined hyperpolarized 3He-enhanced and conventional proton MR imaging in the lung – preliminary observation. Radiology 212:885–889

Driehuys B, Cates GD, Miron E, Sauer K, Walter DK, Happer W (1996) High-volume production of laser-polarized 129Xe. Appl Phys Lett 69:1668–1670

Durand E, Guillot G, Darrasse L, Tastevin G, Nacher PJ, Vignaud A, Vattolo D, Bittoun J (2002) CPMG measurements and ultrafast imaging in human lungs with hyperpolarized helium-3 at low field (0.1 T). Magn Reson Med 47:75–81

Eberle B, Weiler N, Markstaller K et al. (1999) Analysis of intrapulmonary O_2 concentration by MR imaging of inhaled hyperpolarised 3He. J Appl Phys 87:2043–2052

Ebert M, Großmann T, Heil W et al. (1996) Nuclear magnetic resonance imaging on humans using hyperpolarised 3He. Lancet 347:1297–1299

Evans JW, Wagner PD (1977) Limits on V_A/Q distribution from analysis of experimental inert gas elimination. J Appl Physiol 42:588–599

Frahm J, Haase A, Matthaei D (1986) Rapid NMR imaging of dynamic processes using the FLASH technique. Magn Reson Med 3:321–327

Gast KK, Puderback MU, Rodriguez I et al. (2002a) Dynamic ventilation (3)He-magnetic resonance imaging with lung

motion correction: gas flow distribution analysis. Invest Radiol 37:126–134

Gast KK, Viallon M, Eberle B et al. (2002b) MRI in lung transplant recipients using hyperpolarized 3He: comparison with CT. J Magn Reson Imaging 15:268–274

Gentile TR, Jones GL, Thompson AK et al. (2000) Demonstration of a compact compressor for application of metastability-exchange optical pumping of 3He to human lung imaging. Magn Reson Med 43:290–294

Gierada DS, Hakimian S, Slone RM, Yusen RD (1998) MR analysis of lung volume and thoracic dimensions in patients with emphysema before and after lung volume reduction surgery. AJR Am J Roentgenol 170:707–714

Gierada DS, Saam B, Yablonskiy D, Cooper JD, Lefrak SS, Conradi MS (2000) Dynamic echo planar MR imaging of lung ventilation with hyperpolarized [3]He in normal subjects and patients with severe emphysema. NMR Biomed 13:176–181

Goodson BM (2002) Nuclear magnetic resonance of laser-polarized noble gases in molecules, materials, and organisms. J Magn Reson 155:157–216

Guenther D, Eberle B, Hast J et al. (2000) 3-He MRI in healthly volunteers: preliminary correlation with smoking history and lung volumes. NMR Biomed 13:182–189

Heil W, Humblot Otten E W, Schäfer M, Surkau R, Leduc M (1995) Very long nuclear relaxation times of spin polarized helium 3 in metal coated cells. Phys Lett A 201:337–343

Johnson GA, Cates G, Chen XJ et al. (1997) Dynamics of magnetization in hyperpolarized gas MRI of the lung. Magn Reson Med 38:66–71

Johnson GA, Cofer GP, Hedlund LW, Maronpot RR, Suddarth SA (2001) Registered 1H and 3He magnetic resonance microscopy of the lung. Magn Reson Med 45:365–370

Kastler A (1950) Quelques suggestions concernant la production optique et la détection optique d'une inégalité de population des niveaux de quantification spatiale des atomes. Application à l'expérience de Stern et Gerlach et à la resonance magnétique. J Phys Radium 11:255–265

Kauczor HU, Hofmann D, Kreitner KF et al. (1996) Normal and abnormal pulmonary ventilation: visualization at hyperpolarized He-3 MR imaging. Radiology 201:564–568

Kauczor HU, Ebert M, Kreitner KF et al. (1997) Imaging of the lungs using 3He MRI: preliminary clinical experience in 18 patients with and without lung disease. J Magn Reson Imaging 7:538–543

Kauczor HU, Chen XJ, Beek EJR van, Schreiber W (2001) Pulmonary ventilation imaged by MRI: at the doorstep of clinical application. Eur Respir J 17:1–16

MacFall JR, Charles HC, Black RD et al. (1996) Human lung air spaces: potential for MR imaging with hyperpolarized He-3. Radiology 200:553–558

Mai VM, Liu B, Li W et al. (2002) Influence of oxygen flow rate on signal and T(1) changes in oxygen-enhanced ventilation imaging. J Magn Reson Imaging 16:37–41

Mata J, Altes T, Christopher J, Mugler J, Brookeman J, Lange E de (2001) Positional dependence of small inferior ventilation defects seen on hyperpolarized helium-3 MR of the lung. Proc ISMRM 9:949

McAdams HP, Palmer SM, Donnelly LF, Charles HC, Tapson VF, MacFall JR (1999) Hyperpolarized 3He-enhanced MR imaging in lung transplant patients: preliminary results. AJR Am J Roentgenol 173:955–959

Middleton H, Black RD, Saam B et al. (1995) MR imaging with hyperpolarized 3He gas. Magn Reson Med 33:271–275

Miron E, Schaefer S, Schreiber D, Wijngaarden WA van, Zeng X, Happer W (1984) Polarisation of the nuclear spins of noble-gas atoms by spin exchange with optically pumped alkali-metal atoms. Phys Rev A 29:3092–3110

Mugler JP, Driehuys B, Brookeman JR et al. (1997) MR imaging and spectroscopy using hyperpolarized 129Xe gas: preliminary human results. Magn Reson Med 37:809–815

Mugler JP, Brookeman JR, Knight-Scott J, Maier T, Lange EE de, Bogorad PL (1998) Interleaved echo-planar imaging of the lungs with hyperpolarized ^{3}He . Proc ISMRM 6:448

Mugler JP, Salerno M, Lange EE de, Brookeman JR (2002) Optimized TrueFISP hyperpolarized 3He MRI of the lung yields a 3-fold SNR increase compared to FLASH. Proc ISMRM 10:2019

Nacher PJ, Tastevin G, Maître X, Dollat X, Lemaire B, Olejnik J (1999) A peristaltic compressor for hyperpolarized helium. Eur Radiol 9:B18

Nathanson I, Conboy K, Murphy S, Afshani E, Kuhn JP (1991) Ultrafast computerized tomography of the chest in cystic fibrosis: a new scoring system. Pediatr Pulmonol 11:81–86

Paradis I (1998) Bronchiolitis obliterans: pathogenesis, prevention and management. Am J Med Sci 315:161–178

Piepsz A, Wetzburger C, Spehl M (1980) Critical evaluation of lung scintigraphy in cystic fibrosis. J Nucl Med 21:909–913

Quanjer PH, Tammeling GJ, Cotes JE, Pedersen OF, Peslin R, Yernault JC (1993) Lung volumes and forced ventilatory flows: report of the working party standardization of lung function tests. European community for steel and coal – official statement of the European respiratory society. Eur Respir J 6 [Suppl 16]:5–20

Riederer SJ, Tasciyan T, Farzaneh F, Lee JN, Wright RC, Herfkens RJ (1988) MR fluoroscopy: technical feasibility. Magn Reson Med 8:1–15

Rizi RR, Baumgardner JE, Saha PK et al. (2001) Regional lung compliance by hyperpolarized 3helium magnetic resonance imaging. Proc ISMRM 9:944

Roberts DA, Rizi RR, Lipson DA et al. (2000) Detection and localization of pulmonary air leaks using laser-polarized ^{3}He MRI. Magn Reson Med 44:379–382

Roca J, Wagner PD (1994) Principles and information content of the multiple inert gas elimination technique. Thorax 49:815–824

Rodriguez-Roisin R (1994) Contribution of multiple inert gas elimination technique to pulmonary medicine. Thorax 49:813–814

Ruppert K, Brookeman JR, Mugler JP (1998) Real-time MR imaging of pulmonary gas-flow dynamics with hyperpolarized ^{3}He. Proc ISMRM:1909

Saam B, Happer W, Middleton H (1995) Nuclear relaxation of 3He in the presence of O$_2$. Phys Rev A 52:862–865

Saam B, Yablonskiy DA, Gierada DS, Conradi MS (1999) Rapid imaging of hyperpolarized gas using EPI. Magn Reson Med 42:507–514

Sakai K, Bilek AM, Oteiza E et al. (1996) Temporal dynamics of hyperpolarized 129Xe resonances in living rats. J Magn Reson B 111:300–304

Salerno M, Altes TA, Brookeman JR, Lange EE de, Mugler JP III (2001) Dynamic spiral MR imaging of pulmonary gas flow using hyperpolarized 3He: preliminary studies in healthy and diseased lungs. Magn Reson Med 46:667–677

Schreiber WG, Weiler N, Kauczor HU et al. (2000) Ultraschnelle MRT der Lungenventilation mittels hochpolarisiertem Helium-3. Rofo Fortschr Geb Rontgenstr Neuen Bildgeb Verfahr 172:129–133

Tseng CH, Wong GP, Pomeroy VR et al. (1998) Low-field MRI of laser polarized noble gas. Phys Rev Lett 81:3785–3788

Van Vaals JJ, Brummer ME, Dixon WT et al. (1993) "Keyhole" method for accelerating imaging of contrast agent uptake. J Magn Reson Imaging 3:671–675

Viallon M, Cofer GP, Suddarth SA et al. (1999) Functional MR microscopy of the lung using hyperpolarized 3He. Magn Reson Med 41:787–792

Viallon M, Berthezene Y, Callot V et al. (2000) Dynamic imaging of hyper-polarised (3)He distribution in rat lungs using interleaved-spiral scans. NMR Biomed 13:207–213

Wagner PD, Rodriguez-Roisin R (1991) Clinical advances in pulmonary gas exchange. Am Rev Respir Dis 143:883–888

Wagner PD, Smith CM, Davies NJH, McEvoy RD, Gale GE (1985) Estimation of ventilation-perfusion inequality by inert gas elimination without arterial sampling. J Appl Physiol 59:376–383

Wagshul ME, Button TM, Li HF et al. (1996) In vivo MR imaging and spectroscopy using hyperpolarized 129Xe. Magn Reson Med 36:183–191

Walker TG, Happer W (1997) Spin-exchange optical pumping of noble-gas nuclei. Rev Mod Phys 69:629–642

Wild JM, Paley MNH, Viallon M, Schreiber WG, Beek EJR van, Griffiths PD (2002a) K-Space filtering in 2D gradient echo breath-hold hyperpolarized 3He MRI: spatial resolution and signal to noise considerations. Magn Reson Med 47:687–695

Wild JM, Schmiedeskamp J, Paley MNJ et al. (2002b) MR imaging of the lungs with hyperpolarized 3-helium transported by air. Phys Med Biol 47:N185–190

Wild JM, Fichele S, Woodhouse N, Paley MNJ, Swift AJ, Kasuboski L, Beek EJR van (2003a) Assessment and compensation of susceptibility artifacts in gradient echo MRI of hyperpolarized ^{3}He gas. Magn Reson Med 50:417–422

Wild JM, Paley MNJ, Kasuboski L, Swift AJ, Fichele S, Woodhouse N, Griffiths PD, Beek EJR van (2003b) Dynamic radial projection MRI of inhaled hyperpolarized ^{3}He gas. Magn Reson Med 49:991–997

Wong GP, Tseng CH, Pomeroy VR et al. (1999) A system for low field imaging of laser-polarized noble gas. J Magn Reson 141:217–227

Woodhouse N, Wild JM, Fichele S et al. (2002) Measurement of a reduction in the effective lung volume of a smoker as measured by combined proton single shot fast spin echo/hyperpolarized helium-3 MRI. Proc Br Chap ISMRM 58

Zaporohan J, Ley S, Gast K et al. (2002) Aerated versus ventilated lung: Value of the ^{3}He MRI and CT. Proc ISMRM 10:132

7.2 CT for Assessment of Ventilation Distribution

EDWIN J. R. VAN BEEK

CONTENTS

7.2.1 Introduction

The assessment of ventilation distribution is an important clinical parameter for many respiratory disorders. It yields additional information compared to pulmonary function tests, as morphological changes and regional distribution of disease can be visualized. Furthermore, dynamic assessment using paired inspiratory and expiratory imaging is capable of predicting lung function. Thus, ventilation computed tomography (CT) aims to determine the distribution of inspired air in a quantitative, time-resolved and regional fashion. The amount of air in the distal airways will result in changes in density, allowing quantification of air distribution on a regional basis and the assessment of small airway diseases as inspired air becomes trapped at expiration. It is this additional information that makes imaging of ventilation distribution so valuable.

Computed tomography (CT) is the current work-horse of clinical radiology and is described more extensively in Chap. 3–6. Ventilation CT has benefited from the introduction of the slip-ring technology, which made it feasible to speed up the data acquisition process and allowed volumetric data recording (helical or spiral CT). The availability of faster and more powerful computers has allowed the development of image reconstruction algorithms that helped speed up image acquisition or alternatively improve image clarity. The development of multiple detector rows (multi-detector or multi-slice CT) has made it possible to acquire larger volumetric data sets and further increase the speed of imaging or alternatively decrease slice thickness. Thus, present day CT allows a range of diagnostic techniques to be employed, and several strategies may be used for imaging of the lungs.

Imaging of the lung may consist of anatomical (parenchymal) imaging and functional imaging. At present, routine clinical imaging of the lung parenchyma is performed using chest radiography and computed tomography (CT). CT is the gold standard imaging modality for the morphological assessment of the pulmonary parenchyma with high spatial resolution (slice thickness 1 mm or less). Dedicated strategies are applied to estimate lung function (KRAMER and HOFFMAN 1995). By using paired inspiratory and expiratory scans, hypoventilated lung areas caused by expiratory obstruction, i.e. air trapping can be detected (KAUCZOR et al. 2000). In addition, dynamic CT (spiral CT without table movement) is feasible. A temporal resolution of 100 ms can be reached (MARKSTALLER et al. 2001a). However, this approach is limited to one slice. Larger coverage can be achieved by the use of multi-detector ("multi-slice") CT. The whole lung can be scanned within 5 seconds. This temporal resolution is not suited to derive functional parameters. Attempts of direct visualization of ventilation using CT consist of Xenon-enhanced CT and the use of an aerosolized contrast agent. Both techniques aim at an increase of lung density reflecting regional ventilation. The increase in density is rather small and further post-processing is required (THIELE and KLÖPPEL 1995; SIMON et al. 1998). Thus, the use of CT for functional assessment of ventilation is improving using a variety of newly developed techniques.

E. J. R. VAN BEEK, MD, PhD, FRCR
Clinical Senior Lecturer/Honorary Consultant Radiology, Academic Unit of Radiology, Floor C, Royal Hallamshire Hospital, Glossop Road, Sheffield S10 2JF, UK

7.2.2
High Resolution Computed Tomography

High resolution CT (HRCT) is controlled by three main factors (WEBB et al. 2001). First, a narrow beam collimation of 1–1.5 mm is used in order to reduce volume averaging and increase spatial resolution (NAIDICH et al. 1985; NAKATA et al. 1985; ZERHOUNI et al. 1985; MAYO et al. 1987). Secondly, the high spatial frequency reconstruction algorithm used takes advantage of the inherent high contrast between air and soft tissues, thus allowing increased sharpness of the structures at high spatial resolution (MAYO et al. 1987; MURATA et al. 1988). Finally, the application of a small field-of-view (FOV) increases the spatial resolution further. This can be achieved by image targeting after the initial data set has been acquired. The reduction of the FOV results in a decrease in pixel size with corresponding increase in spatial resolution (MAYO et al. 1987; MURATA et al. 1989).

The introduction of volumetric CT scan protocols has the advantage of greater lung coverage, more complete assessment of the lung and the potential to reconstruct images in different (non-axial) planes. Several studies have used maximum intensity projection images to demonstrate that the identification of pulmonary nodules may be improved in general (BHALLA et al. 1996; REMY-JARDIN et al. 1996), although there is a potential problem in very extensive disease with superimposition of opacities (REMY-JARDIN et al. 1996). The usefulness of multidetector CT on workflow related issues was demonstrated in a recent study in 50 patients with suspected or confirmed interstitial lung disease (REMY-JARDIN et al. 2002a). This study showed that the number of images to be interpreted could be reduced by 40%, without affecting diagnostic accuracy, if a coronal plane was reconstructed rather than the routine axial plane.

HRCT is usually performed during suspended full inspiration. This improves the inherent air-soft tissue contrast as the alveoli and airways are fully distended, revealing the interstitial tissues more clearly (Fig. 7.2.1a). Furthermore, full inspiration by its very nature reduces (gravity dependent) transient collapse of lung segments, which will take place in a physiological setting due to the weight of the lung tissue if insufficient insufflation has taken place (Fig. 7.2.1b). In addition to the inspiratory images, suspended forced post-expiratory HRCT scans are useful in patients with obstructive lung diseases (WEBB 1987). Post-expiratory HRCT will enhance the obstruction, leading to air being trapped due to early collapse or obstruction of small airways, thus enhancing the diagnostic potential of HRCT (ARAKAWA and WEBB 1998). Air trapping may be focal or diffuse, and is shown by a relative decrease in density compared to normal lung tissues. Air trapping corresponds reasonably well with pulmonary function tests (ARAKAWA and WEBB 1998; CHEN et al. 1998; LUCIDARME et al. 1998) and can help to distinguish between increased lung opacity as seen in infiltrative lung disease and air trapping as seen in obstructive lung disease (ARAKAWA et al. 1998).

Motion artifacts becomes an issue in the lung zones adjacent the heart. One way of minimizing motion effects is to obtain images in sub-second scan

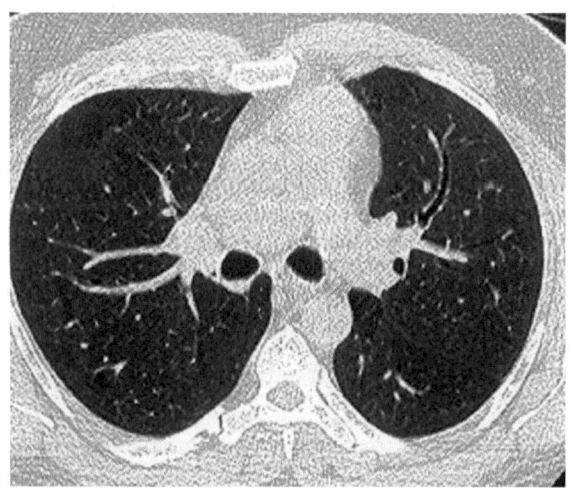 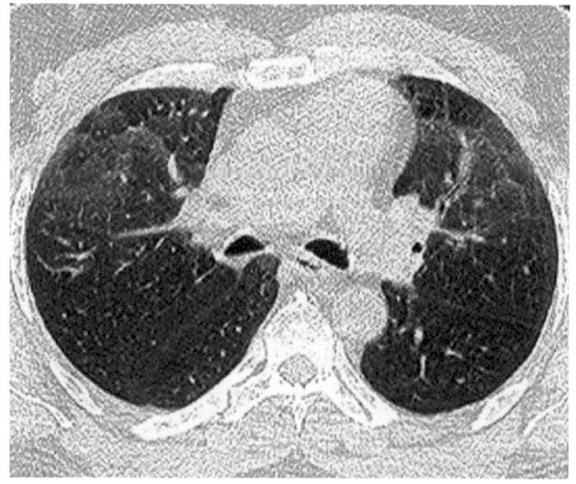

a   b

Fig. 7.2.1. a High-Resolution CT at full inspiration with patient supine. **b** High-Resolution CT at expiration shows gravity dependent increased attenuation of dependent lung tissue (patient supine). Note the change in calibre of the main bronchi during expiration

rotations, as is now feasible with current CT systems (FLOHR et al. 2002). This can be enhanced by the use of prospective ECG gating, where data acquisition takes place during the same phase of the cardiac cycle (SCHOEPF et al. 1999). An alternative is to perform retrospective ECG gating, which allows omission of data within a given period of the cardiac cycle after the entire data set has been acquired (FLOHR et al. 2002). This latter technique is especially suited for multi detector CT systems with sub-second scan rotation times. However, although these techniques improved the image quality in both studies, it did not change the diagnostic value in a series of 35 patients (SCHOEPF et al. 1999).

Correction for respiratory motion is also required. Usually, a breath-hold is sufficient. However, there may be instances where either a patient is too breathless or where cooperation is difficult to achieve. In these situations, spirometrically triggered expiratory HRCT is an option, which allows for reproducible HRCT images as well as quantitative assessment of the airways. In this setting, the spirometric data of the patient's vital capacity are ascertained and a trigger level is set. The patient keeps breathing through the spirometer during CT scanning, and the data acquisition is triggered according to preset lung volumes, when the inflow valve is closed (KALENDER et al. 1990, 1991; LAMERS et al. 1994). As a result of the standardized lung volume, quantitative assessment of lung attenuation is feasible and reproducible.

HRCT is considered the standard work-up for a variety of lung diseases, such as emphysema (KNUDSON et al. 1991; STERN EJ et al. 1994), chronic

airways disease (WEBB 1997; LUCIDARME et al. 1998), interstitial lung diseases (BHALLA M et al. 1996; REMY-JARDIN et al. 1996, 2002a), asthma (PARK et al. 1997), bronchiolitis obliterans (PADLEY et al. 1993; ARAKAWA and WEBB 1998; LEUNG et al. 1998), bronchiectasis (Fig. 7.2.2) (HANSELL et al. 1994) and alveolitis (SMALL et al. 1996). Expiratory HRCT should be obtained in any patient in whom bronchiolitis obliterans, asthma, small airways obstruction or sarcoidosis is considered (GLEESON et al. 1996; ARAKAWA and WEBB 1998; BARTZ and STERN 2000). Furthermore, expiratory HRCT can help distinguish between air trapping as a cause for inhomogeneous attenuation and that seen as a result of lung perfusion abnormalities (mosaic perfusion, Fig. 7.2.3) (WORTHY et al. 1997).

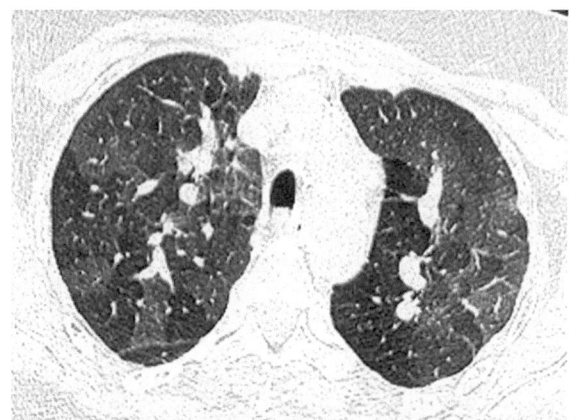

a

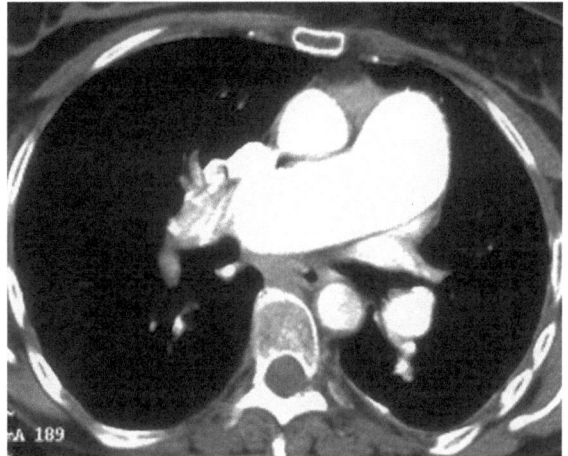

b

Fig. 7.2.3. Patient with primary pulmonary hypertension. a High Resolution CT showing inhomogeneous attenuation of parenchyma (mosaic perfusion pattern). b CT pulmonary angiography shows dilated pulmonary artery (PA) when compared to the aorta (Ao); chronic thromboembolic disease was excluded using this investigation

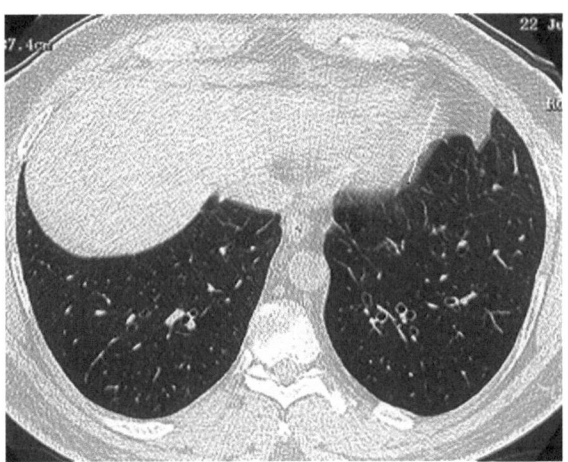

Fig. 7.2.2. High-Resolution CT of lower lung lobes showed "signet ring" (arrow) of dilated bronchi when compared to corresponding pulmonary arteries. This is the classic image of bronchiectasis

Emphysema is a disease, which is characterized by destruction of lung parenchyma, loss of alveolar walls and reduction of elastic recoil of lung tissue. At HRCT, emphysematous areas showed decreased attenuation values due to the reduction of interstitial tissue and the increase in airspace sizes (Fig. 7.2.4a). At expiratory HRCT scans, emphysematous areas remain relatively hyperinflated, leading to air trapping and an increase in the difference between the normal lung tissue (which undergoes an increase in density, Fig. 7.2.4b) (KNUDSON et al. 1991; STERN EJ et al. 1994).

In lung diseases such as peripheral airways disease or asthma, the main causes for abnormal findings are related to inflammation with increased mucus production and irritability of airways with increased bronchiolar muscular tension (PADLEY et al. 1993; PARK et al. 1997; ARAKAWA and WEBB 1998; LEUNG et al. 1998; HANSELL et al. 1994; SMALL et al. 1996). Thus, a combination of airway collapse (due to mucus obstruction) and hyperinflation with air trapping (due to early collapse of distal airways during expiration) may be seen.

Finally, diseases which cause fibrosis show a combination of coarsening of the interstitium, parenchymal distortion and traction effects (including bronchiectasis). The inspiratory scans will demonstrate increased attenuation, while some air trapping may be seen in bronchiectasis (BHALLA M et al. 1996; REMY-JARDIN et al. 1996, 2002a).

A more detailed description of findings in these disease states may be found in Chap. 3–6.

7.2.3
Lung Volume Measurement

Pulmonary function tests (pulmonary spirometry) are routinely used for determination of lung volumes in a clinical setting and consist of static inspiratory volumes, static expiratory volumes and dynamic expiratory volumes. A summary was described in Chap. 2 above. The main problem of standard spirometry is the lack of determination of which parts of the lungs contribute. As a result, the volumes are calculated with the assumption that both lungs contribute equally. CT is well suited to calculate lung volumes for individual lungs and (with additional reconstruction) lobes of each lung.

Several methods can be employed to enhance the information on regional contribution to lung function. First, as described above, a global assessment of attenuation changes on HRCT correlate with lung function tests. The addition of post-expiratory HRCT imaging has a high predictive value for obstructive pulmonary function tests if air trapping is demonstrated (KNUDSON et al. 1991; PADLEY et al. 1993; HANSELL et al. 1994; GLEESON et al. 1996; ARAKAWA and WEBB 1998; LUCIDARME et al. 1998). Second, spirometrically controlled HRCT can be used to quantify lung attenuation within lung regions at standardised lung volumes (LAMERS et al. 1994). A difference was observed between densitometric measurements at 90% and 10% vital capacity between patients with emphysema and chronic bronchitis in a study involving 20 patients in each group (LAMERS et al. 1994).

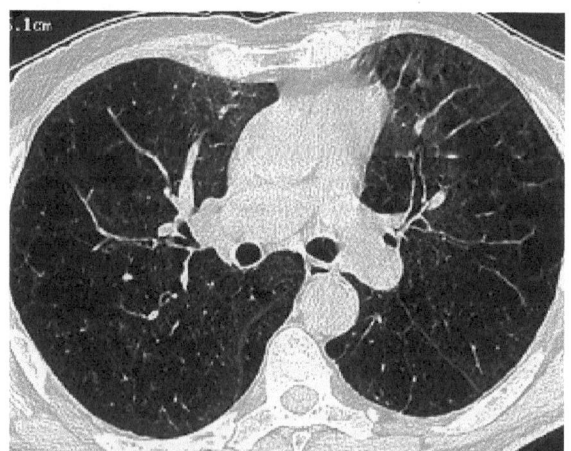
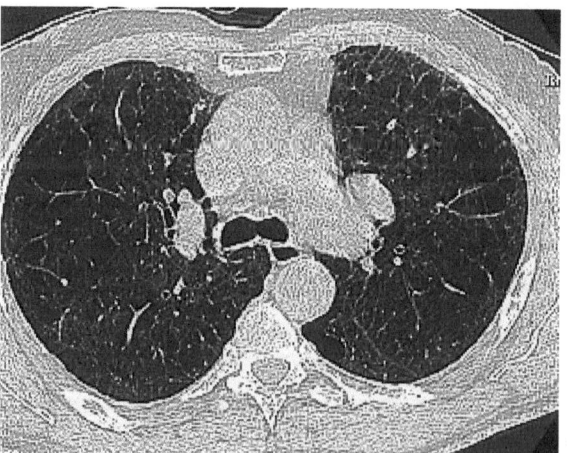

Fig. 7.2.4. Patient with emphysema. **a** High Resolution CT shows parenchymal destruction as areas with relatively low attenuation, while normal lung tissue is perceived as having relatively high attenuation. **b** High Resolution CT during expiration shows increased difference in attenuation values due to air trapping

The introduction of spiral CT has made it possible to obtain volumetric data sets. Two separate studies deserve further exploration in this context (KAUCZOR et al. 1998; MERGO et al. 1998). The first study performed full inspiration CT in 100 patients (and full expiration CT in 53 of them), which were compared with spirometry in 79 patients (MERGO et al. 1998). Abnormal low attenuation volumes were assessed and compared with forced vital capacity, FEV$_1$ and the ratio of FEV$_1$ and FVC. Interestingly, inspiratory low attenuation volumes correlated slightly better with spirometry indicating emphysema than expiratory volumes. Furthermore, a correlation was shown with diffusion capacity, which confirmed a study in patients who underwent HRCT (GEVENOIS et al. 1996).

The second study performed a similar investigation in 72 patients with a variety of lung diseases (KAUCZOR et al. 1998). The thresholds used were slightly higher than in the previous study, and segmentation was performed using a semi-automatic algorithm. A consistent 10–15% underestimation of total lung capacity was demonstrated using inspiratory helical CT, but correlation was very good. At expiratory helical CT there was a systematic overestimation of residual volume by nearly 1 litre (KAUCZOR et al. 1998). Finally, a comparison between 2D and 3D post processing techniques revealed that the 3D technique showed consistently higher volumes, which were in better agreement with spirometry data.

An automated segmentation technique has also been applied successfully (HU et al. 2001). This showed that following lung extraction from the 3D dataset and automated segmentation algorithms were capable of producing reproducible results of lung volumes with an error margin of less than 3%. Thus, volumetry using CT is not only accurate, but can also yield differential results of the individual lungs (or even lung regions when a region of interest is applied).

7.2.4
Dynamic CT

Dynamic CT imaging relies heavily on both hardware and software of the CT system. Ideally, one needs a rapid rotation time and interpolation algorithms for reconstruction of the images. Therefore, initial attempts used electron beam CT, as this technology does not require any rotating parts, but rather relies on rapidly changing electron beams. These beams can be controlled and swiped using electromagnetic (rapid!) pulses, similar to those observed in a tele-

vision set. The beams are aimed at Tungsten target rings, which causes generation of X-rays (BONGAERTS and SHEEDY 1999). This technique is capable of producing slices at less than 100 ms (BONGAERTS and SHEEDY 1999). Initial applications were in the field of cardiac imaging, but dynamic imaging of the airways has been performed as well (WEBB et al. 1993; STERN and WEBB 1993; STERN et al. 1994).

The term dynamic ultrafast HRCT was used to describe electron beam CT with a series of 10 scans during a 6-second period, in which a single level was imaged during inhalation and forced expiration (WEBB et al. 1993). With the movement of the diaphragm, the level will image slightly different anatomical areas as the lungs will move cranially during expiration. However, it has been shown that the application at 3 levels (aortic arch, carina and lung bases) showed reliable changes in attenuation that were not significantly affected by lung motion (WEBB et al. 1993). The regional attenuation increases with a decrease in cross sectional lung area (WEBB et al. 1993). This technique has been employed in obstructive lung diseases (WEBB et al. 1993), cystic lung diseases (STERN et al. 1992), pediatric lung diseases (LYNCH et al. 1990) and bronchopulmonary sequestration (STERN et al. 1991). Furthermore, assessments of the trachea and changes in its diameter and orientation have been evaluated using this technique (STERN et al. 1993). In 10 normal volunteers, the tracheal airway surface changed from 280 mm^2 to 178 mm when subjects went from full inspiration to end expiration (a 35% decrease). In one patient with tracheomalacia a decrease of 82% was demonstrated after forced expiration.

A study in patients with obstructive sleep apnea showed that the oro- and nasopharyngeal airways were four-fold smaller than weight-matched controls (GALVIN et al. 1989). Furthermore, there was greater collapse during normal tidal volume breathing by up to 75% diameter, compared to 27% in the normal controls.

Ultrafast CT was also evaluated in 25 children with suspected airway obstruction (BRASCH et al. 1987). The technique was able to depict dynamic airway caliber changes, small intraluminal polyps, focal tracheal atresia, compressive mediastinal masses, and foreign body obstructions of the major bronchi. In another pediatric study, 17 children underwent both ultrafast CT and endoscopy (BRODY et al. 1991). Cine CT results agreed with endoscopy in 10 cases, but overcalled tracheomalacia in 5 (these patients in fact had a focal stenosis). If the results of HRCT were incorporated, agreement with endoscopy was 10 of 11 cases. The authors conclude that a combined approach for diagnosis of airway abnormalities is required, consisting of both HRCT and cine CT.

Largely due to the limited availability of electron beam CT and the improved design of the newer generations spiral CT systems, dynamic forced expiratory CT was developed for more widespread use. The first attempt was performed in a patient population with obstructive airways disease (IM et al. 1996). Another early attempt took place in a pediatric setting using a simple protocol of quiet breathing, fixed table position and 5 second acquisition time for 0.7 second temporally overlapping scans (JOHNSON et al. 1998). The technique was well tolerated in relatively young children.

Another study compared continuous dynamic expiratory HRCT with suspended expiratory HRCT in 49 patients with a variety of airways diseases (LUCIDARME et al. 2000). A CT scanner with a scanning time of 1 second was used. The cross sectional areas of air trapping were assessed semi-quantitatively using a grid system (CHEN et al. 1998). Air trapping was seen in 80% of the patients, and was significantly more extensive on dynamic (continuous) expiratory HRCT scans (LUCIDARME et al. 2000). One of the possible mechanisms is the fact that patients are able to exhale deeper on forced (dynamic) expiration, thus increasing the potential for detection of air trapping. Motion artifacts are a problem for relatively slow CT scanners, but even at 1-second rotation this did not negatively affect interpretation of images (LUCIDARME et al. 2000).

Most imaging has been performed with patients supine, but an alternative in more difficult patients is the lateral decubitus position (GOTWAY et al. 2000). In this situation, the gravitational forces change position by 90 degrees, effectively running horizontally through the chest. Furthermore, movement of the lowest hemithorax is more restricted, thus allowing for changes in attenuation to become more pronounced.

Quantification of air trapping has been shown to correlate well with routine pulmonary function tests (HEREMANS et al. 1992). The extent of air trapping can be expressed in a four point scale, from no air trapping (0), 1–25% air trapping (1), 26–50% (2), 51–75% (3) to 76–100% air trapping (4), based on cross-sectional area of lung affected in each slice (STERN et al. 1993). An automated technique, which uses a predetermined threshold value for pixel density, can display an index of densities in regions of interest in the lung (NEWMAN et al. 1994). The computed density outperformed macroscopic morphometry in one study in patients with emphysema (GEVENOIS et al. 1995). This technique is of interest, as it takes out the potential influence of

interobserver variability, albeit that this is already quite low in the semi-quantitative technique (NG et al. 1999). Some have advocated the use of a standardized level of lung volume (vital capacity 90% and 10%) to reduce inter- and intra-subject variation based on inequalities of inspiration and expiration (LAMERS et al. 1994). A more recent study in 29 patients with emphysema, who were referred for lung reduction surgery, compared standard vs spirometrically gated HRCT (GIERADA et al. 2001). In this study, it was shown that repeatability of quantitative CT indexes of emphysema was similar for both techniques, and that spirometrically gated CT did not offer any advantages (GIERADA et al. 2001). Intra-individual variation in quantitative CT results was predominantly explained by intraindividual differences in lung volume, which suggests that consistent coaching of patient inspiratory efforts will help to maximize test precision. However, no value could be demonstrated for use of spirometrically gated HRCT.

The preoperative CT investigations were compared with the effects of lung volume reduction surgery on pulmonary function tests in a study in 39 patients (HUNSAKER et al. 2002). Using various visual and computer-based scoring systems, the study showed that the preoperative CT scan correlated well with the improvement of postoperative FEV_1 in these patients (HUNSAKER et al. 2002). There was no obvious difference between qualitative or quantitative scoring systems, while CT did as well as perfusion-ventilation scintigraphy.

Initial attempts to perform quantitative studies relied on 2-D analysis, where a density mask had to be applied to each individual slice. However, this technique is rather time consuming. More recently, a 3D analysis of lung densitometry was compared with the 2-D technique, visual emphysema scores and pulmonary function tests (PARK et al. 1999). In this study, CT scans of 35 normal and 25 emphysematous patients were reconstructed as a 3D model and compared with 2D densitometry. The results were comparable, but reconstruction time was reduced and the slices were less influenced by level of inspiration, as the entire thorax was scanned during a single breath-hold rather than using individual slices.

In a recent study in 155 patients with and without lung diseases, paired inspiratory and expiratory HRCT was compared with pulmonary function tests in an attempt to characterize different types of ventilatory impairment (KAUCZOR et al. 2002). Mean lung density was measured and showed good correlation with static and dynamic lung volumes. The mean lung density and emphysema index correlated better for scans obtained at full expiration than at full inspiration. Furthermore, differentiation between obstructive and restrictive ventilatory impairment could be distinguished from normal subjects, again with the best results obtained from full expiratory scans (KAUCZOR et al. 2002).

HRCT may also be employed for longitudinal quantitative assessment of patients, such as smokers at risk of developing emphysema. A study compared serial HRCT scans over a 5-year period in 57 persistent current smokers, 31 persistent non-smokers, 13 persistent ex-smokers and 10 subjects who stopped smoking during the study. The findings of HRCT of persistent smokers revealed excellent correlation between development of emphysema and lung function tests (REMY-JARDIN et al. 2002b).

True temporal dynamics of lung aeration may be determined using continuous rapid CT acquisitions, however, in humans this would be difficult due to the inherent high radiation dose. A study in a porcine model of ARDS used 250 ms/image continuous HRCT to evaluate the changes of ARDS and response to ventilatory intervention (MARKSTALLER et al. 2001b).

7.2.5
Xenon CT

Xenon (Xe) is a non-radioactive noble gas with higher density than air. Thus, in a CT environment Xe-gas will exhibit higher Hounsfield units (HU), and the change in HU is closely correlated with concentration (Fig. 7.2.5) (CULLEN and CROSS 1951; FOLEY et al. 1978). Furthermore, a reduction in kilovoltage increases this change in density measurements (CULLEN and CROSS 1951). It has been shown that a concentration of approximately 40–50% will enhance the lung parenchyma in the order of 100 HU. Wash-in and wash-out of Xe can be assessed using spirometrically controlled repeated CT imaging, resulting in a local exponential density curve. This curve can be drawn for any region of interest. The greater the density, the better the ventilation in that particular lung region.

The technique of Xe-CT was recognised over 30 years ago (FOLEY et al. 1978; WINKLER et al. 1987; MARCUCCI et al. 2001). Side-effects due to the anaesthetic properties of Xenon were apparent in concentrations over 30% (CULLEN and CROSS 1951; FOLEY et al. 1978; WINKLER et al. 1987). Thus, initial work mainly involved animal studies (GUR et al. 1979; GUR et al. 1981; TOMIYAMA et al. 1993; MARCUCCI et al. 2001), normal subjects (FOLEY et al. 1978; HERBERT et al. 1982) and patients who were critically ill in intensive care under sedation and on respirators (HERBERT et al. 1982; SNYDER et al. 1984). Nevertheless, initial attempts proved difficult as a result of technical and interpretation difficulties. Only relatively recently was work carried out to determine standardization and accurate parameter estimates (including 95% confidence intervals) (SIMON et al. 1988).

Much of the early work used slow CT scanners, but with the introduction of electron beam CT true

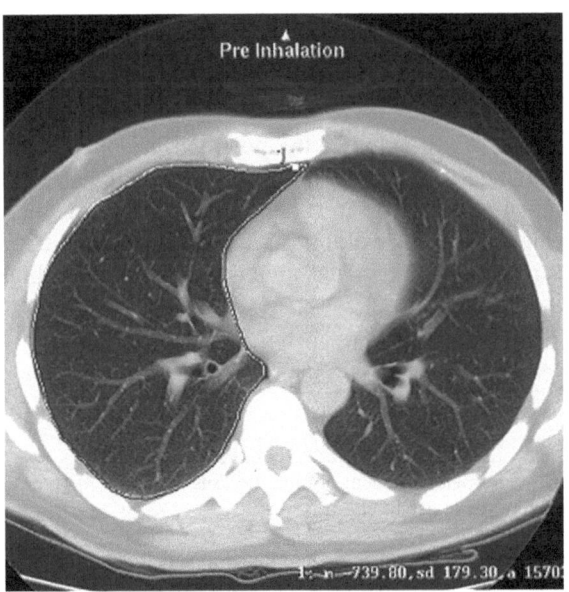
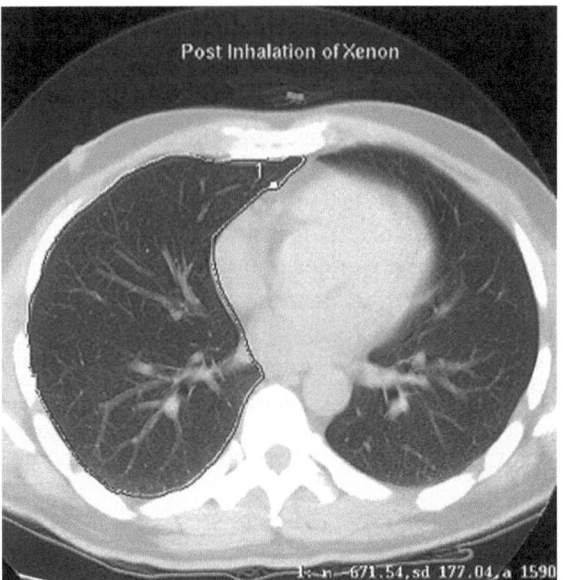

a   b

Fig. 7.2.5. Normal volunteer breathing room air (**a**) and after breathing of Xenon (**b**). The average attenuation of the room air image is –740 HU, while the attenuation after Xenon is –670 HU

dynamic imaging became possible at increased speed (TAJIK et al. 1996) and making it possible to use a single-breath protocol with subtraction technique (TAJIK et al. 2002). With the introduction of faster standard spiral CT scanners, the technique is currently experiencing new interest, not only for lung imaging, but also for assessment of lung perfusion and brain perfusion studies (KRECK et al. 2001; FURUKAWA et al. 2002; SASE et al. 2002). Thus, a combination of CT techniques, including Xe-CT could offer a completely non-invasive assessment of lung structure, pulmonary ventilation and its distribution, pulmonary perfusion and even perfusion in more distant organ systems like the brain (HOFFMAN et al. 1995, TAJIK et al. 2002).

One of the main problems of any lung imaging modality is the difficulty of imaging small airways. Mapping of pulmonary ventilation is possible, even if results of single breath-hold techniques must be interpreted with caution as gravity dependent mechanisms play a role in the distribution patterns (TAJIK et al. 2002). More recently, Xenon has been employed in an animal experiment involving monochromatic synchotron radiation CT (BAYAT et al. 2001). By obtaining simultaneous images at two different energy levels, absolute Xe concentrations can be obtained via a subtraction mechanism. The spatial distribution of Xe concentrations in airspaces, dynamic of airspace filling and images of airways with a diameter in the range of 1 mm could be obtained (BAYAT et al. 2001). Thus, this technology is capable of studying small airways physiology as well as the influences of pathology and treatment interventions.

A recent experiment in ventilated sheep used a combination of Xenon ventilation and standard CT to assess radiodensity changes (KRECK et al. 2001). The radiodensity changes are a function of regional ventilation (inspiration will increase radiodensity changes) and perfusion (good perfusion will result in a decrease of radiodensity). The study demonstrated vertically oriented and isogravitational heterogeneity of ventilation-perfusion measurements. It would seem that sequential CT imaging using Xenon breathing is feasible to determine regional ventilation and perfusion with high spatial resolution.

Finally, not only Xenon may be useful for determination of physiological lung parameters. A recent study in 28 non-smoking controls and 47 patients with smoking-induced COPD examined lung density changes and compared findings during breathing of a mixture of 21% Oxygen in Sulfur Hexofluorane (SF_6) or a mixture of 21% Oxygen in Helium (YAMAGUCHI et al. 2001). Continuous breathing was performed, and end-inspiration and end-expiration images were obtained during the second (early phase) and 60th (equilibrium state) breaths, and mean lung density and relative area with low attenuation were determined. The authors demonstrated that the differences increased from early phase to equilibrium state, and that this increase was more marked in smokers and in the lower lung fields (YAMAGUCHI et al. 2001). These differences were only detected at end-inspiration. Thus, this technique may be helpful for predicting acinar gas distribution abnormalities in patients with COPD.

7.2.6
Conclusions

Computed tomography is helpful for the assessment of (regional) ventilation distribution and this influences management of a wide range of lung diseases. Table 7.2.1 gives an overview of the available techniques and their relative merits and disadvantages. The introduction of more rapid multi-detector systems will enhance the potential use of dynamic CT, albeit that radiation exposure remains an issue at present. However, the main advantages of imaging-based lung volume parameters are their demonstration of regional abnormalities and the high

Table 7.2.1. Computed Tomography techniques for assessment of ventilation of the lung

Technique	Reproducibility	Spatial Resolution	Temporal Resolution	Clinical Applications
HRCT in/exp	++	++++	0	routine assessment of lung diseases
Spirometric HRCT	+++	++++	0	possible use in children
Dynamic HRCT	++	+++	++	only one slice
Multidetector CT	+++	++++	++	combination of static and dynamic imaging
Xenon CT	++	++++	0	improved visualisation of air trapping (?)
Ultrafast CT (+/− Xe)	++	++++	++	research tool, limited availability

level of reproducibility. Thus, they incorporate morphology and functional information. This will very likely lead to additional information, and therefore, it is expected that the use of these technologies will increase in areas such as physiology (to aid the understanding of the normal lung in different conditions), respiratory medicine (quantification of disease, assessment of therapeutic response) and thoracic surgery (lung cancer respectability assessment and planning of lung volume reduction surgery).

The use of inert gases, such as Xenon and Helium, are potentially of additional value for physiological analysis of lung function. However, there are only limited clinical applications at present, as most of the work thus far was carried out using ultrafast CT modalities.

References

Arakawa H, Webb WR (1998) Air trapping on expiratory high-resolution CT scans in the absence of inspiratory scan abnormalities: correlation with pulmonary function tests and differential diagnosis. Am J Roentgenol 206:89–94

Arakawa H, Webb WR, McCowin M, et al. (1998) Inhomogenous lung attenuation at thin-section CT: diagnostic value of expiratory scans. Radiology 206:89–94

Bartz RR, Stern EJ (2000) Airways obstruction in patients with sarcoidosis. Expiratory CT scan findings. J Thorac Imaging 15:285–289

Bayat S, Le Duc G, Porra L, et al. (2001) Quantitative functional lung imaging with synchrotron radiation using inhaled xenon as contrast agent. Phys Med Biol 46:3287–3299

Bhalla M, Naidich DP, McGuinness G, et al. (1996) Diffuse lung disease: bassessment with helical CT – preliminary observations of the role of maximum and minimum intensity projection images. Radiology 200:341–347

Bongaerts AHH, Sheedy II PF (1999) Electron beam tomography and pulmonary thromboembolism. In: Oudkerk M, Beek EJR van, Cate JW ten (eds) Pulmonary embolism. Blackwell, Berlin, pp 226–249

Brasch RC, Gould RG, Gooding CA, Ringertz HG, Lipton MJ (1987) Upper airway obstruction in infants and children: evaluation with ultrafast CT. Radiology 165:459–466

Brody AS, Kuhn JP, Seidel FG, Brodsky LS (1991) Airway evaluation in children with use of ultrafast CT: pitfalls and recommendations. Radiology 178:181–184

Chen D, Webb WR, Storto ML, et al. (1998) Assessment of air trapping using postexpiratory high-resolution computed tomography. J Thorac Imaging 13:135–143

Cullen SC, Gross EG (1951) Anesthetic properties of Xenon in animals and human beings with additional observation on Krypton. Science 113:580–581

Flohr T, Prokop M, Becker C, et al. (2002) A retrospectively ECG-gated multislice spiral CT scan and reconstruction technique with suppression of heart pulsation artifacts for cardio-thoracic imaging with extended volume coverage. Eur Radiol 12:1497–1503

Foley WD, Haughton VM, Schmidt J, Wilson CR (1978) Xenon contrast enhancement in computed body tomography. Radiology 129:219–220

Furukawa M, Kashiwagi S, Matsunaga N, Suzuki M, Kishimoto K, Shirao S (2002) Evaluation of cerebral perfusion parameters measured by perfusion CT in chronic cerebral ischemia: comparison with xenon CT. J Comput Assist Tomogr 26:272–278

Galvin JR, Rooholamini SA, Stanford W (1989) Obstructive sleep apnea: diagnosis with ultrafast CT. Radiology 171:775–778

Gevenois PA, Martelaer V de, De Vuyst P, Zanen J, Yernault JC (1995) Comparison of comuputed density and macroscopic morphometry in pulmonary emphysema. Am J Respir Crit Care Med 152:653–657

Gevenois PA, De Vuyst P, Sy M, et al. (1996) Pulmonary emphysema: quantitative CT during expiration. Radiology 199:825–829

Gierada, DS, Yusen RD, Pilgram TK, et al. (2001) Repeatability of quantitative CT indexes of emphysema on patients evaluated for lung volume reduction surgery. Radiology 220: 448–454

Gleeson FV, Traill ZC, Hansell DM (1996) Evidence on expiratory CT scans of small airway obstruction in sarcoidosis. Am J Roentgenol 166:1052–1054

Gotway MB, Lee ES, Reddy GP, Golden JE, Webb WR (2000) Low-dose, dynamic, expiratory thin-section CT of the lungs using a spiral CT scanner. J Thorac Imaging 15: 168–172

Gur D, Drayer BP, Borovetz HS, Griffith BP, Hardesty RL, Wolfson SK (1979) Dynamic computed tomography of the lung: regional ventilation measurements. J Comput Assist Tomogr 3:749–753

Gur D, Shabason L, Borovetz HS, et al. (1981) Regional pulmonary ventilation measurements by xenon enhanced dynamic computed tomography: an update. J Comput Assist Tomogr 5: 678–683

Hansell DM, Wells AU, Rubens MB, et al. (1994) Bronchiectasis: functional significance of areas of decreased attenuation at expiratory CT. Radiology 193:369–374

Herbert DL, Gur D, Shabason L, et al. (1982) Mapping of human local pulmonary ventilation by xenon enhanced computed tomography. J Comput Assist Tomogr 6:1088–1093

Heremans A, Verschakelen JA, Van Fraeyenhoven L, Demedts M (1992) Measurement of lung density by means of quantitative CT scanning: a study of correlations with pulmonary function tests. Chest 102: 805–811

Hoffman, EA, Tajik JK, Kugelmass SD (1995) Matching pulmonary structure and perfusion via combined dynamic multislice CT and thin-slice high-resolution CT. Comput Med Imaging Graph 19:101–112

Hu S, Hoffman EA, Reinhardt JM (2001) Automatic lung segmentation for accurate quantitation of volumetric X-ray CT images. IEEE Trans Med Imaging 20:490–498

Hunsaker AR, Ingenito IP, Reilly JJ, Costello P (2002) Lung volume reduction surgery for emphysema: correlation of CT and V/Q imaging with physiologic mechanisms of improvement in lung function. Radiology 222:491–498

Im JG, Kim SH, Chung MJ, Koo JM, Han MC (1996) Lobular low attenuation of the lung parenchyma on CT: evaluation of 48 patients. J Comput Assist Tomogr 20:752–762

Johnson JL, Kramer SS, Mahboubi S (1998) Air trapping in children: evaluation with dynamic lung densitometry with spiral CT. Radiology 206:95–101

Kalender WA, Rienmuller R, Seissler W, et al. (1990) Measurement of pulmonary parenchymal attenuation: use of spirometric gating with quantitative CT. Radiology 175:265–268

Kalender WA, Fichte H, Bautz W, et al. (1991) Semiautomatic evaluation procedures for quantitative CT of the lung. J Comput Assist Tomogr 15:248–255

Kauczor HU, Heussel CP, Fischer B, Klamm R, Mildenberger P, Thelen M (1998) Assessment of lung volumes using helical CT at inspiration and expiration: comparison with pulmonary function tests. Am J Roentgenol 171:1091–1095

Kauczor HU, Hast J, Heussel CP, Schlegel J, Mildenberger P, Thelen M (2000) Focal airtrapping at expiratory high-resolution CT: comparison with pulmonary function tests. Eur Radiol 10:1539–1546

Kauczor HU, Hast J, Heussel CP, Schlegel J, Mildenberger P, Thelen M (2002) CT attenuation of paired HRCT scans obtained at full inspiratory/expiratory position: comparison with pulmonary function tests. European Radiology 12: 2757–2763

Knudson RJ, Standen JR, Kaltenborn WT, et al. (1991) Expiratory computed tomography for assessment of suspected pulmonary emphysema. Chest 99:1357–1366

Kramer SS, Hoffman EA (1995) Physiologic imaging of the lung with volumetric high-resolution CT. J Thorac Imag 10:280–290

Kreck TC, Krueger MA, Altemeier WA, et al. (2001) Determination of regional ventilation and perfusion in the lung using xenon and computed tomography. J Appl Physiol 91: 1741–1749

Lamers RJ, Thelissen GR, Kessels AG, et al. (1994) Chronic obstructive pulmonary diseases. Evaluation with spirometrically controlled CT lung densitometry. Radiology 193: 109–113

Leung AN, Fisher K, Valentine V, et al. (1998) Bronchiolitis obliterans after lung transplantation: detection using expiratory HRCT. Chest 113:365–370

Lucidarme O, Coche E, Cluzel P, et al. (1998) Expiratory CT scans for chronic airway disease: correlation with pulmonary function test results. Am J Roentgenol 170:301–307

Lucidarme O, Grenier PA, Cadi M, Mourey-Gerosa I, Benali K, Cluzel P (2000) Evaluation of air trapping at CT: comparison of continuous versus suspended expiration CT techniques. Radiology 216:768–772

Lynch DA, Brasch RC, Hardy KA, et al. (1990) Pediatric pulmonary disease: assessment with high-resolution ultrafast CT. Radiology 176:243–248

Marcucci C, Nyhan D, Simon BA (2001) Distribution of pulmonary ventilation using Xe-enhanced computed tomography in prone and supine dogs. J Appl Physiol 90:421–430

Markstaller K, Arnold M, Döbrich M, et al. (2001a) A software tool for automatic image-based ventilation analysis using dynamic chest CT-scanning in healthy and ARDS lungs. Fortschr Rontgenstr 173:830–835

Markstaller K, Eberle B, Kauczor HU, et al. (2001b) Temporal dynamics of lung aeration determined by dynamic CT in a porcine model of ARDS. Br J Anaesth 87:459–468

Mayo JR, Webb WR, Gould R, et al. (1987) High-resolution CT of the lungs: an optimal approach. Radiology 163:507–510

Mergo PJ, Williams WF, Gonzalez-Rothi R, et al. (1998) Three-dimentional volumetric assessment of abnormally low attenuation of the lung from routine helical CT: inspiratory and expiratory quantification. Am J Roentgenol 170: 1355–1360

Murata K, Khan A, Rojas KA et al. (1988) Optimization of computed tomography technique to demonstrate the fine structure of the lung. Invest Radiol 23:170–175

Murata K, Khan A, Herman PG (1989) Pulmonary parenchnymal disease: evaluation with high-resolution CT. Radiology 170:629–635

Naidich DP, Zerhouni EA, Hutchins GM, et al. (1985) Computed tomography of the pulmonary parenchyma: part 1, distal air-space disease. J Thorac Imaging 1:39–53

Nakata H, Komoto T, Nakayama T, et al. (1985) Diffuse peripheral lung disease: evaluation by high-resolution computed tomography. Radiology 157:181–185

Newman KB, Lynch DA, Newman LS, Ellegood D, Newell JD Jr (1994) Quantitative computed tomography detects air trapping due to asthma. Chest 106:105–109

Ng CS, Desai SR, Rubens MB, Padley SPG, Wells AU, Hansell DM (1999) Visual quantitation and observer variation of signs of small airways disease at inspiratory and expiratory CT. J Thorac Imaging 14:279–285

Padley SPG, Adler BD, Hansell DM, et al. (1993) Bronchiolitis obliterans: high-resolution CT findings and correlation with pulmonary function tests. Clin Radiol 47:236–240

Park CS, Müller NL, Worthy SA, et al. (1997) Airway obstruction in asthmatic and healthy individuals: inspiratory and expiratory thin-section CT findings. Radiology 203:361–367

Park KJ, Bergin, CJ, Clausen, JL (1999) Quantitation of emphysema with three-dimensional CT densitometry: comparison with two-dimensional analysis, visual emphysema scores, and pulmonary function test results. Radiology 211541–547

Remy-Jardin M, Remy J, Artaud D, et al. (1996) Diffuse infiltrative lung disease: clinical value of sliding-thin-slab maximum intensity projection CT scans in the detection of mild micronodular patterns. Radiology 200:333–339

Remy-Jardin M, Campistron P, Amara A, et al. (2002a) Workflow issue with multislice CT (MSCT) of the thorax. Usefulness of multiplanar reformations in the diagnostic approach of infiltrative lung disease. Eur Radiol 12 (Suppl 1):134 (abstract)

Remy-Jardin M, Edme JL, Boulenguez C, Remy J, Mastora I, Sobaszek A (2002b) Longitudinal follow-up study of smoker's lung with thin-section CT in correlation with pulmonary function tests. Radiology 222:261–270

Sase S, Honda M, Kushida T, Seiki Y, Machida K, Shibata I (2002) Quantitative cerebral blood flow calculation method using white matter lambda in xenon CT. J Comput Assist Tomogr 26:471–478

Schoepf UJ, Becker CR, Bruening RD, et al. (1999) Electrocardiographically gated thin-section CT of the lung. Radiology 212:649–654

Simon B, Marcucci C, Fung M, Lele S (1998) Parameter estimation and confidence intervals for Xe-CT ventilation studies: a Monte Carlo approach. J Appl Physiol 84:709–716

Small JH, Flower CD, Traill ZC, et al. (1996) Air trapping in extrinsic allergic alveolitis on computed tomography. Clin Radiol 51:684–688

Snyder JV, Pennock B, Herbert D, et al. (1984) Local lung ventilation in critically ill patients using nonradioactive xenon-enhanced transmission computed tomography. Crit Care Med 12:46–51

Stern EJ, Webb WR (1993) Dynamic imaging of lung morphology with ultrafast high-resolution computed tomography. J Thorac Imaging 8:273–282

Stern EJ, Webb WR, Warnock ML, et al. (1991) Broncho-pulmonary sequestration: dynamic, ultrafast, high-resolution CT evidence of air trapping. Am J Roentgenol 157: 947–949

Stern EJ, Webb WR, Golden JA, et al. (1992) Cystic lung disease associated with eosinophilic granuloma and tuberous sclerosis: air trapping at dynamic ultrafast high-resolution CT. Radiology 182:325–329

Stern EJ, Graham CM, Webb WR, Gamsu G (1993) Normal trachea during forced expiration: dynamic CT measurements. Radiology 187:27–31

Stern EJ, Webb WR, Gamsu G (1994) Dynamic quantitative computed tomography: a predictor of pulmonary function in obstructive lung diseases. Invest Radiol 29:564–569

Tajik, JK, Tran BQ, Hoffman EA (1996) Xenon enhanced CT imaging of local pulmonary ventilation. Proc SPIE 2709: 40–54

Tajik JK, Chon D, Won C, Tran BQ, Hoffman EA (2002) Sub-second multisection CT of regional pulmonary ventilation. Acad Radiol 9:130–146

Thiele J, Klöppel R (1995) Computertomographische Messung der Lungenventilation durch Inhalation von Isovist 300. Röntgenpraxis 48: 259–260

Tomiyama N, Takeuchi N, Imanaka H, et al. (1993) Mechanism of gravity-dependent atelectasis. Analysis by non-radioactive xenon-enhanced dynamic computed tomography. Invest Radiol 28:633–638

Webb WR (1997) Radiology of obstructive pulmonary disease. Am J Roentgenol 169:637–647

Webb WR, Stern EJ, Kanth N, et al. (1993) Dynamic pulmonary CT: findings in normal adult men. Radiology 186:117–124

Webb WR, Müller NL, Naidich PD (2001) Technical aspects of HRCT. In: Webb WR, Müller NL, Naidich PD. High-resolution CT of the lung, 3rd edn. Lippincott Williams & Wilkins, Philadelphia, pp 1–47

Winkler S, Nielsen A, Mesina J (1987) Respiratory depression in goats by stable Xenon: implication for CT studies. J Comput Assist Tomogr 11:496–498

Worthy SA, Müller NL, Hartman TE, Swensen SJ, Padley SPG, Hansell DM (1997) Mosaic attenuation pattern on thin-section CT scans of the lung: differentiation among infiltrative lung, airway and vascular diseases as a cause. Radiology 205:465–470

Yamaguchi K, Soejima K, Koda E, Sugiyama N (2001) Inhaling gas with different CT densities allows detection of abnormalities in the lung periphery of patients with smoking-induced COPD. Chest 120:1907–1916

Zerhouni EA, Naidich DP, Stitik FP, et al. (1985) Computed tomography of the pulmonary parenchyma: part 2, interstitial disease. J Thorac Imaging 1:54–64

7.3 Lung Scintigraphy

Edwin J. R. van Beek

CONTENTS

7.3.1
Introduction

Ventilation lung scintigraphy was developed during the 1950s and 1960s. One of the first compounds to be used was 133-xenon gas, which was first used in 1955 (Knipping et al. 1955) and was developed to ultimately be used in one of the largest studies ever, the PIOPED study (PIOPED Investigators 1990). However, in conjunction with the use of 133-xenon gas, other tracers were developed. Another noble gas, 81m-krypton, which is derived from the decay of 81-rubidium, came into use in the late 1970s (Goris et al. 1977) (Fig. 7.3.1). Both gases produce photons at 81 keV, xenon by beta decay with a half-life of 5.3 days and krypton by isomeric transition with a half-life of 13 s.

Aerosols were also tried, such as 99m-technetium diethylene triamine pentacetate (99m-Tc-DTPA), which was developed in the early 1980s, and has particle sizes between 0.1 and 3 µm (Dolovich et al. 1993) (Fig. 7.3.2). Finally, an ultrafine dispersion of 99m-Tc-labeled carbon particles (Technegas), came into use in the late 1980s with a particle size of approximately 0.005 µm (Burch et al. 1986). Technetium produces photons at 140 keV.

All these compounds have their own individual characteristics with advantages and disadvantages.

E. J. R. van Beek, MD, PhD, FRCR
Clinical Senior Lecturer/Honorary Consultant Radiology, Academic Unit of Radiology, Floor C, Royal Hallamshire Hospital, Glossop Road, Sheffield S10 2JF, UK

The main advantage of noble gases is that they offer homogeneous distribution throughout the lungs, which is not necessarily dependent on the state of health of the bronchial mucosa. In contrast, aerosols and dispersions are dependent on particle size homogeneity, and central deposition is a common problem in patients with inflamed mucosa or increased mucus secretion. If one used aerosols for ventilation, it is neither possible to perform pseudodynamic studies nor simultaneous perfusion studies, whereas both are feasible with the use of one of the inert gases. Thus, these properties can be exploited depending on the type of information one needs to obtain (Figs. 7.3.1, 7.3.2).

Apart from dedicated diagnostic ventilation agents, novel pharmaceutical compounds can also be labeled with radioisotopes to allow the study of deposition and pharmacokinetics. An example of this technology is 99m-Tc-labeled triamcinolone acetonide (Hirst et al. 2001b). Alternatively, novel compounds can be tested by mixing the new compounds with 99m-Tc-DTPA aerosols, where one assumes that the deposition of the aerosolized particles will be similar (Hirst et al. 2001a). More recently, positron emission tomography (PET) has been used in a few centers.

7.3.2
Technical Developments

Standard gamma cameras have been used for many years. More recently, multiple detector systems and the availability of simultaneous imaging at different energy spectra has allowed improvements in spatial and temporal resolution. Furthermore, it has opened the possibility of co-registration of ventilation and perfusion imaging.

Single photon emission tomography (SPECT or SPET) uses a revolving gamma camera system. With the availability of double-headed and triple-headed gamma camera systems, the imaging time for full volumetric data capture is reduced to under 20 min (Fig. 7.3.3). Several studies have demonstrated that

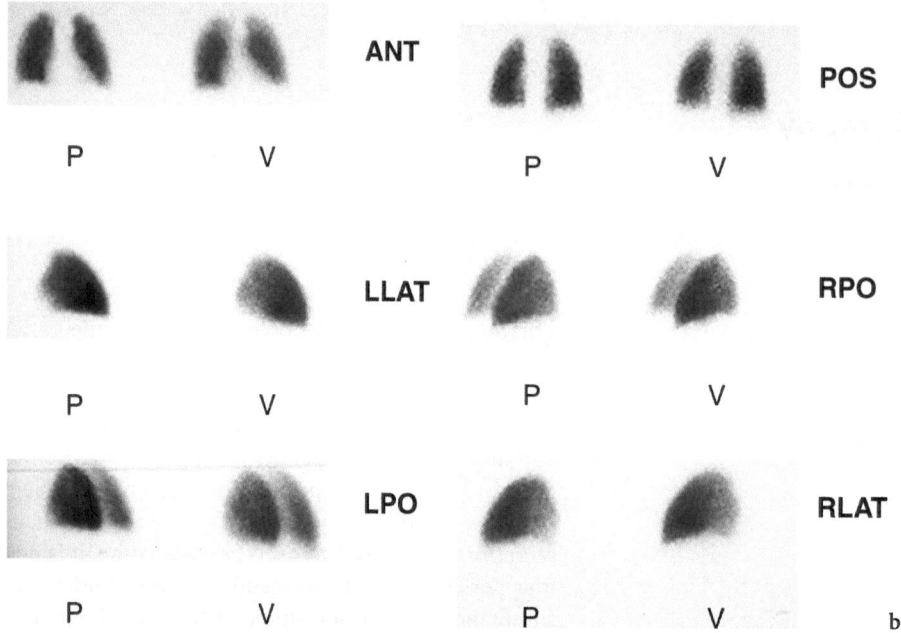

Fig. 7.3.1. Normal combined scintigraphy using 81m-krypton for ventilation (*V*) and 99m-technetium-labeled macro-aggregates of albumin for perfusion (*P*). Homogeneous matching distribution of ventilation and perfusion is normally demonstrated. Images in six projections are normally obtained: **a** anterior, left lateral, and left posterior oblique; **b** posterior, right posterior oblique and right lateral

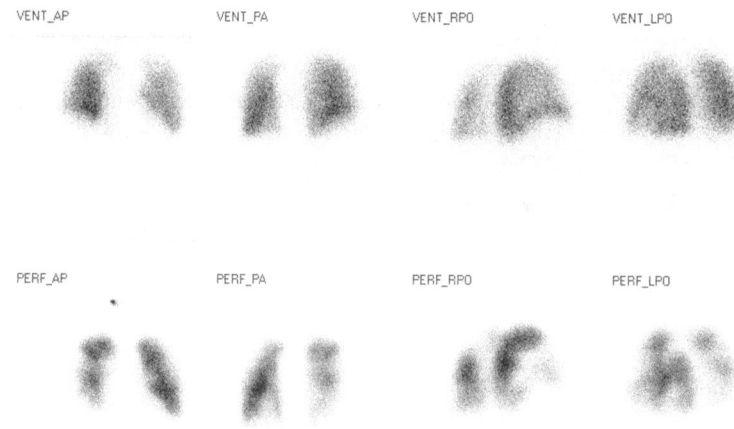

Fig. 7.3.2. Normal ventilation scintigraphy using 99m-technetium-DTPA (*Vent, upper row*) and abnormal perfusion scintigraphy using 99m-technetium-labeled macro-aggregates of albumin (*Perf, lower row*) with multiple wedge-shaped defects. Note the mismatch between the two corresponding to perfusion defects caused by pulmonary embolism and unaffected ventilation

full quantitation is feasible, allowing for true ventilation, perfusion and V/Q ratio analysis (ALMQUIST et al. 1999; SANCHEZ-CRESPO et al. 2002). Furthermore, respiratory-gated techniques now allow a reduction in motion artifacts, thus allowing better image fusion capability for ventilation-perfusion SPECT data (SUGA 2002).

Ventilation scintigraphy using inert noble gases, like 133-xenon or 81m-krypton, will give the best estimate of actual ventilation of the lungs. The reason that these gases are preferred is that they do not suffer from central deposition, like the aerosols and dispersions. Although another form of xenon, 127-xenon, has been used in the past, it is less widely available and has therefore been abandoned. Scintigraphy with 133-xenon uses a mixture of 133-xenon and oxygen, which the patient inhales over a period of several minutes. The initial phase (wash-in) requires the patient to take a single breath, at which point an image is obtained. Subsequently, the patient keeps breathing the mixture and two 90 s images are obtained (equilibrium phase). Finally, the patient breathes normal room air and several 45 s images are obtained while the exhaled air is captured. Usu-

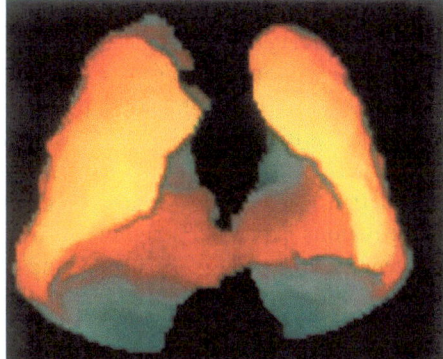

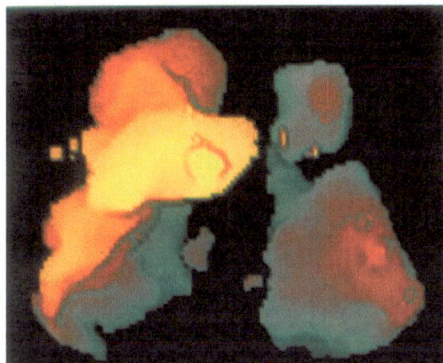

Fig. 7.3.3. Normal ventilation SPECT using 81m-krypton (*top*) and abnormal perfusion SPECT using 99m-technetium-labeled macro-aggregates of albumin (*bottom*). A triple head gamma camera was used to acquire the full volumetric data set within 20 min. Note the mismatch corresponding to perfusion defects caused by pulmonary embolism and unaffected ventilation

ally anterior and posterior views only are obtained, although 45° oblique images may also be obtained during the equilibrium phase. Xenon scintigraphy allows a "dynamic" assessment of the inflow and outflow of gas, but the temporal resolution is poor as many counts are required to obtain a single image. Thus, it is capable of giving some information on in-flow restriction of the gas during the initial phase or out-flow obstruction with air trapping during the final phase of the study. When compared to the sub-second acquisition of present-day CT systems and the millisecond data acquisition of dynamic hyper-polarized 3-helium MRI sequences, it is obvious that the scintigraphy technique is becoming outdated.

Ventilation images with 81m-krypton are much faster due to the keV setting and the possibility of high count rates. As this gas has a very short half-life, simultaneous imaging with perfusion scintigraphy is feasible in 4–8 directions. However, it does require a cyclotron within a few hours travel time from the imaging department. Thus, availability may be a prob-

lem. Aerosol ventilation scintigraphy requires homogeneous particle nebulizers to avoid central deposition in the airways. This is of particular importance in patients with inflamed airways, as increased mucus production and/or airway narrowing causes turbulent flow and increases the deposition of particles.

In one comparative study in 92 patients with suspected pulmonary embolism, 81m-krypton gas demonstrated better ventilation images with less central airway deposition than 99m-technetium Technegas (HARTMANN et al. 2001).

7.3.3
Ventilation Distribution

Ventilation with inert gases will demonstrate homogeneous distribution in the normal lungs (Fig. 7.3.1). Airway obstruction, such as that due to infection, infarction or even reactive physiological processes, will result in ventilation defects. These defects tend to have a segmental distribution (Fig. 7.3.2), although the diffusion of gas through anastomoses and pores can result in ventilation of segments with centrally obstructed airways. This latter observation is generally not observed with the use of particulates or aerosols.

In patients with severe emphysema, who were potential candidates for lung volume reduction surgery, ventilation-perfusion SPECT was not uniformly useful. One study applied 99m-Tc-DTPA aerosol in 50 patients and demonstrated that 23 patients had such severe central airway deposition that quantitation was not technically feasible (JAMADAR et al. 1999). However, in a study of 28 patients with emphysema, who underwent 133-xenon scintigraphy, abnormal wash-out and retention of 133-xenon was demonstrated (SUGA et al. 2000). In this latter study, nine patients subsequently underwent lung volume reduction surgery based on the imaging findings, and this resulted in significant improvement of both 133-xenon scintigraphy and pulmonary function tests.

Semi-quantitative V/Q imaging has been applied for dynamic 133-xenon scintigraphy in the post-operative assessment of lung volume reduction surgery. A study on 8 patients who were imaged prior to surgery and post-reduction surgery, demonstrating improved wash-out of the 133-xenon, suggesting that lung volume reduction surgery improves lung physiology (SUGI et al. 2000). In another study of 20 patients, who underwent bilateral lung volume reduction surgery, similar findings were obtained (TRAVALINE et al. 2000). A study of 39 patients who underwent lung volume reduction surgery compared preoperative CT with V/Q scintigraphy with postoperative outcome measurement of spirometry. Computer-based assessment and visual scoring of both modalities were almost equivalent in predicting the postoperative improvement in FEV(1) measurements (HUNSAKER et al. 2002).

In a study involving a total of 11 asthmatic subjects, 99m-Tc-technegas scintigraphy demonstrated increased deposition in central airways while expiratory flow limitation remained undetectable at the mouthpiece. The authors concluded that expiratory flow limitation in asthma occurs asynchronously across the lung and that methods based on mouth flow measurements are insensitive (PELLEGRINO et al. 2001). Another study in 12 patients with oral corticosteroid-dependent bronchial asthma demonstrated significant 99m-Tc-technegas ventilation defects in spite of the patients being clinically in remission (FUJITA et al. 1999). Others studies have used V/Q scintigraphy to evaluate regional lung ventilation in asthmatics (AGNEW et al. 1984; VERNON et al. 1986). However, radionuclide scanning is not routinely used in the clinical assessment of this disease because of its poor spatial resolution and exposure to ionizing radiation. In this context the use of hyperpolarized 3-helium MR imaging is of interest, as similar findings have been demonstrated in greater spatial and temporal detail in exercise-induced asthma, following medical challenging with methacholine and before and after bronchodilator therapy (ALTES et al. 2001; SAMEE et al. 2003).

A study investigated the role of 133-xenon ventilation scintigraphy in nine military recruits who presented with hyperlucent lung (indicative of Swyer-James-MacLeod syndrome) on the screening chest radiograph (ARSLAN et al. 2001). Air-trapping was demonstrated on the wash-out phase, while perfusion scintigraphy also demonstrated perfusion defects of the affected lungs.

Ventilation scintigraphy using 133-xenon and in combination with aerosol deposition imaging has been used for early detection of bronchiolitis obliterans following lung transplantation (UCHIYAMA et al. 1998). Two cases demonstrated delayed wash-out and increased deposition or aerosols, whereas respiratory function tests and high resolution CT were still normal. Functional imaging of ventilation using either radionuclides or hyperpolarized 3-helium MRI is more sensitive than global lung function measurements or morphologic imaging using CT (GAST et al. 2002).

Finally, PET has been used in a few studies to evaluate ventilation and associated physiological processes. One study used 13-nitrogen intravenously to study perfusion and ventilation of six healthy, spontaneously breathing subjects in prone and supine positions to study the gravity-dependent effects on V/Q matching. They found that 1) vertical gradients favoring dependent lung regions explained a significant fraction of heterogeneity, especially of perfusion, and 2) although perfusion did not seem to be systematically more homogeneous in the prone position, differences in individual behavior may make the prone position advantageous, in terms of V-to-Q matching in selected subjects (MUSCH et al. 2002).

7.3.4
Perfusion Scintigraphy

It is difficult to discuss ventilation scintigraphy without a few words about perfusion lung scintigraphy, as these tests are usually performed during a single diagnostic sitting. Perfusion scintigraphy is centered around the intravenous injection of macro-aggregates of albumin (MAA), which are labeled with 99m-technetium (MILLER and O'DOHERTY 1992). The injection takes place with the patient supine to remove the gravitational effects on the pulmonary circulation. During the injection, the patient is asked to inhale deeply, thus leading to a homogeneous distribution of particles throughout the lungs. The particles are of a size in the order of 200 µm, and they become trapped in the capillaries of the pulmonary arteries, giving rise to temporary obstruction in approximately 1% of the total pulmonary capillary bed (DALEN et al. 1967).

Imaging takes place simultaneously with ventilation imaging, resulting in the potential to subtract areas to lead to ventilation-perfusion mismatch assessment. The technique has been particularly useful for the diagnosis of pulmonary embolism, which is largely based on the assumption that pulmonary emboli give rise to obstruction of the pulmonary vascular bed without obstruction of the airways (VAN BEEK et al. 1993). However, the technique has also been used in the assessment of other lung diseases, such as the pre-operative and postoperative evaluation of various surgical procedures (lung cancer, lung reduction surgery and lung transplantation) and limited asthma studies. Generally, the main data include the use of differential perfusion ratios to determine resectability and the assessment of ventilation-perfusion matching.

7.3.5
Conclusion

Ventilation scintigraphy still has a (limited) role in the functional assessment of the lungs. However, as this book demonstrates, newer techniques are now capable of more detailed and more rapid evaluation of functional parameters that enhance our knowledge overall. It is possible that PET techniques will be capable of producing data on physiology that are of general interest. However, the clinical use of PET for ventilation imaging in humans has to be regarded as limited.

References

Agnew JE, Bateman JR, Pavia D et al. (1984) Radionuclide demonstration of ventilatory abnormalities in mild asthma. Clin Sci 66:525–531

Almquist H, Jonson B, Palmer J et al. (1999) Regional VA/Q ratios in man using 133Xe and single photon emission computed tomography (SPECT) corrected for attenuation. Clin Physiol 19:475–481

Altes TA, Powers PL, Knight-Scott J et al. (2001) Hyperpolarized ^{3}He MR lung ventilation imaging in asthmatics: preliminary findings. J Magn Reson Imaging 13:378–384

Arslan N, Ilgan S, Ozkan M et al. (2001) Utility of ventilation and perfusion scan in the diagnosis of young military recruits with an incidental finding of hyperlucent lung. Nucl Med Commun 22:525–530

Burch WM, Sullivan PJ, McLaren CJ (1986) Technegas – a new ventilation agent for lung scanning. Nucl Med Commun 7:865–871

Dalen JE, Haynes FW, Hoppin FG Jr et al. (1967) Cardiovascular responses to experimental pulmonary embolism. Am J Cardiol 20:3–9

Dolovich MB, Cockcroft DW, Coates G (1993) Aerosols in diagnosis: ventilation, airway penetrance, epithelial permeability, mucociliary transport and airway responsiveness. In: Moren F, Colovitch MB, Newhouse MT, Newman SP (eds) Aerosols in medicine, principles, diagnosis and therapy. Elsevier Science, Amsterdam, pp 195–234

Fujita J, Takahashi K, Satoh K et al. (1999) Tc-99m Technegas scintigraphy to evaluate the lung ventilation in patients with oral corticosteroid-dependent bronchial asthma. Ann Nucl Med 13:247–251

Gast K, Viallon M, Eberle B et al. (2002) MRI in lung transplant recipients using hyperpolarized ^{3}He: comparison with CT. J Magn Reson Imaging 15:268–274

Goris ML, Daspit SG, Walter JP et al. (1977) Applications of ventilation lung imaging with 81m Krypton. Radiology 122:399–403

Hartmann IJ, Hagen PJ, Sokkel MP et al. (2001) Technegas versus (81m)Kr ventilation-perfusion scintigraphy: comparative study in patients with suspected acute pulmonary embolism. J Nucl Med 42:393–400

Hirst PH, Bacon RE, Pitcairn GR et al. (2001a) A comparison of the lung deposition of budesonide from Easyhaler, Turbohaler and pMDI plus spacer in asthmatic patients. Respir Med 95:720–727

Hirst PH, Pitcairn GR, Richards JC et al. (2001b) Deposition and pharmacokinetics of an HFA formulation of triamcinolone acetonide delivered by pressurized metered dose inhaler. J Aerosol Med 14:155–165

Hunsaker AR, Ingenito EP, Reilly JJ et al. (2002) Lung volume reduction surgery for emphysema: correlation of CT and V/Q imaging with physiologic mechanisms of improvement in lung function. Radiology 222:491–498

Jamadar DA, Kazerooni EA, Martinez FJ et al. (1999) Semiquantitative ventilation/perfusion scintigraphy and single photon emission tomography for evaluation of lung volume reduction surgery candidates: description and prediction of clinical outcome. Eur J Nucl Med 26:734–742

Knipping HW, Bolt W, Ventrath H et al. (1955) Eine neue Methode zur Prüfung der Herz- und Lungenfunktion die regionale Funktionsanalyse in der Lungen und Herzklinik mit Hilfe des Radioaktiven Edelgases Xenon 133 (isotopenthorakographie). Dtsch Med Wochenschr 80:1146–1147

Miller RF, O'Doherty MJ (1992) Pulmonary nuclear medicine. Eur J Nucl Med 19:355–368

Musch G, Layfield JD, Harris RS et al. (2002) Topographical distribution of pulmonary perfusion and ventilation, assessed by PET in supine and prone humans. J Appl Physiol 93:1841–1851

Pellegrino R, Biggi A, Papaleo A (2001) Regional expiratory flow limitation studied with Technegas in asthma. J Appl Physiol 91:2190–2198

PIOPED Investigators (1990) Value of the ventilation-perfusion scan in acute pulmonary embolism. Results of the prospective investigation of pulmonary embolism diagnosis (PIOPED). JAMA 263:2753–2759

Samee S, Altes TA, Powers P et al. (2003) Imaging the lungs in asthmatics using hyperpolarized helium-3 MR: assessment of response to methacholine and exercise challenge. J Allergy Clin Immunol 111:1205–1211

Sanchez-Crespo A, Petersson J, Nyren S et al. (2002) A novel quantitative dual-isotope method for simultaneous ventilation and perfusion SPET. Eur J Nucl Med Mol Imaging 29:863–875

Suga K (2002) Technical and analytical advances in pulmonary ventilation SPECT with xenon-133 gas and Tc-99m-Technegas. Ann Nucl Med 16:303–310

Suga K, Tsukuda T, Awaya H et al. (2000) Interactions of regional respiratory mechanics and pulmonary ventilatory impairment in pulmonary emphysema: assessment with dynamic MRI and xenon-133 single-photon emission CT. Chest 117:1646–1655

Sugi K, Kaneda Y, Suga K et al. (2000) Improvement of contralateral pulmonary function after unilateral lung volume reduction surgery (LVRS). Ann Thorac Cardiovasc Surg 6:363–368

Travaline JM, Maurer AH, Charkes ND et al. (2000) Quantitation of regional ventilation during the washout phase of lung scintigraphy: measurement in patients with severe COPD before and after bilateral lung volume reduction surgery. Chest 118:721–727

Uchiyama H, Uchiyama M, Shishikura A et al. (1998) Bronchiolitis obliterans after bone marrow transplantation: evaluation with lung scintigraphy. Int J Hematol 68:213–220

Van Beek EJR, Tiel-van Buul MMC, Büller HR et al. (1993) The value of lung scintigraphy in the diagnosis of pulmonary embolism. Eur J Nucl Med 20:173–181

Vernon P, Burton GH, Seed WA (1986) Lung scan abnormalities in asthma and their correlation with lung function. Eur J Nucl Med 12:16–20

8 Oxygen-Sensitive Imaging, Gas Exchange

Balthasar Eberle and Hans-Ulrich Kauczor

CONTENTS

8.1 Physiology and Pathophysiology of Alveolar Gas Exchange

8.1.1 Physiological Considerations

Respiration in mammalians is a cyclic tidal phenomenon. Each breath replenishes a portion of alveolar gas containing humidified dead space gas with fresh gas by exchanging gas back and forth through common conducting airways and by diffu-

B. Eberle, MD
Institut für Anästhesiologie der Universität und des Inselspitals Bern, Inselspital, 3010 Bern, Switzerland
H.-U. Kauczor, MD
Professor of Radiology, Innovative Krebsdiagnostik und Therapie, Deutsches Krebsforschungszentrum (DKFZ), Im Neuenheimer Feld 280, 69120 Heidelberg, Germany

sive mixing in the very distal air spaces. Pulmonary perfusion, on the other hand, is a pulsatile and continuous process. Oxygen diffuses from the alveolar gas space across the alveolocapillary membrane into the oxygen-depleted mixed venous blood, whereas carbon dioxide diffuses in the opposite direction. This process is extremely efficient: partial pressures of oxygen and carbon dioxide of pulmonary capillary blood and alveolar gas nearly equilibrate within one-third of the red blood cell average transit time (~1 s) through pulmonary capillaries.

The main reason for this efficiency is the minute physical diffusion barrier between gas and blood: the alveolocapillary membrane itself is less than 0.5 μm thick, and during their passage through the sheet-like pulmonary capillary network, the red blood cells are forced into close contact with the pulmonary capillary endothelium. In addition, hemoglobin oxygenates faster with higher O_2 saturation, which compensates for a decreasing diffusion gradient. The structural alveolocapillary diffusion barrier is estimated to result in a PO_2 difference of merely 4 mmHg between mixed alveolar gas and arterial blood.

Thus, in the normal lung, exchange of the respiratory gases is not limited by alveolocapillary diffusion but rather by perfusion (Lumb 2000). In order to quantify the diffusive barrier better, a very low concentration of carbon monoxide (CO) is substituted as a test gas for O_2. CO has a similar molecular size, but much higher hemoglobin affinity than O_2. Thus CO transfer is only limited by diffusion and not by uptake into the erythrocytes. The pulmonary diffusing capacity for CO (DL_{CO}) is therefore a parameter used clinically to assess both barrier thickness and surface area which is available for gas exchange. This is an important parameter in lung fibrosis and is associated with an increase in the diffusive barrier thickness as well as in emphysema which is associated with a decrease of the alveolar surface area.

Within a model alveolus, the gas composition is therefore determined 1) by the balance between oxygen supplied by ventilation and oxygen removed by perfusion (Fig. 8.1) and 2) by carbon dioxide

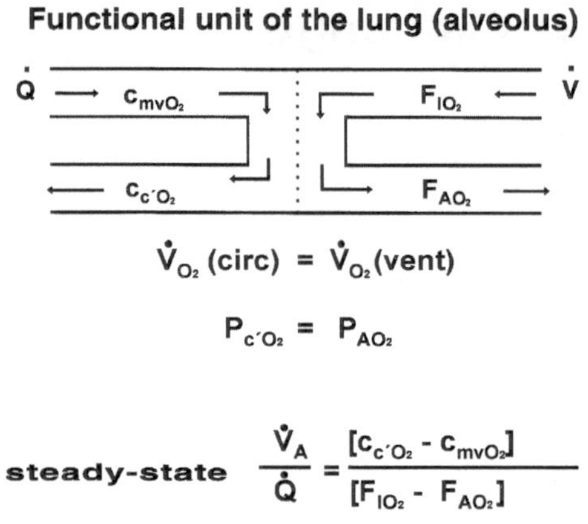

Functional unit of the lung (alveolus)

$$\dot{V}_{O_2}(\text{circ}) = \dot{V}_{O_2}(\text{vent})$$

$$P_{c'O_2} = P_{AO_2}$$

$$\text{steady-state} \quad \frac{\dot{V}_A}{\dot{Q}} = \frac{[c_{c'O_2} - c_{mvO_2}]}{[F_{IO_2} - F_{AO_2}]}$$

Fig. 8.1. During steady states of ventilation, perfusion and inert gas concentrations, respiratory gas composition within a functional lung unit is determined by the equilibrium between ventilatory supply ($\dot{V}_{O_2}$) of O_2 and its removal by perfusion ($\dot{Q}_{O_2}$), and accordingly, by perfusion-driven CO_2 delivery and its expiratory elimination. Since nearly complete diffusion equilibrium can be assumed in non-diseased functional lung units, alveolar PO_2 (P_AO_2) equals end-capillary PO_2, ($P_{c'}O_2$). Using a mass balance relationship between end-capillary-to-mixed venous difference in O_2 content ($C_{c'}O_2$, $C_{mv}O_2$) and inspired-to-alveolar difference of fractional O_2 (F_IO_2, F_AO_2), this PO_2 allows theoretically derivation of the local ratio of alveolar ventilation to perfusion $\dot{V}_A/\dot{Q}$ (WAGNER 1998)

delivery by perfusion, and its elimination by expiration. At a given ventilation and perfusion, and in a steady state of inspired gas mixture and whole-body metabolism, alveolar gas composition is assumed to be fairly constant and quite homogeneous throughout a functional unit of the lungs. The latter is comprised, in reality, not of an individual alveolus but actually of a secondary lobule which represents a group of roughly 2,000 alveoli supplied by a respiratory bronchiole, with a size of about 3.5 mm in diameter (WEIBEL 1970). In each individual functional lung unit with its diffusion equilibrium, the alveolar PO_2 (P_AO_2) equals the end-capillary PO_2 and according to the law of mass balance, both are uniquely defined by the ratio of this individual unit alveolar ventilation to its perfusion ($\dot{V}_A/\dot{Q}$ ratio) (WAGNER 1998).

In the normal individual breathing room air (20.9% O_2 in nitrogen), approximately 130,000 of such functional lung units transfer their individual alveolar gas partial pressures across their alveolocapillary membranes to the pulmonary capillary blood. If each of these units had the same $\dot{V}_A/\dot{Q}$ ratio, the transfer of oxygen and carbon dioxide across the entire lung would demonstrate that the normal physical diffusion barrier is indeed negligible, and mixed alveolar gas partial pressures would closely match arterial blood

gases. Even in healthy lungs, however, there is a non-uniform distribution of $\dot{V}_A/\dot{Q}$ ratios with some scatter around the normal value of 0.8. Effects of gravitation on pulmonary blood flow distribution, and of lung biomechanics on the distribution of ventilation have been invoked to explain this scatter, which also increases with age even in healthy lungs.

8.1.2
Pathophysiological Considerations

In many types of pulmonary disease, a significant number of lung units with abnormal $\dot{V}_A/\dot{Q}$ ratios contribute to mixed alveolar and arterial gas composition. Alveolar-arterial PO_2 differences widen and hypoxemia ensues with or without hypercapnia. Particularly during early stages of disease, however, compensatory mechanisms are activated to counteract $\dot{V}_A/\dot{Q}$ maldistribution to some degree. Most prominently, hypoxic pulmonary vasoconstriction (HPV) reduces blood flow to hypoventilated lung regions very efficiently (von Euler-Liljestrand reflex). This reflex compensates regional hypoventilation, e.g., in emphysema, atelectasis, cystic fibrosis or during one-lung ventilation, and supports oxygenation for quite a long time. Another reflectory mechanism, termed hypocapnic bronchoconstriction, shifts alveolar ventilation away from acutely embolized lung regions towards perfused areas, again redressing $\dot{V}_A/\dot{Q}$ maldistribution to a measurable degree (VIDAL MELO et al. 2002).

8.1.3
Methods of Analysis

A global assessment of gas exchange in the lung is conventionally done by measuring O_2 and CO_2 partial pressures in inspired, end-expiratory and mixed expiratory gas, either by mass spectrometry, or by polarographic or paramagnetic O_2 measurement, and by infrared CO_2 analysis. In mixed venous and arterial blood, electrochemical detection and multi-wavelength spectrometry are employed for blood gas analysis and oximetry. Using such global data, gas exchange is conventionally modeled to occur in one "ideal" alveolar compartment of the whole lung. A ventilated but non-perfused (dead space, $\dot{V}_A/\dot{Q} = \infty$) and a non-ventilated but perfused compartment (shunt, $\dot{V}_A/\dot{Q} = 0$) complete this clinically still convenient three-compartment model (RILEY and COURNAND 1949).

The seminal concepts of $\dot{V}_A/\dot{Q}$ distributional analysis were initially developed by WEST and

DOLLERY (1960) using radioisotope ventilation and perfusion scanning. They went on to develop a methodology which describes the distribution of ventilation and perfusion throughout 50 virtual lung compartments, which are characterized by typical $\dot{V}_A/\dot{Q}$ ratios. This multiple inert gas elimination technique (MIGET) uses an infusion of six biologically inert gases with conveniently different blood/gas solubilities. Their steady-state distribution between arterial blood and mixed expired gas is analyzed using gas chromatography (WAGNER et al. 1974) or recently, mass spectrometry (BAUM-GARDNER et al. 2000). The matching of ventilation and perfusion, the scatter and distribution of $\dot{V}_A/\dot{Q}$ ratios, the size of the intrapulmonary shunt (perfused compartment without ventilation) and alveolar dead space (ventilated compartment without perfusion) and the resultant oxygenation are assessed. Although MIGET does not allow for any topographical allocation of functional lung compartments, the technique has contributed much to the understanding of pulmonary physiology and pathophysiology.

Distribution of ventilation can be assessed, indirectly and without topographical allocation, by time-resolved expiratory inert gas analysis. Spatial information is conveyed by radioisotope ventilation using positron emission tomography (PET) of the lung with a resolution of 3–5 mm^3 (VENEGAS 1998). Superior temporal resolution is possible with dynamic high-resolution computed tomography (HOFFMAN and OLSON 1998; MARKSTALLER et al. 2001) and hyperpolarized noble gas MRI (SCHREIBER et al. 2000b; SALERNO et al. 2001).

Regional $\dot{V}_A/\dot{Q}$ analysis is currently performed in clinical routine, by ventilation and perfusion scanning using radioisotope SPECT. Its spatial resolution is far from even approaching the dimensions of a terminal lobule (3–4 mm). Also, acquisition times of many minutes do not allow for breath-hold imaging. Scans therefore represent a cluster of lung regions averaged during continuous movement over many respiratory cycles.

During the last several years, PET with radioactive $^{13}N_2$ as a tracer has evolved into a potent, if still experimental, technique to acquire $\dot{V}_A/\dot{Q}$ maps of the lung (VENEGAS 1998). Images of pulmonary perfusion are acquired during a short breath-hold following injection of the dissolved tracer. After equilibration of the tracer, ventilation mapping is derived from washout kinetics of the inert gas during respiration. Spatial resolution is today in the order of 2–3 mm, and maps convey three-dimensional information. The technique has been used to date in large-animal

experimentation and in healthy human volunteers (VIDAL MELO et al. 2002).

8.2
Rationale of Oxygen-Sensitive Imaging

Alveolar gas composition within a functional lung unit is unequivocally defined by its $\dot{V}_A/\dot{Q}$ ratio (WAGNER 1998). Spatially resolved and direct measurement of alveolar or end-capillary gas concentrations has not been possible up to now. Information about topographical and temporal PO_2 distributions within the pulmonary biocompartments would (1) enhance the understanding of $\dot{V}_A/\dot{Q}$ distribution and heterogeneity in the normal lung, e.g., with respect to effects of species differences, maturation, gravity and positioning; (2) contribute to diagnostic assessment and follow-up in obstructive, interstitial, inflammatory and thromboembolic pulmonary disease, and 3) help to monitor drug efficiency which, e.g. influence the pulmonary vasculature in a non-invasive way. Several agents and MR techniques which are currently undergoing research and development in this field are presented in the following sections.

8.3
Oxygen-Sensitive Hyperpolarized ^{3}He-MRI

Magnetic resonance imaging of the bronchoalveolar space has its inherent physical limitations, first, in the low proton density of the respiratory gas mixture, and second, due to the lungs sponge-like structure. The latter gives rise to a vast and geometrically complex gas-tissue interface, with a very inhomogeneous distribution of magnetic susceptibilities depending on the strength of B_0. This causes rapid dephasing of spin signal and hence, a very short T2*. However, the introduction of certain highly spin-polarized noble gases, i.e. hyperpolarized ^{3}He and ^{129}Xe, as inhaled imaging agents has opened up a new route to achieve satisfactory spin density imaging of lung air spaces, and even to perform functional ventilation MRI. The non-equilibrium polarization ("hyperpolarization") of ^{3}He, generated by exposure to circularly polarized laser light, allows an increase of spin density in the alveolar space by a factor of 10^5 in comparison to proton MR. This artificially strong polarization may be used to obtain superior spatial or temporal resolution, and that even at lower B_0 field strength than in current clinical 1.5 T scanners.

Hyperpolarization is not a stable state but one that returns in time towards thermal equilibrium ("Boltzmann") polarization. The rate at which longitudinal spin relaxation occurs is governed by several factors: relaxation due to magnetic field gradients, bulk relaxation from ^{3}He-^{3}He dipolar interaction, wall relaxation from surface contacts of the ^{3}He atoms, and O_2-induced relaxation (Table 8.1). Under clinical conditions (1.5 Tesla), field and bulk relaxation processes have longitudinal time constants (T_1) of 2,000 h and more. Surface relaxation within lung parenchyma occurs at T_1 in the range of 4–5 min (DENINGER et al. 1999), whereas the destructive effect of paramagnetic O_2 molecules at 1 bar shortens T_1 to 2.6 s, and even to 16 s at a typical alveolar O_2 concentration of 16% (SAAM et al. 1995).

Molecular oxygen is the most abundant paramagnetic molecule in inspired and alveolar gas. The rapid decay of ^{3}He hyperpolarization in the lungs is thus dominated by ambient alveolar oxygen partial pressure. During early experiments in the field, this rapid destruction of polarization by alveolar O_2 appeared to be an obstacle when developing feasible ^{3}He-MR imaging protocols. Therefore, deoxygenation of the lungs was attempted by inhaling pure nitrogen prior to a ^{3}He breath. This was not an entirely benign maneuver since temporary hypoxemia will be provoked even in healthy volunteers. Soon enough, however, the exquisite O_2 sensitivity of ^{3}He polarization was employed to use the gas as an indicator substance to measure instantaneous and regional intraalveolar PO_2 (EBERLE et al. 1997).

8.3.1
Theory

O_2 reduces the strong magnetization – and hence signal intensity – of inhaled hyperpolarized ^{3}He rapidly, i.e., within the time frame of a long breath-hold, towards its very weak equilibrium magnetization inside the scanner's B_0 field. The exponential decay characteristics of ^{3}He hyperpolarization in the presence of O_2 and the dependency of the O_2-induced relaxation rate (GO_2, s^{-1}) on temperature (T, Kelvin) and molecular density of O_2 ([O_2], Amagat) have been described based upon in vitro NMR spectroscopy studies:

$$\Gamma O_2 = 1/T_1 = 0.45 \times [O_2] \times (299/T)^{0.42} \ s^{-1}/amagat$$

for $T \approx 200$–400 K (SAAM et al. 1995).

Translated to in vivo conditions, the rate of signal decay R (t) depends on the ambient partial pressure of oxygen (PO_2), according to the proportional relationship $R(t) = PO_2/\xi$, with a so-called longitudinal relaxation time constant ξ of 2.61 sbar at 37°C (SAAM et al. 1995).

Furthermore, magnetic resonance imaging itself "consumes" the non-equilibrium polarization of ^{3}He, since each of the n repetitive radio frequency (RF) pulses required for image data acquisition reduces the non-renewable polarization P by a fixed factor towards thermal equilibrium. Thus, following each RF pulse that imposes a flip angle α (with α being approximately proportional to RF voltage) upon a ^{3}He spin population of polarization P_n, the residual polarization P_{n+1} of this spin ensemble will be $P_{n+1} = P_n \cos \alpha$.

Technically, the contribution of the PO_2 effect to total signal decay rate within the intraalveolar ^{3}He/alveolar gas mixture can be isolated from that of the imaging process, and can be quantified. For instance, the kinetics of signal decay may be obtained from a series of repetitive fast low angle shot (FLASH) images performed during breath-hold. During a second series of breath-hold images, one of the two major determinants of relaxation, i.e. flip angle α (~RF voltage) or exposure time to O_2 (interval between consecutive images τ) is varied methodically. The two series may be obtained during two separate – but well reproduced – breath-holds (DENINGER et al. 2000), or during one and the same breath-holding maneuver (DENINGER et al. 2002). Comparison of the two image series and their regional signal kinetics allows the relative contributions of local PO_2 and flip angle α to relaxation to be mathematically extracted (EBERLE et al. 1997; DENINGER et al. 1999, 2002). If prolonged breath-holding is tolerated, the time-dependent decrease of alveolar PO_2 can be observed (DENINGER et al. 2000), which is the net result of both alveolocapillary O_2 transfer and non-ventilatory mass movement of fresh gas.

8.3.2
Methods

If hyperpolarized ^{3}He gas is used as an indirect reporter of alveolar PO_2, the smallest possible disturbance of alveolar gas composition by the inert noble gas is desirable. Hence, for this particular technique of functional imaging, triggered administration of only small amounts of highly polarized ^{3}He followed by normal inspiratory gas appears preferable to inhalation of full breaths of hypoxic ^{3}He/N_2 mixtures. The other alternative, i.e. quantitative replacement of a fraction of the inspired N_2 by hyperpolarized ^{3}He, is technically more difficult due to the immediate O_2-induced relaxation effect upon the tracer gas, and has been accomplished so far only in small animal set-ups under controlled ventilation (MÖLLER et al. 2001).

Therefore, in current PO$_2$-sensitive ^{3}He MRI studies of large animals and humans, approximately 3–4 ml/kg predicted body weight of ^{3}He, at 50% polarization measured at the filling station, are administered at the front of normal or vital capacity inspirations. Gas dosage and administration require dedicated PC-controlled applicator devices (EBERLE et al. 1997; DENINGER et al. 2002). Applicator devices used in clinical research allow insertion of measured bolus amounts (between 20 and 500 ml) of ^{3}He into any phase of the inspiratory cycle, followed by the normal respiratory gas mixture. The system is connected into the inspiratory branch of a valved one-way breathing circuit. The tracer gas may be administered during spontaneous breathing or, if the applicator is backed up by a respirator machine, also during a variety of respiratory support modes (Fig. 8.2). The patient inhales the tracer gas and the subsequent inspiratory volume via a nasal continuous positive airway pressure (CPAP) mask or mouthpiece, and then the breath is held. Respiratory flows, volumes and gas concentrations are measured and recorded. Supplemental O_2, inspiratory pressure support or CPAP can be provided for patients with significantly impaired lung function. Throughout the procedure, heart rate and arterial oxygenation of the patient are monitored by finger pulse oximetry.

8.3.3
Results and Current Status

Image-based determination of intrapulmonary PO$_2$ has been validated in phantom experiments, in animals and human volunteers (EBERLE et al. 1997; DENINGER et al. 2000, 2002). Following equilibration with various inspired O_2 concentrations, the alveolar O_2 fractions measured in peripherally located lung regions of ventilated animals correlated closely with O_2 fractions measured in the end-expiratory gas. A technique to derive intrapulmonary PO$_2$, its temporal evolution during breath-holding as well as local flip angle calibration from paired serial image acquisitions, has been developed theoretically. This has been tested in a phantom model of a "single alveolus" and in human volunteer subjects. At an in-plane resolution of about 1 cm^2, this double-acquisition technique permits determination of regional PO$_2$ in the lung with a relative error of 3%, and of apneic PO$_2$ decrease rate with a relative error of 7% during controlled ventilation (DENINGER et al. 1999, 2000).

Meanwhile, using spectroscopy and imaging techniques with repetitive RF pulse series, other researchers have confirmed in animal experiments that spin-lattice relaxation kinetics of inspired ^{3}He are dominated by alveolar PO$_2$. In guinea pigs ventilated with a mixture of 79% ^{3}He and 21% O_2, T_1 dropped from 19.6 s in vivo to 14.6 s after cardiac arrest, i.e. termination of O_2 uptake. The initial difference in oxygen concentration between inspired and alveolar air, and the temporal decay during apnea were related to functional parameters like pulmonary blood flow. Plausible approximations of functional residual capacity, oxygen uptake from the lungs and cardiac output were possible from such data (MÖLLER et al. 2001).

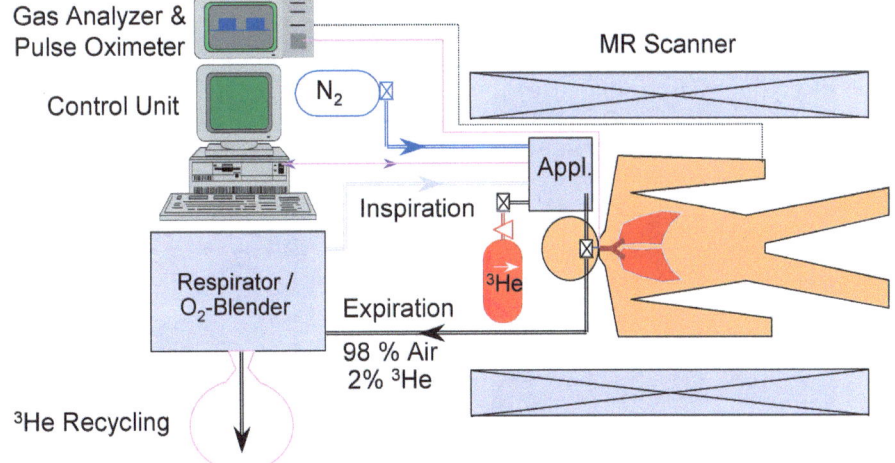

Fig. 8.2. Set-up for ^{3}He-MRI patient studies (*appl.*, applicator, i.e., microprocessor-controlled and pneumatically driven gas dosage and administration device)

In order to improve accuracy of PO$_2$ measurements, determinants of ^{3}He signal attenuation other than RF pulses and oxygen must be taken into account or minimized, such as the relaxation properties of the surfaces that come into contact with the ^{3}He. However, it has been found that the surface relaxation effect of bronchoalveolar epithelia in healthy porcine lungs is, at T$_1 \geq 4.3$ min, surprisingly weak and comparable to most uncoated glass surfaces (Table 8.1) (DENINGER et al. 1999).

Also, convective and diffusive movement of ^{3}He in and out of the imaged field or slice, affects measurement accuracy in O$_2$-sensitive signal analysis. Incomplete breath-holding, particularly in dyspneic patients, as well as cardiac oscillations or diaphragmatic movements induce measurement artifacts. In order to circumvent the problem of reproducibility of breath-holding maneuvers, a single-acquisition imaging sequence for O$_2$-sensitive ^{3}He-MRI has been developed, together with a computer routine for PO$_2$ mapping of the lungs (Fig. 8.3). With this technique, static PO$_2$ determination at the onset of breath-hold with an accuracy comparable to the double-acquisition method is possible, whereas the rate of PO$_2$

Table 8.1. Longitudinal relaxation time constants T$_1$ of hyperpolarized ^{3}He due to various relaxing mechanisms

Mechanism	Conditions	Time constant T$_1$	Reference
Field gradient	25 mT/m (B$_0$ 1.5 T), P=1 bar	>2000 h	SCHEARER and WALTERS (1965)
Dipole relaxation or self-relaxation	T=21°C, P$_{He}$ <0.2 bar	>3700 h	NEWBURY et al. (1993)
Surface relaxation	Supremax glass (h=55 h/cm)	71 h	SURKAU et al. (1997)
	Porcine alveoli (h>22 h/cm)	>4.3 min	DENINGER et al. (1999)
Molecular oxygen	20.9% O$_2$, 21°C, 760 mmHg	12.5 s	SAAM et al. (1995)

B_0 magnetic field strength (T, Tesla); gas pressure (bar, mmHg); T temperature (°C); h surface-specific relaxation coefficient (h/cm)

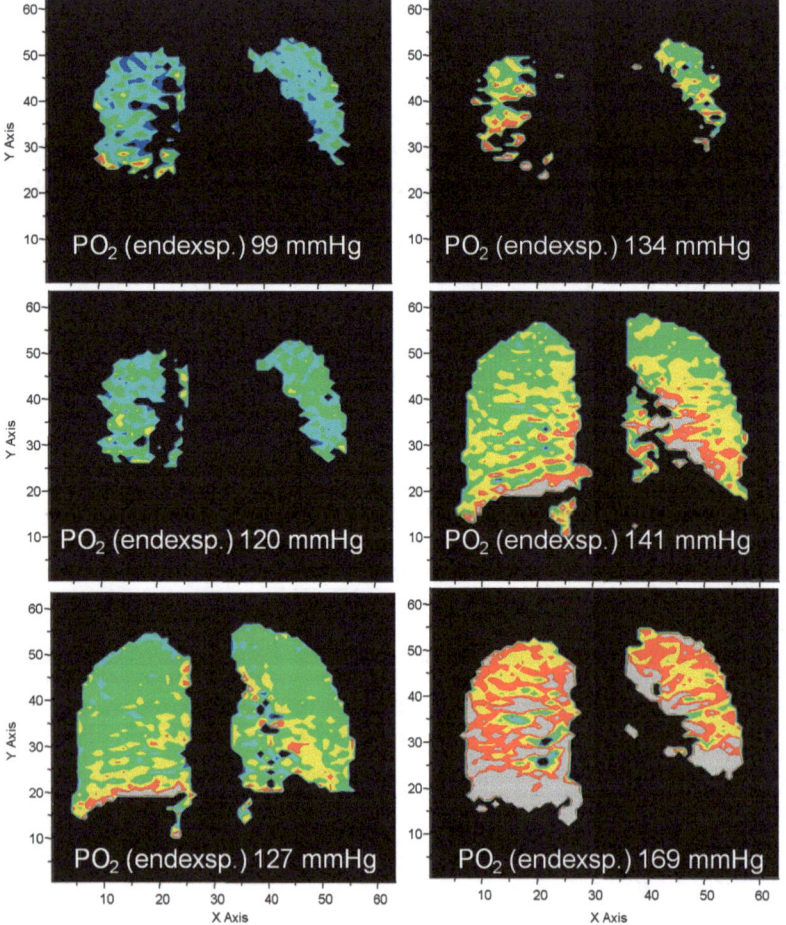

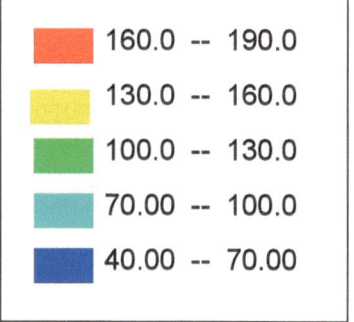

	Regional intrapulmonary PO$_2$ (mm Hg)
160.0 -- 190.0	
130.0 -- 160.0	
100.0 -- 130.0	
70.00 -- 100.0	
40.00 -- 70.00	

Regional intrapulmonary PO$_2$ (mm Hg)

Fig. 8.3. Projection mapping of the PO$_2$ distribution at the end of inspiration in the lungs of a healthy volunteer. Each breath-hold imaging period was preceded by steady-state breathing at a different F$_I$O$_2$. Amounts of approximately 250 ml ^{3}He were then included in an inspiratory tidal volume, and were imaged during a single breath-hold (single acquisition technique). The mean of the intrapulmonary PO$_2$ distribution is in good agreement with end-expiratory PO$_2$ values (*PO$_2$ endexp.*) measured at the mouth using conventional respiratory gas analysis

decrease over time was not as well reproduced due to limitations from low signal-to-noise ratio at the end of breath-holding. A major advantage of a single-acquisition technique, however, remains the reduction of ^{3}He dose requirements per study (DENINGER et al. 2002).

Apparent diffusion coefficients (ADC) of ^{3}He inhaled into the lungs range from 0.2 cm^2/s in healthy subjects to 0.9 cm^2/s in severe emphysema (SAAM et al. 2000). This particularly limits accuracy of thin two-dimensional slice-selective measurements, since during repetitive acquisitions, less-depolarized ^{3}He diffuses into the imaged partition from adjacent non-imaged regions. Two-dimensional projection imaging, as it is used in most current studies, is robust against this diffusional error, but provides low spatial resolution only. Use of three-dimensional imaging sequences can circumvent these problems by simultaneous depolarization of all imaged partitions.

First human applications of the aforementioned techniques demonstrated a very homogeneous PO$_2$ distribution in healthy young subjects, the mean being in good agreement with end-tidal PO$_2$ measured at the mouth (Fig. 8.4). PO$_2$ decrease rates of approximately -1.6 mmHg/s during breath-holding at an inspiratory reserve capacity of the lungs ($\sim$3,000 ml) are well compatible with an oxygen consumption of about 400 ml O$_2$/min derived for such subjects using standard formulae. In patients with pulmonary fibrosis after unilateral lung transplantation, image-based alveolar

PO$_2$ was found to be quite heterogeneously distributed; specifically, it was higher in functioning grafts than in native fibrotic lungs. Mean MRI-determined alveolar PO$_2$ of these patients ranged quite consistently between the PO$_2$ levels found in end-tidal gas and arterial blood (EBERLE et al. 2000). More recent work of the same group demonstrates that the alveolar dead space generated by pulmonary arterial obstruction is topographically correctly detected by MRI-determined alveolar PO$_2$ measurement (EBERLE et al. 2002). Meanwhile, intrapulmonary PO$_2$ measurement using low-field MR imaging has been also shown to be feasible in small animals (OLSSON et al. 2002).

In perspective, functional ^{3}He-MRI including techniques to utilize the oxygen-sensitivity of intraalveolar spin hyperpolarization may develop into a quick, non-invasive and easily repeatable method to assess the regional matching of ventilation with perfusion, at least in those parts of the lung reached by inhaled ^{3}He, and to estimate local oxygen uptake from selected regions within lung parenchyma.

8.4
Oxygen-Enhanced ^{1}H-MRI

O$_2$ molecules are paramagnetic due to the two unpaired orbital electrons. This effect is weak but

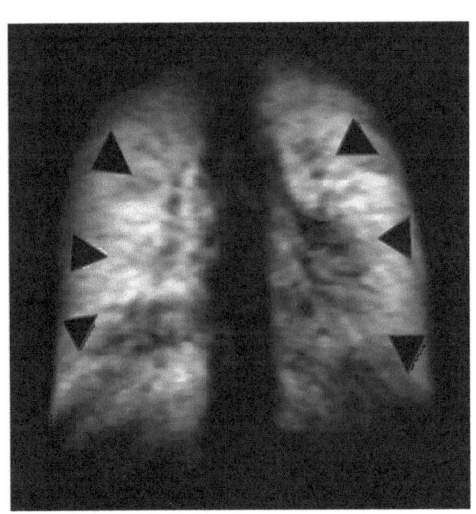

Healthy nonsmoking male, 25 y

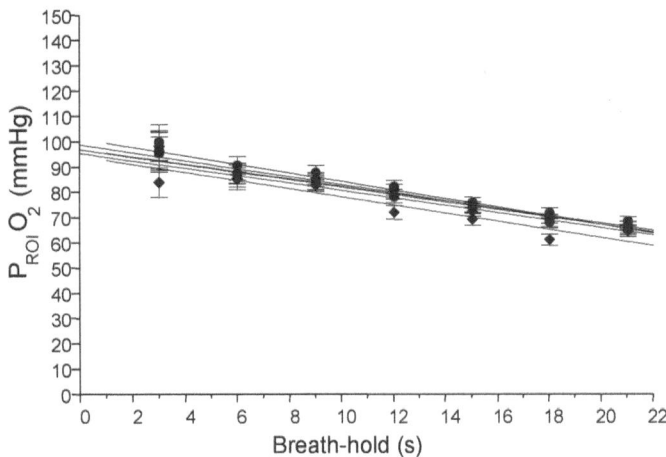

Initial P$_{ROI}$O$_2$ = 97 ± 3mmHg
Rate of decrease RO$_2$ = −1.6 ± 0.1mmHg

Fig. 8.4. Intrapulmonary PO$_2$ kinetics during breath-holding in the lungs of a healthy 25-year-old non-smoker for room air (projection image with 299 ml ^{3}He; ROI area 165 mm^2). Again, a quite homogeneous PO$_2$ distribution between the alveolar regions-of-interest (three black triangles in the periphery of each lung) is apparent, which varies closely around 97 mmHg at end-inspiration. A PO$_2$ decrease rate of −1.6 mmHg/s during breath-holding at an inspiratory reserve capacity of the lungs ($\sim$3,000 ml) would correspond to an oxygen uptake of at least 380 ml O$_2$/min

substantially influences the proton signal in the lungs because of their extremely large surface area. Therefore, EDELMAN et al. (1996) proposed another novel and non-invasive approach for ventilation scanning, i.e., using ¹H-MRI combined with inhaled molecular oxygen as a contrast agent. Theoretically, the method should be able to depict transfer of oxygen increments across the alveolus into lung tissues and capillaries.

8.4.1
Theory

Aerated lung parenchyma is not well visualized by ¹H-MRI with conventional pulse sequences due to abundant tissue-air interfacing. This causes extremely short T2* and, in consequence, decreases signal intensity. Only air-free structures like blood vessels and atelectasis are depicted somewhat better.

The T1 relaxation time constant, on the other hand, is shortened in the presence of paramagnetic molecules, which in turn can be utilized to increase MR signal intensity. The lungs contain two major populations of such molecules, i.e., molecular oxygen (gaseous and dissolved) and deoxyhemoglobin (in red blood cells). The latter is confined within the erythrocyte compartment, and therefore not in close enough proximity to the main population of tissue and plasma water protons to induce useful T1 shortening. Nevertheless, blood oxygenation level-dependent (BOLD) effects of hemoglobin upon T2* within red blood cells can be utilized for regional blood flow and oxygenation studies (OGAWA et al. 1990).

Inhaled supplemental O_2 produces T1 shortening in the ventilated lung tissue (by 9–12% at 100% O_2 (EDELMAN et al. 1996; LÖFFLER et al. 2000). Furthermore, relaxivity (1/T1) of lung parenchyma correlates excellently with arterial oxygen partial pressure (PaO_2) (HATABU et al. 2001). O_2-induced signal intensity enhancement thus originates from O_2 dissolved in tissues and fluids. This has also been demonstrated by the ability of oxygen-enhanced MRI to follow tumor and muscle cell PO_2 after exposure to hyperbaric oxygen (KINOSHITA et al. 2000).

Thus, rather than simply representing alveolar wash-in of O_2 increments by pulmonary ventilation, the method more likely reflects incremental O_2 uptake from ventilated air spaces into lung parenchyma, and finally its diffusive transfer into pulmonary capillaries and veins. At a change from room air breathing to a fraction of inspired O_2 (F_IO_2) of 1.0, the step-up of dissolved O_2 between mixed venous

(PO_2 40 mmHg) and pulmonary venous plasma (PO_2 650 mmHg) should reduce T1 of plasma even more than that of lung parenchyma. Consequently, marked signal enhancement is observed also in pulmonary veins and left heart chambers, and even in adjacent well-perfused organs like the aorta, spleen and kidneys (Fig. 8.5) (HATABU et al. 2001).

8.4.2
Technical Considerations

Technically, step changes in F_IO_2 are induced and act as a T1 contrasting principle for non-invasive assessment of regional wash-in and uptake of O_2 by ventilation and alveolocapillary diffusion. The technique produces images with more inherent noise than hyperpolarized ³He-MRI, but this can be overcome by sequence optimization and signal averaging (MCADAMS et al. 2000). A short effective echo time (TE) and short inter-echo spacing, e. g. in a centrally reordered single-shot rapid acquiring relaxation enhancement (RARE) sequence, yields good signal-to-noise ratios. The contrast-to-noise ratio can be improved using an optimized inversion delay time (CHEN et al. 1998).

Appropriate sequences have been successfully used in animal models and healthy subjects (CHEN et al. 1998; OHNO et al. 2001a; MAI et al. 2001) (Fig. 8.6).

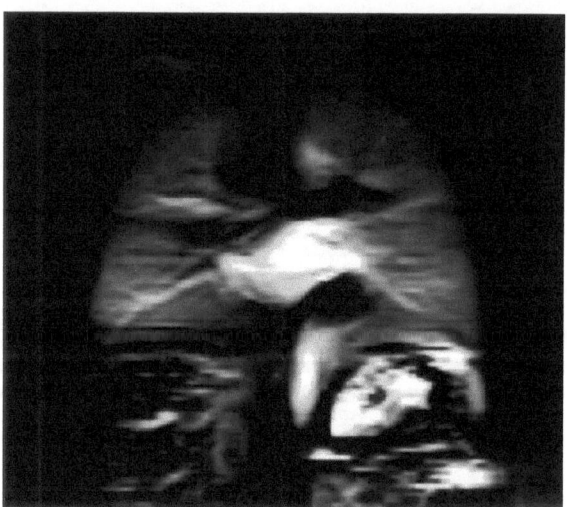

Fig. 8.5. Oxygen-enhanced ¹H-MRI of a healthy volunteer's lungs. Signal change maps were calculated from images obtained both during room air breathing and after changing the fractional oxygen concentration from 0.21 to 1.0. The marked signal enhancement of lung parenchyma is clearly associated with enhancement of pulmonary veins, aorta and spleen, but not pulmonary arteries. Reprinted with permission from HATABU et al. (2001)

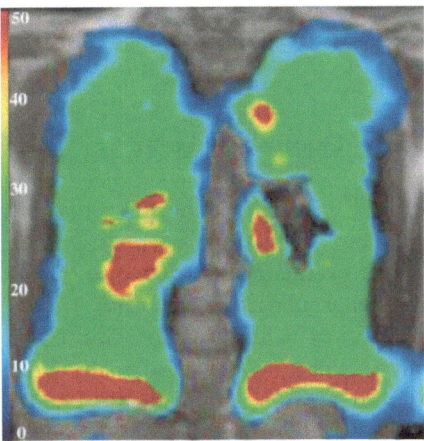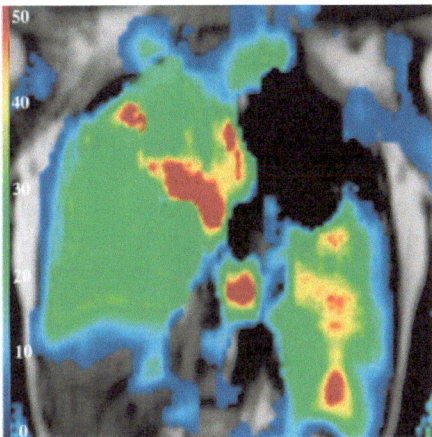

Fig. 8.6. Oxygen-enhanced 1H-MRI of the lungs of a healthy 28-year-old subject (*left*), and of a 65-year-old male patient with a left upper lobe mass (*right*). Color-coded oxygen enhancement maps show marked and homogeneous oxygen uptake in the healthy subject, whereas in the patient the tumor prevents enhancement locally, and even reduces it in adjacent lung parenchyma. Reprinted with permission from OHNO et al. (2001a)

In volunteers, breathing 100% O_2 enhances signal intensity by up to 18% compared to breathing room air , and reduces mean T1 in the lungs from 1.22 s to 1.07 s (LÖFFLER et al. 2000). Also, the expected dose-dependency between step-ups in F_IO_2, mean T1 decrease and average signal intensity has been demonstrated (MAI et al. 2002a).

The feasibility of O_2-enhanced ^{1}H-MRI on open 0.2 T MR scanners has also been shown. A SNR of ~10 was achieved on the 0.2 T MR system, compared to ~70 at 1.5 T (STOCK et al. 1999). Moreover, to carry out fast pulmonary imaging at low field (0.2 T), an inversion recovery true fast imaging with steady precession (true FISP) pulse sequence has been developed. Here, lungs of healthy volunteers exhibited signal enhancement of about 12% after equilibration with an F_IO_2=1.0 (MÜLLER et al. 2001).

8.4.3
Results

Since signal enhancement is due to O_2 dissolved in tissue and plasma, regional interruption of ventilation causes signal voids (Fig. 8.7). In regions with interrupted perfusion, venous stasis settles in, but lung tissue and intravascular water still equilibrate with the increased alveolar dead space PO_2. Hence, O_2-enhanced MRI in animals and patients clearly visualizes ventilation defects due to bronchial obstruction, tumor and bullous emphysema, but not perfusion defects from pulmonary embolism. The depiction of a ventilation-perfusion mismatch in pulmonary embolism requires the additional use of perfusion MRI techniques (EDELMAN et al. 1996; CHEN et al. 1999). A combination of O_2-enhanced

Fig. 8.7. Oxygen-enhanced functional ventilation imaging during balloon bronchial obstruction in a pig. (*left*) T1-weighted, coronal, inversion-recovery single-shot turbo spin echo image of the lungs with the inflated balloon. (*right*) Oxygen-enhanced coronal RARE MR image delineates a signal void which corresponds to nearly absent ventilation to the right lower lobe. Reprinted with permission from CHEN et al. (1998)

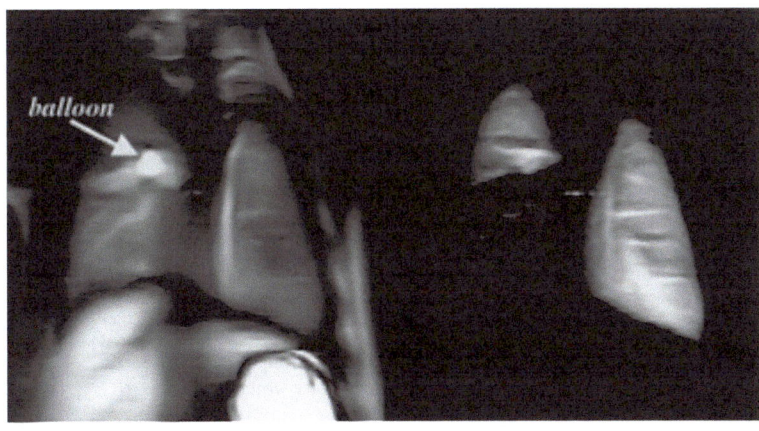

MRI and arterial spin labeling techniques even allows a distribution of ventilation-perfusion signal intensity ratios to be generated . In healthy volunteers, the range and distribution of such ventilation-perfusion signal intensity ratios have been found to resemble those obtained by the multiple inert gas elimination technique (MAI et al. 2001, 2002b).

Another variant of this combined approach to pulmonary ventilation-perfusion assessment is to combine O_2-enhanced MRI with first-pass contrast-enhanced perfusion MRI. With this approach, all of a group of patients with pulmonary embolism were shown to have regional perfusion deficits without ventilation abnormality (NAKAGAWA et al. 2001).

Since O_2-enhanced MRI appears to rely on diffusive oxygen uptake into proton-rich lung compartments, it has been compared with the pulmonary diffusing capacity for CO (DL_{CO}) as a conventional reference test (OHNO et al. 2001b, 2002; MÜLLER et al. 2002). MÜLLER et al. (2002) demonstrated that signal intensity time kinetics during breathing of 100% O_2 are indeed correlated with DL_{CO}. Signal intensity slopes differed significantly between healthy and diseased lungs, and regional maps of this parameter showed good agreement with findings on radiographs and CT scans. Similarly, OHNO et al. (2001b) found that maximum mean relative enhancement ratio was significantly decreased in emphysema patients compared to healthy volunteers, and that correlation with DL_{CO} was excellent. The same parameter also correlated strongly with the [81m]Kr distribution ratio. The mean slope of relative enhancement was strongly correlated to forced expiratory volume in 1 s (FEV1).

Therefore, apart from ventilation imaging, O_2-enhanced MRI also appears to allow spatially resolved assessment of the pulmonary diffusing capacity (OHNO et al. 2002; MÜLLER et al. 2002) and, in combination with perfusion MRI, analysis of ventilation-perfusion matching. Although presently the clinical experience is more extensive with hyperpolarized [3]He-MRI, the limited availability of [3]He, the technical sophistication of hyperpolarization and the need to tune the MR system to the [3]He resonant frequency, still hamper widespread proliferation. Oxygen-enhanced MR imaging is performed at the conventional proton frequency. This obviates the need of readjustment of the MR system, and offers some advantages in lower cost and ready availability (McADAMS et al. 2000).

8.5
^{19}F-MRI

8.5.1
Oxygen-Sensitive Fluorinated Gases

Natural fluorine (^{19}F) has the highest gyromagnetic ratio of all elements besides hydrogen (^{1}H). Therefore, ^{19}F-MRI is a promising alternative to [1]H-MRI and MRI using hyperpolarized noble gases. In addition, as the portion of natural fluorine in the body (and especially in the lungs) is very small, it does not interfere with the measurement. Morphological imaging of the ventilated bronchial and alveolar space has already been tested using highly fluorinated compounds. Such compounds are, e.g., ^{19}F-containing fluids like the oxygen-carrying perfluorocarbons (PFC) (THOMAS et al. 1986; LAUKEMPER-OSTENDORF et al. 2002; HEUSSEL et al. 2003) or biologically inert fluorinated gases such as tetrafluoromethane (CF_4) (RINCK et al. 1984), hexafluoroethane (C_2F_6) (KUETHE et al. 1998), and sulfur hexafluoride (SF_6) (SCHREIBER et al. 2000a) (Fig. 8.8). Costly and time-consuming polarization of the substances is not necessary, and there are no biological side-effects reported when sufficient oxygen is provided. PFC is being studied in liquid ventilation (Figs. 8.9, 8.10) (QUINTEL et al. 1998), e.g., in acute respiratory distress syndrome (ARDS) and respiratory insufficiency in preterm neonates (GREENSPAN et al. 1989) (review in KAISERS et al. 2003).

Since in these compounds, a large number of ^{19}F atoms per molecule resonate at the same Larmor frequency, a strong MR signal is emitted. However, PFC

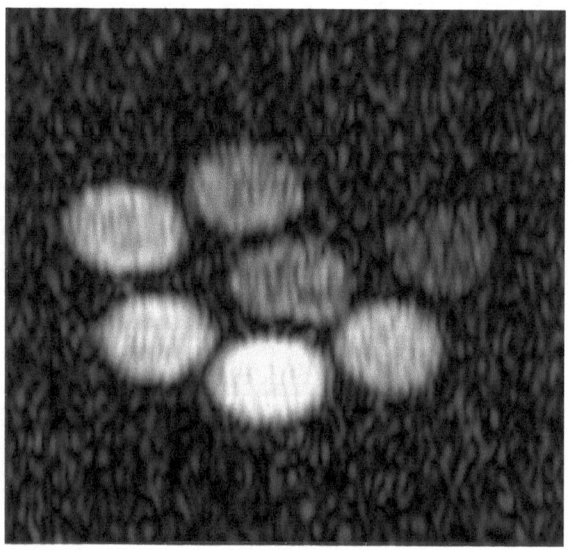

Fig. 8.8. Phantoms containing different concentrations of SF_6 ranging from 40–100% showing the signal intensity which can be obtained in ^{19}F-MRI of gases

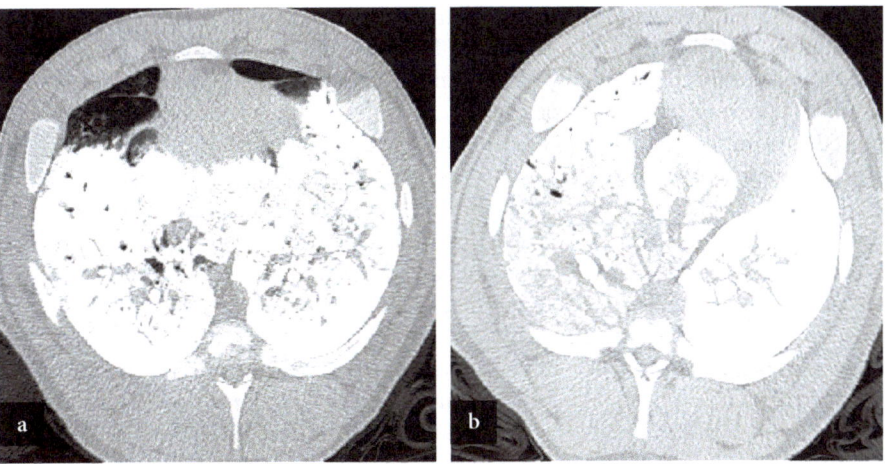

Fig. 8.9. CT of a porcine lung after induction of ARDS and (**a**) partial liquid ventilation (10 ml·kg^{-1}·bw^{-1}) and (**b**) total liquid ventilation (20 ml·kg^{-1}·bw^{-1})

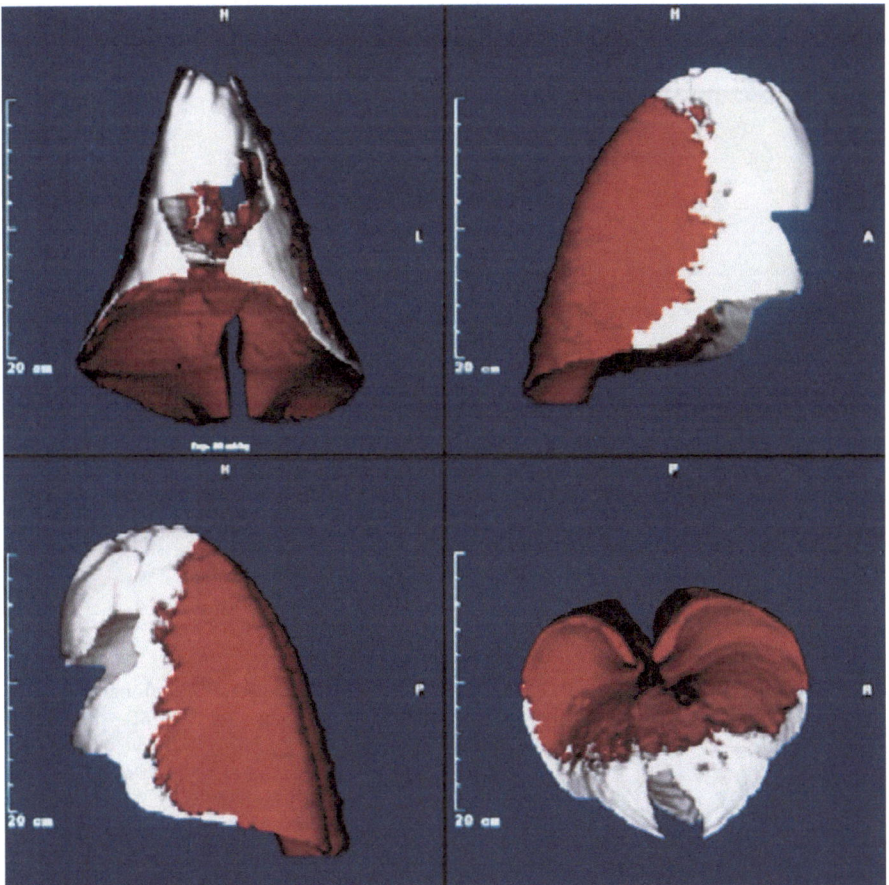

Fig. 8.10. Three-dimensional image reconstruction from electron-beam computed tomograms of a healthy sheep lung undergoing partial liquid ventilation with perfluorooctyl bromide (30 ml/kg body weight; Perflubron/LiquiVent®, Alliance Pharm., San Diego, CA). In the supine position there is preferential distribution of the perfluorocarbon in the dorsal (dependent) lung regions. *Red* mostly perfluorocarbon-filled regions (+400 to +2000 HU). *White* mostly gas-containing regions (−1000 to −200 HU). *Left upper panel* ventral lung aspect, *right upper* right lateral aspect, *left lower* left lateral aspect, *right lower* diaphragmatic aspect in expiration. Reprinted with permission from QUINTEL et al. (1998). Reconstruction from 140 EBCT slices; thickness 2 mm, collimation 3 mm, 130 kV, 128 mAs, scan time 200 ms, Perflubron ~+2000 HU

consists of different fluorinated carbon groups (e.g. CF_2, CF_3, $CBrF_2$), which resonate at slightly different Larmor frequencies. This requires countermeasures (e.g. CHESS presaturation) to prevent chemical shift artifacts (LAUKEMPER-OSTENDORF et al. 2002). Nevertheless, the signal in fluorinated gases is low compared to ¹H-MRI of solid organs. Fortunately, the longitudinal relaxation times T1 of these fluorinated molecules are in the order of milliseconds. This allows rapid RF pulse repetition in appropriate gradient power and hence a large number of signal averages, which compensates for the low spin density of ¹⁹F-containing gaseous tracers.

Early animal work showed that ¹⁹F-MR image quality achievable with fluorinated gases required either integration of signal from large gas volumes (RINCK et al. 1984), or acquisition times in the order of several hours. Therefore it was found appropriate for imaging steady-state rather than transient gas concentrations (KUETHE et al. 1998, 2000). Even

with older gradient hardware (rise time 24 mT/m, slew rate 80 mT/m ms⁻¹) imaging of, e.g., SF_6 gas within animal lungs during a single breath-hold (acquisition times, 9 s/2D-image or to 49 s/3D-stack) (Fig. 8.11) was feasible. Actual systems (rise time 58 mT/m, slew rate 200 mT/m ms⁻¹) can further shorten acquisition time or improve spatial resolution. Linearity of the relationship between SF_6 concentration and ¹⁹F signal intensity in such lung images has been established (SCHREIBER et al. 2000a). Thus, not only morphological ventilatory SF_6 distribution within porcine lungs during breath-holds but also techniques to describe alveolar wash-in and wash-out kinetics of this gas have been demonstrated (SCHREIBER et al. 2001) (Fig. 8.11). Alternatively, cyclical lung volume measurements are possible using inhaled SF_6 signal strength. Rats with elastase-induced emphysema showed increased lung volumes and abnormally slow exhalation times with this technique (KUETHE et al. 2002).

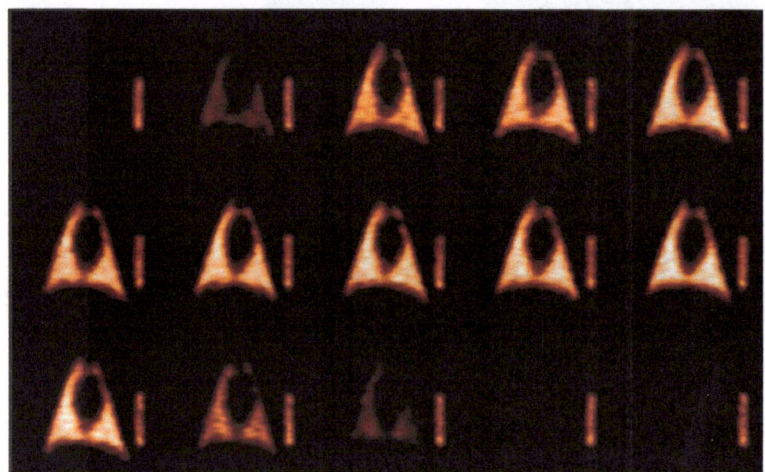

a

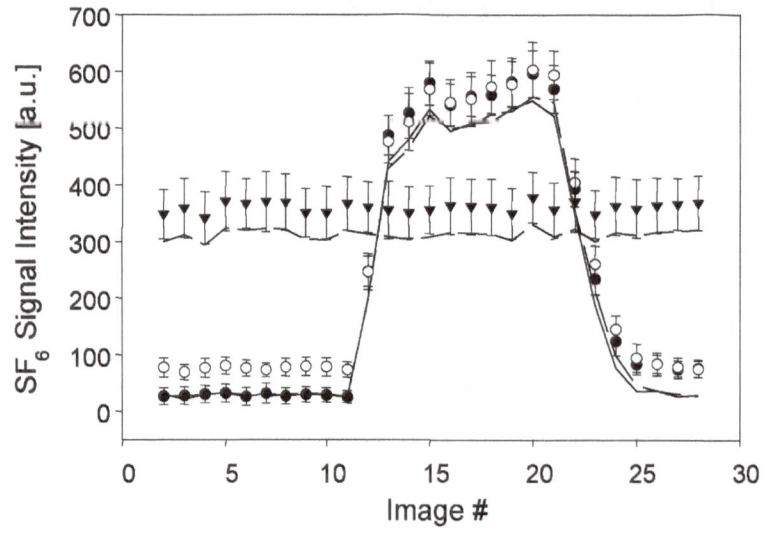

b

Fig. 8.11a. Series of coronal 2D projection images of the SF_6 distribution obtained during wash-in of 100% SF_6 (upper two rows, images 11–23 in graph b), followed by wash-out (lower row, images 24–28 in graph b). Wash-out was performed by breathing 100% O_2. Lateral to the left lung, a phantom containing 100% SF_6 is visible. b Shows the signal time curves in ROIs in the left lung (filled circles), in the right lung (open circles), and in the phantom (filled diamonds). The kinetics of the SF_6 distribution are similar in right and left lung. Reprinted with permission from SCHREIBER et al. (2001).

Oxygen-sensitive MRI using fluorinated gases has, as yet, not made use of the shortening of T1 of ^{19}F-nuclei by O_2. Instead, the effect of insoluble inert gas (SF_6) concentration in the alveoli has been employed, which is due to oxygen uptake into pulmonary capillary blood during breath-holding. At high inspired fractions of O_2 (F_IO_2), inhaled SF_6 gas concentrates preferentially in lung regions with low ventilation-to-perfusion ratios ($\dot{V}_A/\dot{Q}$), and ^{19}F-MR signal increases in contrast to lung regions with high $\dot{V}_A/\dot{Q}$. If a reference image is acquired at low F_IO_2 and high F_ISF_6 (e.g. 30% O_2, 70% SF_6), the concentration effect due to oxygen uptake is much less relevant. From the quotient of images taken at different F_IO_2 concentrations, a lung map of $\dot{V}_A/\dot{Q}$ distribution can be generated. With this technique, low $\dot{V}_A/\dot{Q}$ areas produced by partial bronchial obstruction have been demonstrated in rats (KUETHE et al. 2000) (Fig. 8.12).

8.5.2
Oxygen-Sensitive Perfluorocarbon Imaging

Perfluorocarbons (PFC) are biologically inert, organic fluid compounds whose hydrogen atoms have been replaced by halogens (mostly fluoride). They have the ability to take up and transport molecular oxygen as well as carbon dioxide in physical solution. The compounds are studied as CT and as MRI contrast agents (LOWE 1999), and are investigated in emulsified preparation for use as heme-free oxygen-carrying blood substitutes at the level of phase III trials. Specific advantages are a viscosity similar to water and lower than blood, which is of interest for applications during coronary interventions, the lack of untoward effects of heme-containing solutions (e.g., hypertension), safety from biological hazards and long-term storability.

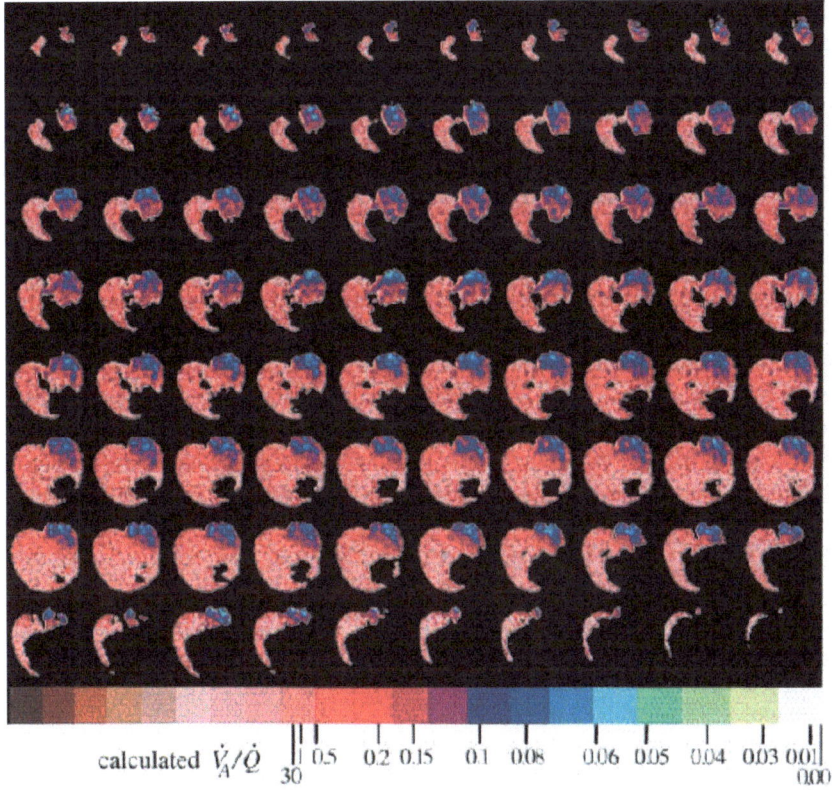

Fig. 8.12. This image of obstructed ventilation is a quotient of 2 MR images of SF_6 in a rat's lungs. The left main stem bronchus has been partially obstructed by a glass bead. An obstruction-detecting image (requiring 1.5 h of data collection while the animal breathed 25% SF_6/75% O_2) was divided by a reference image (from 1 h of data collection while rat breathed 80% SF_6/20% O_2). *From top left to bottom right,* successive planes (72×74 pixels each) show the lungs in cranio-caudal sequence. Heart, mediastinum, and in the most caudal planes, diaphragm and liver appear dark. The right lung appears pink, and the left lung purple and blue. Pixel values range from 0 (*black*) to 1 (*white*) and are divided into 21 color bands. The width of the color bands is 1 SD for individual pixels of that color. This measurement error is greater for higher pixel values; therefore, right color bands are wider than left ones. The $\dot{V}_A/\dot{Q}$ scale relates colors to $\dot{V}_A/\dot{Q}$ values. Like many laboratory rats, this one has a small lobe of right lung in the posterior left thorax, which accounts for the pink lung portion in the left hemithorax. Reprinted with permission from KUETHE et al. (2000)

In pure or emulsified form, use of a second generation PFC (perflubron) during so-called partial or total liquid ventilation has also been under evaluation recently at a phase III level. Rationales for this latter use were that 1) PFC has low surface tension (LOWE 1997); thus, when introduced into the surfactant-depleted and partially or entirely atelectatic alveolar space of ARDS lungs, it will help during positive-pressure ventilation to re-expand the alveolar surface area more easily towards conditions of functional residual capacity of the lung. 2) Liquid ventilation with PFC transfers oxygen from the gas phase to pulmonary capillary blood and eliminates carbon dioxide from it. 3) In addition, PFC is thought to exert anti-inflammatory effects on lung parenchyma (CROCE et al. 1998). Whereas animal studies have been promising, clinical trials of Perflubron have failed so far to improve the outcome over that of control groups in phase III trials of pediatric patients and an adult population with acute lung injury (review in (KAISERS et al. 2003).

With the use of appropriate MR equipment, the visualization of PFC emulsions in the vascular system, highly perfused organs (e.g., heart, lungs, liver, spleen), the reticular-endothelial system, as well as of PFC instilled into the lungs of animals has been demonstrated (JOSEPH et al. 1985; THOMAS et al. 1986; LAUKEMPER-OSTENDORF et al. 2002; HEUSSEL et al. 2003). Double-resonant $^{19}F/^{1}H$ radiofrequency coils even allow a correlation of anatomic representation from proton images with compartment-specific fluorine distribution (SAMARATUNGA et al. 1994).

Also in ^{19}F-MRI, oxygen reduces the spin-lattice relaxation rate 1/T1 of fluorine contained in PFC by its paramagnetism according to the proportional relationship between PO2 and 1/T1. A measurement of the T1 time again allows in vivo monitoring and mapping of PO2 gradients within the fluorine-tagged physiological compartment (FISHMAN et al. 1987; PRATT et al. 1997).

Using this technique of T1 monitoring and mapping, oxygen-sensitive ^{19}F-MR images of the circulatory system of small and large animals infused with emulsified perfluorochemical blood substitutes have been obtained. Marked organ-specific PO2 increases due to the change from ambient air breathing to respiration at increased F_IO_2 were observed in lungs, liver and spleen (FISHMAN et al. 1987; THOMAS et al. 1994). Using the same technique, abnormally reduced PO2 levels were found within implanted mammary adenocarcinomas after intravenous PFC infusion, indicating that tumor blood flow did not match its oxygen consumption (FISHMAN et al. 1989).

Accordingly, in PFC-filled lungs of living animals, enhancement of MR image signals of up to 90% has been demonstrated after transition from ambient air to 100% oxygen breathing (THOMAS et al. 1986). Oxygen-enhanced ^{19}F-MRI has also been attempted using inhalation of aerosolized PFC compounds (THOMAS et al. 1997). During partial liquid ventilation, in vivo determination of regional PO2 distribution within the PFC phase of porcine lungs has been developed (LAUKEMPER-OSTENDORF et al. 2002). In a subsequent study, marked PO2 gradients within the PFC phase have been found, which decrease from non-dependent towards dependent PFC-filled lung regions (HEUSSEL et al. 2003) (Fig. 8.13). This indicates an abnormal pattern of ventilation-perfusion ratios, which would also decrease from non-dependent towards dependent PFC-filled lung regions. Thus, perfusion is redistributed away from the dependent lung regions with arteriovenous shunting. This results in an improvement of the $\dot{V}_A/\dot{Q}$ ratio in both dependent and non-dependent lung regions.

In summary, availability and handling of both fluorinated gases (SF_6) and PFC compounds is technically much less demanding and also less expensive than that of hyperpolarized ^{3}He. On the other hand, ^{19}F-containing tracers provide much lower spin density in comparison to hyperpolarized noble gas MRI, which currently makes ^{19}F-MRI "noisier" and also "slower" due to protracted signal averaging. This limits the use of ^{19}F-MRI in functional imaging. In particular, accurate PO2 quantitation requires current MR hardware, optimized pulse sequences, chemical shift presaturation, appropriate background noise correction, error analysis of the calibration relating PO2 to measured ^{19}F relaxation rate, and ^{19}F signal independence from other constituents of the imaged biocompartment (THOMAS et al. 1994; LAUKEMPER-OSTENDORF et al. 2002; HEUSSEL et al. 2003).

8.6
Hyperpolarized ^{129}Xe-MRI

8.6.1
Morphological Imaging

Xenon is another inert noble gas that can be imaged using MR techniques. Actually, the first published hyperpolarized gas space MR image of the lungs has been produced using inhaled hyperpolarized ^{129}Xe (ALBERT et al. 1994). Xenon has a 30-fold lower self-diffusion coefficient than He, which would actually

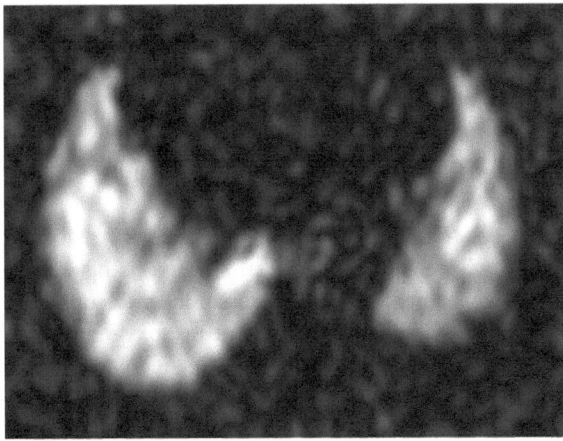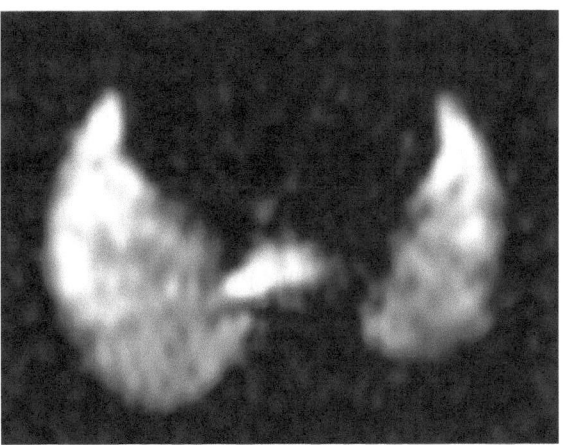

a   b

Fig. 8.13. ^{19}F-MRI of a porcine lung with total liquid ventilation (20 ml·kg^{-1}·bw^{-1}) and ventilation with **a** 40% and **b** 100% oxygen. The homogeneous oxygen distribution in **a** is shifted away from the dependent to the non-dependent lung regions **b** after increase of the oxygen concentration to 100%

favor the heavier noble gas especially for small scale imaging. However, when compared to hyperpolarized ^{3}He-MRI, disadvantages of ^{129}Xe are both its lower gyromagnetic ratio (^{129}Xe, –12 MHz/T *vs.* ^{3}He, –32 MHz/T) and – at least currently – lower polarization levels (^{129}Xe, approximately 30% *vs.* ^{3}He, 60%). These characteristics presently reduce the attainable signal-to-noise ratio of ^{129}Xe-MRI and hence, its spatial and temporal resolution in static and dynamic spin density lung imaging. Static morphological gas-space imaging of lungs and also other gas-filled cavities (e. g., oral, nasal and sinus air spaces) has been shown to be feasible both in animals and in human volunteers, at spatial resolutions to the order of less than 1 cm^{3} (SAKAI et al. 1996; ALBERT et al. 1996; MUGLER et al. 1997). Spatially resolved measurements of the properties of hyperpolarized ^{129}Xe, such as measurement of T2* and the apparent diffusion coefficient, reveal information about pulmonary microstructure. Thus, the apparent diffusion coefficient and T2* times of ^{129}Xe in guinea pigs were 0.068 cm^{2}/s and 40.8 ms in the trachea as well as 0.021 cm^{2}/s and 18.5 ms in the lung parenchyma (CHEN et al. 1999a, CHEN et al. 1999b). Both results reflect the fact that the alveolar space is much smaller than the air-filled trachea. After inhalation, only a small fraction (~2%) of xenon is dissolved into lung parenchyma or blood. Once xenon enters the bloodstream it is distributed throughout the body by the circulation.

8.6.2
Functional Imaging

In the field of functional MR imaging, however, there are several specific and quite prominent advantages of ^{129}Xe, which have the potential to make it a sensitive reporter of its biochemical environment: first, it is more soluble in biological fluids and especially lipid-rich tissues than ^{3}He; second, its resonance spectrum displays a wide-ranging chemical shift upon solution and absorption. ^{129}Xe resonances were observed at 0, 192, 199, and 210 ppm and assigned to xenon in gas, fat, tissue and blood, respectively (SWANSON et al. 1999). The collection of gas-phase and dissolved-phase signals is possible within a single acquisition, and after distribution with the bloodstream, chemical shift imaging can be performed in the lungs, heart, kidneys, and brain. Here, images of the blood resonance show xenon in the lungs and the heart ventricle, whereas images of the tissue resonance reveal xenon in the pulmonary parenchyma and myocardium. Comparing ^{1}H and ^{129}Xe gas images a good correlation was found between the gas-space signal void on the proton images and the gas-space signal on ^{129}Xe images (MUGLER et al. 1997). Combined imaging of the gas-phase and dissolved-phase ^{129}Xe MRI, may lead to simultaneous ventilation-perfusion studies of the lungs. Time-resolved spectroscopy shows that the dynamics of the blood resonance match the dynamics of the gas resonance and demonstrate efficient diffusion of xenon gas to the lung parenchyma and then to the pulmonary blood (SWANSON et al. 1999). Intentional imaging of the dissolved phase is feasible by imaging the gas phase followed by a series of a radiofrequency pulses that selectively destroy the longitudinal magnetization of xenon dissolved in the lung parenchyma. During the delay time between the consecutive radiofrequency pulses, the depolarized xenon rapidly exchanges with the gas phase, thus lowering the gas polarization. The resulting contrast in the ^{129}Xe gas image provides a measure of the absorption of xenon into the tissue and about local tissue density (RUPPERT et al. 2000) (Fig. 8.14). Due to the high temporal resolution, the time course of xenon uptake representing gas exchange can be calculated. Time constants of 61 ms were measured for tissue saturation with xenon and 70 ms for red blood cell saturation. Future developments may provide additional characterizing information about alveolar surface area or lung perfusion.

Third, the T1 of hyperpolarized [129]Xenon's gaseous and dissolved phases are, at about 10–50 s, long enough to trace its spins at least to well-perfused organs like heart and brain (ALBERT et al. 1995; SAKAI et al. 1996). This has already enabled researchers to apply chemical shift-sensitive sequences and to produce dissolved-phase images of the distribution of hyperpolarized [129]Xe within brain, chest and abdomen of animals (SWANSON et al. 1997). An alternate route to introduce hyperpolarized [129]Xe into the macrocirculation and microcirculation for these purposes has also been developed by injecting it dissolved in saline, emulsified lipids or perfluorocarbons (GOODSON et al. 1997; MÖLLER et al. 1998).

A fourth characteristic of [129]Xe useful in this respect is that the presence of oxygen significantly affects its longitudinal relaxation kinetics T1 both in the gas and fluid phases (JAMESON et al. 1988).

8.6.3
Oxygen Sensitivity in Blood and Tissues

The relationship between blood oxygen content and [129]Xe relaxivity has been studied using various approaches, which have led to sometimes contradictory results. In an in vitro set-up, oxygenation of human blood foamed with [129]Xe decreased T1 of [129]Xe from more than 40 s to about 21 s (TSENG et al. 1997). Other researchers, however, have observed shortening of T1 in venous compared to arterial blood and attributed this to the increased content of deoxyhemoglobin, and xenon's interaction with its paramagnetic heme centers (SAKAI et al. 1996; WOLBER et al. 1999). Subsequent T1 measurements in whole blood also found that oxygenation markedly increases T1 of [129]Xe from about 4 s to 13 s, and that it shifts the resonance of erythrocytes to a higher frequency (ALBERT et al. 1999, 2000). Since in quantitative terms, relaxation rate decreased with increasing hemoglobin oxygen saturation in a non-linear fash-

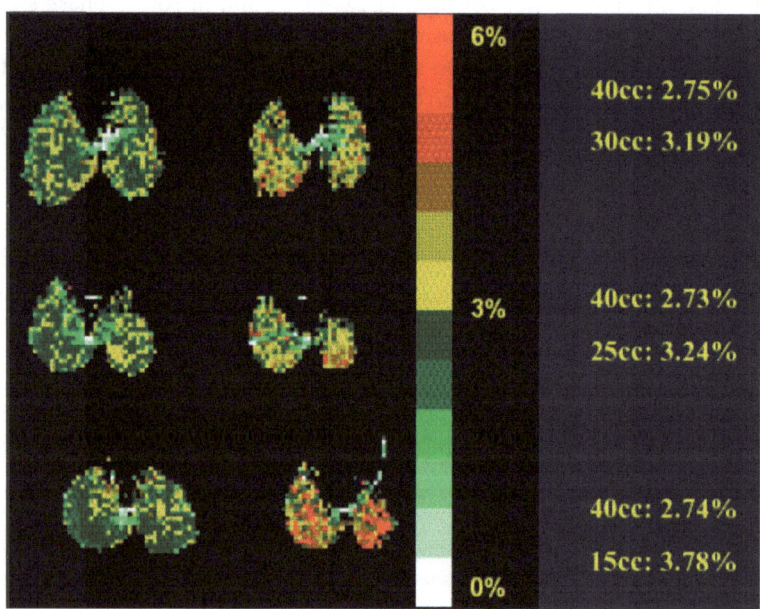

Fig. 8.14. Gas volume-dependent xenon polarization transfer contrast (XTC) MRI. Selective imaging of the dissolved phase is feasible by gas-phase imaging followed by selective inversion of [129]Xe dissolved-phase, i.e., parenchymal, magnetization. During the delay time between the consecutive radiofrequency pulses, depolarized [129]Xe rapidly exchanges with the gas phase and reduces its polarization. The resulting contrast in the [129]Xe gas image provides a measure of [129]Xe absorption into tissue, and also about local tissue density. The latter depends, e.g., on the degree of lung inflation and on gravity-induced compression, but is physiologically also influenced by oxygen uptake in low $\dot{V}_A/\dot{Q}$ areas (absorption atelectasis). A lower gas volume contained within the rabbit lungs shown (volume in cc; left-sided images: higher inflation volumes) increases the lung tissue density, and hence, depolarization levels (given as color-coded map, scale and as mean in %). Courtesy of Kai Ruppert, PhD (RUPPERT et al. 2000)

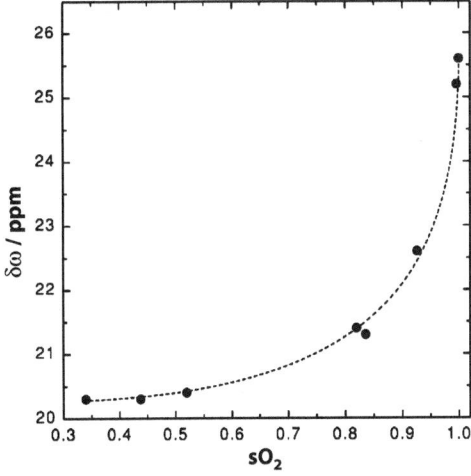

Fig. 8.15. In contrast to the intracellular ^{129}Xe NMR signal, the extracellular ^{129}Xe resonance is insensitive to blood oxygenation. The separation of the two resonance peaks ($\delta\omega$) is plotted as a function of oxyhemoglobin concentration (sO$_2$). The dependence of $\delta\omega$ on sO$_2$ is non-linear, and is more pronounced for higher oxygenation levels. Reprinted with permission from WOLBER et al. (2000)

ion (Fig. 8.15), the oxygenation-dependence of ^{129}Xe T1 may be due to conformational changes within hemoglobin during oxygen binding (WOLBER et al. 2000). At this stage of experimental research, it remains undisputed that blood oxygenation affects T1 of hyperpolarized ^{129}Xe to a measurable extent, and that this effect will quite certainly be exploited in the future for both imaging vascular beds and mapping their oxygenation status.

Finally, at the tissue level this potential of ^{129}Xe is also being explored. T1 determinations of ^{129}Xe in homogenates of rat brain, kidney, liver, and lung at varying oxygenation levels found values ranging from approximately 4 s in deoxygenated lung homogenates to 22 s in deoxygenated brain homogenates. Such relaxation times are long enough to allow accumulation and subsequent MRI of – previously inhaled or injected – ^{129}Xe dissolved in tissues, with a signal-to-noise ratio of approximately 3% of proton SNR. T1 of tissue-phase ^{129}Xe also depends on oxygenation levels of its microenvironment, and on the amount of hemoglobin in the tissue homogenate (WILSON et al. 1999).

In summary, functional MRI of hyperpolarized ^{129}Xe is still less far developed en route to clinical applicability when compared to other MRI techniques. However, easy availability at low cost, various non-invasive access routes to internal organs, and the specific sensitivity of hyperpolarized ^{129}Xe towards internal physiological environments, promise great potential for functional, oxygen-sensitive ^{129}Xe-MRI in the future.

References

Albert MS, Cates GD, Driehuys B et al. (1994) Biological magnetic resonance imaging using laser-polarized 129Xe. Nature 370:199–201

Albert MS, Schepkin VD, Budinger TF (1995) Measurement of 129Xe T1 in blood to explore the feasibility of hyperpolarized 129Xe MRI. J Comput Assist Tomogr 19:975–978

Albert MS, Tseng CH, Williamson D et al. (1996) Hyperpolarized 129Xe MR imaging of the oral cavity. J Magn Reson B 111:204–207

Albert MS, Kacher DF, Balamore D et al. (1999) T(1) of (129)Xe in blood and the role of oxygenation. J Magn Reson 140: 264–273

Albert MS, Balamore D, Kacher DF et al. (2000) Hyperpolarized (129)Xe T (1) in oxygenated and deoxygenated blood. NMR Biomed 13:407–414

Baumgardner JE, Choi IC, Vonk-Noordegraaf A et al. (2000) Sequential V(A)/Q distributions in the normal rabbit by micropore membrane inlet mass spectrometry. J Appl Physiol 89:1699–1708

Chen Q, Jakob PM, Griswold MA et al. (1998) Oxygen enhanced MR ventilation imaging of the lung. Magma 7:153–161

Chen Q, Levin DL, Kim D et al. (1999) Pulmonary disorder: ventilation-perfusion MR imaging with animal models. Radiology 213:871–879

Chen XJ, Moller HE, Chawla MS et al. (1999a) Spatially resolved measurements of hyperpolarized gas properties in the lung in vivo, part I: diffusion coefficient. Magn Reson Med 42:721–728

Chen XJ, Moller HE, Chawla MS et al. (1999b) Spatially resolved measurements of hyperpolarized gas properties in the lung in vivo, part II: T* (2). Magn Reson Med 42:729–737

Croce MA, Fabian TC, Patton JH Jr et al. (1998) Partial liquid ventilation decreases the inflammatory response in the alveolar environment of trauma patients. J Trauma 45:273–280

Deninger A, Eberle B, Ebert M et al. (1999) Quantitation of regional intrapulmonary oxygen partial pressure evaluation during apnoe by 3He-MRI. J Magn Reson 141: 207–216

Deninger A, Eberle B, Ebert M et al. (2000) 3He-MRI-based measurements of intrapulmonary pO2 and its time course during apnea in healthy volunteers: first results, reproducibility and technical limitations. NMR Biomed 13:194–201

Deninger A, Eberle B, Bermuth J et al. (2002) Assessment of a single-acquisition imaging sequence for oxygen-sensitive 3He MRI. Magn Reson Med 47:105–114

Eberle B, Weiler N, Markstaller K et al. (1997) Analysis of regional intrapulmonary O$_2$-concentrations by MR imaging of inhaled hyperpolarized helium-3. J Appl Physiol 87: 2043–2052

Eberle B, Markstaller K, Lill J et al. (2000) Oxygen-sensitive 3He magnetic resonance imaging of the lungs in patients after unilateral lung transplantation. Am J Respir Crit Care Med 161:A718

Eberle B, Markstaller K, Stepniak A et al. (2002) 3Helium-MRI-based assessment of regional gas exchange impairment during experimental pulmonary artery occlusion. Anesthesiology 96:A1309

Edelman RR, Hatabu H, Tadamura E et al. (1996) Noninvasive assessment of regional ventilation in the human lung using oxygen-enhanced magnetic resonance imaging. Nat Med 2:1236–1239

Fishman JE, Joseph PM, Floyd TF et al. (1987) Oxygen-sensitive 19F NMR imaging of the vascular system in vivo. Magn Reson Imaging 5:279–285

Fishman JE, Joseph PM, Carvlin MJ et al. (1989) In vivo measurements of vascular oxygen tension in tumors using MRI of a fluorinated blood substitute. Invest Radiol 24:65–71

Goodson BM, Song Y, Taylor RE et al. (1997) In vivo NMR and MRI using injection delivery of laser-polarized 129 Xe. Proc Natl Acad Sci USA 94:14725–14729

Greenspan JS, Wolfson MR, Rubenstein SD, Shaffer TH (1989) Liquid ventilation of preterm baby (letter). Lancet 2:1095

Hatabu H, Tadamura E, Chen Q et al. (2001) Pulmonary ventilation: dynamic MRI with inhalation of molecular oxygen. Eur J Radiol 37:172–178

Heussel CP, Scholz A, Schmittner M et al. (2003) Measurements of alveolar PO2 using ^{19}F-MRI in partial liquid ventilation. Invest Radiol 10 (in press)

Hoffman EA, Olson LE (1998) Characteristics of respiratory system complexity captured via X-ray computed tomography: image acquisition, display, and analysis. In: Hlastala MP, Robertson HT (eds) Complexity in structure and function of the lung. Dekker, New York, pp 325–378

Jameson CJ, Jameson AK, Hwang JK (1988) Nuclear spin relaxation by intermolecular magnetic dipole coupling in the gas phase. ^{129}Xe in oxygen. J Chem Phys 89:4074–4081

Joseph PM, Yuasa Y, Kundel HL et al. (1985) Magnetic resonance imaging of fluorine in rats infused with artificial blood. Invest Radiol 20:504–509

Kaisers U, Kelly KP, Busch T (2003) Liquid ventilation. Br J Anaesth 91:143–151

Kinoshita Y, Kohshi K, Kunugita N et al. (2000) Preservation of tumour oxygen after hyperbaric oxygenation monitored by magnetic resonance imaging. Br J Cancer 82:88–92

Kuethe DO, Caprihan A, Fukushima E et al. (1998) Imaging lungs using inert fluorinated gases. Magn Reson Med 39:85–88

Kuethe DO, Caprihan A, Gach HM et al. (2000) Imaging of obstructed ventilation with NMR using inert fluorinated gases. J Appl Physiol 88:2279–2286

Kuethe DO, Behr VC, Begay S (2002) Volume of rat lungs measured throughout the respiratory cycle using 19F NMR of the inert gas SF6. Magn Reson Med 48:547–549

Laukemper-Ostendorf S, Scholz A, Bürger K et al. (2002) 19F-MRI of perflubron for measurement of oxygen partial pressure in porcine lungs during partial liquid ventilation. Magn Reson Med 47:82–89

Löffler R, Muller CJ, Peller M et al. (2000) Optimization and evaluation of the signal intensity change in multisection oxygen-enhanced MR lung imaging. Magn Reson Med 43:860–866

Lowe KC (1997) Perfluorochemical respiratory gas carriers: applications in medicine and biotechnology. Sci Prog 80:169–193

Lowe KC (1999) Perfluorinated blood substitutes and artificial oxygen carriers. Blood Rev 13:171–184

Lumb AB (2000) Nunn's applied respiratory physiology, 5th edn. Butterworth Heinemann, Oxford

Mai VM, Bankier AA, Prasad PV et al. (2001) MR ventilation-perfusion imaging of human lung using oxygen-enhanced and arterial spin labeling techniques. J Magn Reson Imaging 14:574–579

Mai VM, Liu B, Li W et al. (2002a) Influence of oxygen flow rate on signal and T(1) changes in oxygen-enhanced ventilation imaging. J Magn Reson Imaging 16:37–41

Mai VM, Liu B, Polzin JA et al. (2002b) Ventilation-perfusion ratio of signal intensity in human lung using oxygen-enhanced and arterial spin labeling techniques. Magn Reson Med 48:341–350

Markstaller K, Eberle B, Kauczor HU et al. (2001) Temporal dynamics of lung aeration determined by dynamic CT in a porcine model of ARDS. Br J Anaesth 87:459–468

McAdams HP, Hatabu H, Donnelly LF et al. (2000) Novel techniques for MR imaging of pulmonary air spaces. Magn Reson Imaging Clin North Am 8:205–219

Möller H, Chawla MS, Chen XJ et al. (1998) Vascular 129Xe MR imaging in live rats. Proceedings of the International Society of Magnetic Resonance Medicine, 6th meeting, p 1910

Möller H, Hedlund L, Chen X et al. (2001) Measurements of hyperpolarized gas properties in the lung, part III: 3He T1. Magn Reson Med 45:421–430

Mugler JP III, Driehuys B, Brookeman JR et al. (1997) MR imaging and spectroscopy using hyperpolarized 129-xenon gas. Preliminary human results. Magn Reson Med 37:809–815

Müller CJ, Löffler R, Deimling M et al. (2001) MR lung imaging at 0.2 T with T1-weighted true FISP: native and oxygen-enhanced. J Magn Reson Imaging 14:164–168

Müller CJ, Schwaiblmair M, Scheidler J et al. (2002) Pulmonary diffusing capacity: assessment with oxygen-enhanced lung MR imaging preliminary findings. Radiology 222:499–506

Nakagawa T, Sakuma H, Murashima S et al. (2001) Pulmonary ventilation-perfusion MR imaging in clinical patients. J Magn Reson Imaging 14:419–424

Newbury N, Barton A, Cates G et al. (1993) Gaseous 3He-3He magnetic dipolar spin relaxation. Phys Rev A 48:4411–4420

Ogawa S, Lee TM, Kay AR et al. (1990) Brain magnetic resonance imaging with contrast dependent on blood oxygenation. Proc Natl Acad Sci USA 87:9868–9872

Ohno Y, Chen Q, Hatabu H (2001a) Oxygen-enhanced magnetic resonance ventilation imaging of lung. Eur J Radiol 37:164–171

Ohno Y, Hatabu H, Takenaka D et al. (2001b) Oxygen-enhanced MR ventilation imaging of the lung: preliminary clinical experience in 25 subjects. AJR Am J Roentgenol 177:185–194

Ohno Y, Hatabu H, Takenaka D et al. (2002) Dynamic oxygen-enhanced MRI reflects diffusing capacity of the lung. Magn Reson Med 47:1139–1144

Olsson L, Magnusson P, Deninger A et al. (2002) Intrapulmonary pO2 measured by low field MR imaging of hyperpolarized 3He. Proceedings of the International Society of Magnetic Resonance Medicine 10:2021

Pratt RG, Zheng J, Stewart BK et al. (1997) Application of a 3D volume 19F MR imaging protocol for mapping oxygen tension (pO2) in perfluorocarbons at low field. Magn Reson Med 37:307–313

Quintel M, Meinhardt J, Waschke KF (1998) Partial liquid ventilation. Anaesthesist 47:479–489

Riley RL, Cournand A (1949) "Ideal" alveolar air and the analysis of ventilation-perfusion relationships in the lungs. J Appl Physiol 1:825–847

Rinck PA, Petersen SB, Lauterbur PC (1984) NMR imaging of fluorine-containing substances. 19-Fluorine ventilation and perfusion studies. ROFO Fortschr Geb Rontgenstr Nuklearmed 140:239–243

Ruppert K, Brookeman JR, Hagspiel KD, Mugler JP III (2000)

Probing lung physiology with xenon polarization transfer contrast (XTC). Magn Reson Med 44:349–357

Saam B, Happer W, Middleton H (1995) Nuclear relaxation of 3He in the presence of O2. Phys Rev A 52:862–865

Saam BT, Yablonskiy DA, Kodibagkar VD et al. (2000) MR imaging of diffusion of (3)He gas in healthy and diseased lungs. Magn Reson Med 44:174–179

Sakai K, Bilek AM, Oteiza E et al. (1996) Temporal dynamics of hyperpolarized [129]Xe resonances in living rats. J Magn Reson Ser B 111:300–304

Salerno M, Altes TA, Brookeman JR et al. (2001) Dynamic spiral MRI of pulmonary gas flow using hyperpolarized (3)He: preliminary studies in healthy and diseased lungs. Magn Reson Med 46:667–677

Samaratunga RC, Pratt RG, Zhu Y et al. (1994) Implementation of a modified birdcage resonator for 19F/1H MRI at low fields (0.14 T). Med Phys 21:697–705

Schearer L, Walters G (1965) Nuclear spin-lattice relaxation in the presence of magnetic field gradients. Phys Rev A 139:1398–1402

Schreiber WG, Markstaller K, Weiler N et al. (2000a) 19F-MRT of pulmonary ventilation in the breath-hold technic using SF6 gas. Rofo Fortschr Geb Rontgenstr Neuen Bildgeb Verfahr 172:500–503

Schreiber WG, Weiler N, Kauczor HU et al. (2000b) Ultraschnelle MRT der Lungenventilation mittels hochpolarisiertem Helium-3. RoFo Fortschr Geb Rontgenstr Neuen Bildgeb Verfahr 172:129–133

Schreiber WG, Eberle B, Laukemper-Ostendorf S et al. (2001) Dynamic (19)F-MRI of pulmonary ventilation using sulfur hexafluoride (SF(6)) gas. Magn Reson Med 45:605–613

Stock KW, Chen Q, Morrin M et al. (1999) Oxygen-enhanced magnetic resonance ventilation imaging of the human lung at 0.2 and 1.5 T. J Magn Reson Imaging 9:838–841

Surkau R, Becker J, Ebert M et al. (1997) Realization of a broad band neutron spin filter with compressed polarized 3He gas. Nucl Instr Methods A 384:444–450

Swanson SD, Rosen MS, Agranoff BW et al. (1997) Brain MRI with laser-polarized [129]Xe. Magn Reson Med 38:695–698

Swanson SD, Rosen MS, Coulter KP et al. (1999) Distribution and dynamics of laser-polarized [129]Xe magnetization in vivo. Magn Reson Med 42:1137–1145

Thomas SR, Clark LC, Ackerman JL et al. (1986) MR imaging of the lung using liquid perfluorocarbons. J Comput Assist Tomogr 10:1–9

Thomas SR, Millard RW, Pratt RG et al. (1994) Quantitative pO2 imaging in vivo with perfluorocarbon F-19 NMR: tracking oxygen from the airway through the blood to organ tissues. Artif Cells Blood Substit Immobil Biotechnol 22:1029–1042

Thomas SR, Gradon L, Pratsinis SE et al. (1997) Perfluorocarbon compound aerosols for delivery to the lung as potential 19F magnetic resonance reporters of regional pulmonary pO2. Invest Radiol 32:29–38

Tseng CH, Peled S, Nascimben L et al. (1997) NMR of laser-polarized 129Xe in blood foam. J Magn Reson 126:79–86

Venegas JG (1998) Noninvasive measurement of local VA, Q, and VA/Q distributions by PET. In: Hlastala M, Robertson HT (eds) Complexity in structure and function of the lung. Lung biology in health and disease. Dekker, New York, pp 483–508

Vidal Melo MF, Harris RS, Layfield D et al. (2002) Changes in regional ventilation after autologous blood clot pulmonary embolism. Anesthesiology 97:671–681

Wagner PD (1998) Ventilation, pulmonary blood flow, and ventilation-perfusion relationships. In: Fishman AP (ed) Pulmonary diseases and disorders, 3rd edn. McGraw-Hill, New York, pp 177–192

Wagner PD, Saltzman HA, West JB (1974) Measurement of continuous distribution of ventilation-perfusion ratios: theory. J Appl Physiol 36:507–514

Weibel ER (1970) Anatomical distribution of air channels, blood vessels, and tissue in the lung. In: Arcangeli P (ed) Normal values for respiratory function in man. Panminerva Medica, Milan

West JB, Dollery CT (1960) Distribution of blood flow and ventilation-perfusion ratio in the lung, measured with radioactive CO2. J Appl Physiol 15:405–410

Wilson GJ, Santyr GE, Anderson ME et al. (1999) Longitudinal relaxation times of 129Xe in rat tissue homogenates at 9.4 T. Magn Reson Med 41:933–938

Wolber J, Cherubini A, Dzik-Jurasz AS et al. (1999) Spin-lattice relaxation of laser-polarized xenon in human blood. Proc Natl Acad Sci USA 96:3664–3669

Wolber J, Cherubini A, Leach MO et al. (2000) Hyperpolarized 129Xe NMR as a probe for blood oxygenation. Magn Reson Med 43:491–496

9 Pulmonary Perfusion

U. Joseph Schoepf, Joachim Ernst Wildberger, Matthias Niethammer, Peter Herzog, Stefan Schaller, Hidemasa Uematsu, Hiroto Hatabu, Christian Fink, Hans-Ulrich Kauczor

9.1 CT Perfusion Imaging of the Lung in Pulmonary Embolism

U. Joseph Schoepf, Joachim Ernst Wildberger, Matthias Niethammer, Peter Herzog, Stefan Schaller

CONTENTS

U. J. Schoepf, MD
Department of Radiology, Brigham and Women's Hospital, Harvard Medical School, 75 Francis Street, Boston, MA 02115, USA
J. E. Wildberger, MD
Department of Diagnostic Radiology, University Hospital, RWTH Aachen, Pauwelsstrasse 30, 52074 Aachen, Germany
M. U. Niethammer, PhD
Siemens Medical Solutions, Computed Tomography, Siemensstrasse 1, 91301 Forchheim, Germany
P. Herzog, MD
Department of Radiology, University Hospital Grosshadern, University Munich, Marchioninistrasse 15, 81377 Munich, Germany
S. Schaller, PhD
Siemens Medical Solutions, CT Concepts, Siemensstrasse 1, 91301 Forchheim, Germany

9.1.1 Introduction

Recent years have seen an increasing importance of computed tomography (CT) in the diagnosis of pulmonary embolism (PE), mainly brought about by the advent of fast CT image acquisition techniques (Kauczor et al. 1999; Remy-Jardin and Remy 1999; Schoepf et al. 2000a, b). Competing imaging modalities are in decline: nuclear scanning allows functional assessment of lung ventilation and perfusion but lacks spatial resolution (PIOPED-Investigators 1990). Once the first line of defense in the diagnostic algorithm of PE, this modality is currently withdrawing to diagnostic niches due to limited availability, poor inter-observer correlation (Blachere et al. 2000), and notorious lack of specificity (PIOPED-Investigators 1990). Pulmonary angiography, the one-time gold standard for the diagnosis of PE, is becoming increasingly tarnished (Diffin et al. 1998; Stein et al. 1999). Its ability to detect isolated peripheral emboli does not seem to exceed the accuracy of computed tomography (Diffin et al. 1998; Stein et al. 1999). Magnetic resonance imaging may be a promising tool for the diagnosis of PE (Meaney et al. 1997; Roberts et al. 1999; Oudkerk et al. 2002) in the future and allows for functional analyses (Roberts et al. 1999). To date, however, magnetic resonance imaging has not found widespread use in emergency medicine mainly due to its long examination times and difficulties in patient monitoring. In contrast, CT has become established as a widely available, safe, cost-effective (van Erkel et al. 1996), and accurate modality for the quick and comprehensive diagnosis of the pulmonary circulation and the deep venous system. The evident advantages of CT for the diagnosis of PE have become further enhanced by the introduction of multislice CT technology (Klingenbeck-Regn et al. 1999; McCollough and Zink 1999; Schoepf et al. 1999; Hu et al. 2000). It is now feasible to acquire scans with sub-millimeter resolution of the entire thorax within one breath-hold. Perceived limitations of CT for an accurate morphological, structural diagnosis of central and peripheral emboli are thus overcome.

However, to date CT has not permitted the functional evaluation of pulmonary microcirculation during pulmonary embolism. Yet, the choice of the adequate therapeutic regimen critically hinges on an accurate evaluation of the functional effect of the embolic event on lung perfusion. If large percentages of the lung parenchyma are affected by embolic occlusion, imminent right heart failure (WINTERSPERGER et al. 1999) warrants a more aggressive regimen, such as thrombolysis, that carries a small but definite risk (GOLDHABER 1997; KONSTANTINIDES et al. 1997). Thus the quantitative assessment of the effect of PE on tissue perfusion may bear more important information for patient management than the direct visualization of emboli by CT angiography alone.

It has been shown that with the advent of fast CT scanning techniques functional parameters of lung perfusion can be non-invasively assessed by means of CT imaging (HOFFMAN et al. 1995a; HOFFMAN and McLENNAN 1997; GROELL et al. 1999; SCHOEPF et al. 2000a). In the following we would like to discuss different experimental approaches for visualization and quantification of pulmonary perfusion, based on various CT techniques. We anticipate these methods to evolve in a valuable adjunct to CT pulmonary angiography by providing both structural and functional information using the same modality. The well-established accuracy of CT for the depiction of emboli and thoracic anatomy is thus supplemented by an effective means to quantitatively assess the functional effect of the embolic event on lung perfusion. This way, a comprehensive diagnosis is feasible within few minutes, without having to subject a patient to multiple expensive and time-consuming tests requiring transportation and advanced logistics.

9.1.2
Electron Beam CT

9.1.2.1
Functional EBCT Scan Protocol

A unique feature of Electron Beam CT (EBCT) is that it can be used both for volume scanning for the depiction of structure (STANFORD et al. 1992) and for functional analyses by acquiring high temporal resolution data sets simultaneously on multiple sections of an organ. EBCT has successfully been used for perfusion measurements

in the heart (RUMBERGER et al. 1987; WOLFKIEL et al. 1987; BRUNDAGE 1995), the brain (GOBBEL et al. 1991) and the kidneys (LERMAN et al. 1995). The feasibility of pulmonary blood flow measurements with EBCT has been validated in a number of controlled animal studies (WOLFKIEL and RICH 1992; HOFFMAN et al. 1995b). In a recent study (SCHOEPF et al. 2000a) we were able to demonstrate the usefulness of EBCT as a single modality to image both thoracic structure and function in patients with suspected acute PE.

The technical design of the electron beam scanner is described in detail elsewhere (RUMBERGER 1992; McCOLLOUGH and MORIN 1994). In the multislice mode of the scanner, 8 slices in a 7.6-cm volume at 20 consecutive time-points can be acquired without patient table movement to monitor the passage of a contrast material bolus through the lung parenchyma. To improve the quality of the data, scans can be ECG-triggered to the quiet diastolic phase of the heart cycle. For measuring pulmonary perfusion, contrast material is intravenously injected with a flow rate of 10 cc/s for 4 s.

9.1.2.2
Functional EBCT Analysis

For dynamic blood flow evaluation we use an approach that comprises a qualitative analysis by selectively coding lung pixel attenuation in a color-coded cold-to-hot spectrum. This way maps can be generated for visualization of parameters such as peak Hounsfield Unit (HU) change, time to peak or mean transit time of contrast material. In our experience peak HU change is most suitable for identification of flow deficits. Using this parameter a qualitative analysis of lung perfusion can be performed by generating a color-coded map for the eight scan levels that are simultaneously acquired by EBCT. On color-coded maps, flow deficits are defined by predominance of cold-spectrum colors with segmental distribution. Guided by color-coded maps, a quantitative analysis for the assessment of regional pulmonary blood flow can be performed. To this end, time-density curves (TDCs) are generated by manually tracing regions of interest (ROIs) over lung segments showing flow-deficits. A ROI over the main pulmonary artery or the right ventricle can be used as input function. Regional pulmonary blood-flow in each segment can then be assessed according to the indicator dilution theory (THOMPSON et al. 1964; RUMBERGER et al. 1991) using a basic flow equation (WOLFKIEL et al. 1987; WOLFKIEL and RICH 1992) where $PBF/V = Pul / \int C_{RC} \, dt$ (PBF/V = pulmonary blood flow per unit volume of lung tissue, DPul = peak HU change during contrast injection, $\int C_{RC}\, dt$ = area under a variate-fit of a TDC of the right side of the circulation).

9.1.2.3
Advantages and Disadvantages of Functional EBCT Perfusion Imaging of the Lung

Once a suspicion of pulmonary embolism arises, crucial questions need to be answered: is it indeed PE that causes the patient's symptoms, or are there other reasons for the patient's discomfort? If it is PE, where are the emboli located and how extensive is the disease? Does anticoagulation suffice, or are thrombolysis or invasive measures warranted? And are there conditions that prohibit thrombolysis? Where does the disease originate? Contrast-enhanced CT provides high spatial resolution and allows the objective, non-invasive visualization of thoracic anatomy. Sources of chest pain other than PE can be identified. The location of pulmonary emboli and the extent of the disease can be assessed to determine the need for and feasibility

of anticoagulation, thrombolysis or more invasive measures. However, to date CT has not permitted the functional evaluation of pulmonary microcirculation during pulmonary embolism. CT perfusion measurements as described in this chapter may represent a valuable adjunct to CT pulmonary angiography by providing both structural and functional information using the same modality. The well-established accuracy of CT for the depiction of emboli and thoracic anatomy is thus supplemented by an effective means to quantitatively assess the functional effect of thromboemboli on lung perfusion. In a recent study (SCHOEPF et al. 2000a) we were able to show that EBCT can successfully be used for the functional analysis of pulmonary blood flow and therefore allows a differentiation between segments with normal and reduced capillary perfusion. This allows the percentage of lung parenchyma with impaired microcirculation to be estimated. Thus, a decision whether anticoagulation or thrombolysis is warranted is facilitated. During the course of treatment, the effect of therapy may be monitored by comparing pre- and post-therapy blood-flow values by repetitive scanning (Fig. 9.1.1). Up to two-

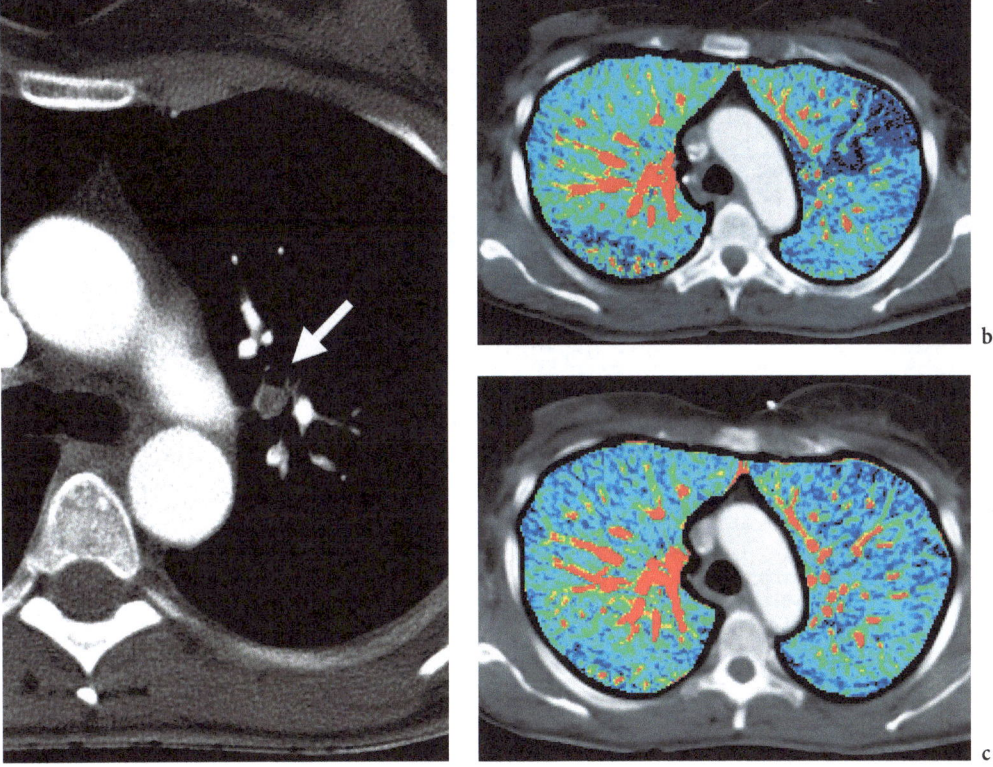

Fig. 9.1.1. a 58-year-old female patient with upper extremity deep vein thrombosis and acute PE after hand surgery. EBCT volume scanning shows thromboembolic material in the apicoposterior segmental artery of the left upper lobe (*arrow*). **b** Dynamic EBCT scanning. Color-coded map at the level of the upper lobes of the lung in the same patient. Cold spectrum colors delineate the perfusion deficit in the apicoposterior segment. **c** Dynamic scanning after heparin therapy at the same level as in **b**. The flow void in the apicoposterior segment has resolved

thirds of patients with initially suspected PE receive other diagnoses including unknown malignancies or life-threatening conditions such as aortic rupture or dissection. By initially performing a contrast-enhanced thin-slice volume study, the presence of pulmonary embolism can be verified and other or additional underlying diseases are readily recognized. The patient with previously unknown small-cell lung cancer in our patient group (Fig. 9.1.2) illustrates the importance of a thorough analysis of thoracic structure. A major limitation is that the 7.6 -cm scan volume of the dynamic study does not cover the entire chest. The additional radiation dose for the dynamic study amounts to an effective dose equivalent of 7.2 mSv, which about equals the radiation exposure usually applied during nuclear scanning (RHODES et al. 1989).

While there is good correlation between volume and functional scans for the detection of segmental emboli, a potential limitation of this technique arises from partial occlusion of vessels with maintained blood flow on a capillary level despite the presence of thrombi. Such findings may be regarded as "false negative" results. However, valuable information can be gained by assessing the actual effect of small emboli on lung microcirculation. Completely or partially maintained perfusion despite the presence of emboli revealed by functional scanning, may influence the decision whether or not to start thrombolytic therapy in a patient, the latter carrying a small but definite risk (GOLDHABER 1997; KONSTANTINIDES et al. 1997). Thus the quantitative assessment of the effect of PE on tissue perfusion may bear more important information for patient management than the direct visualization of emboli by CT angiography alone. Non-occluding emboli are also a well-known problem in lung scintigraphy, where even extensive central emboli frequently go undiagnosed if they are only partially occluding and do not cause localized perfusion defects. Similar

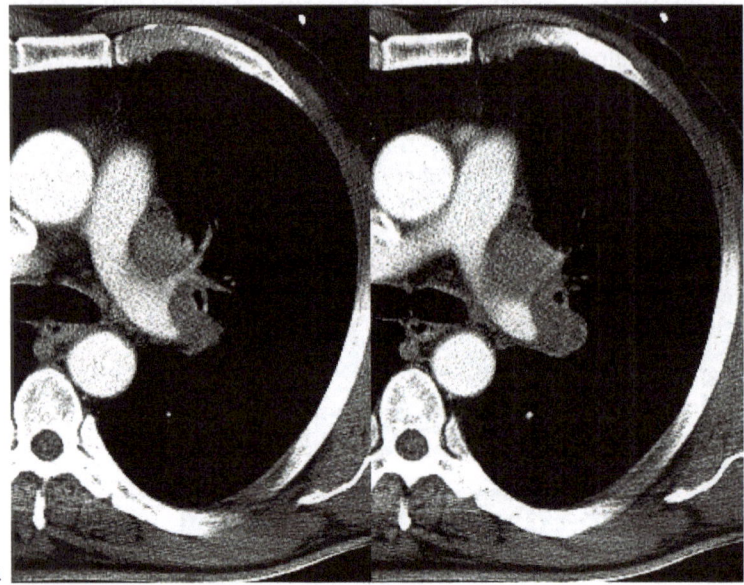

a

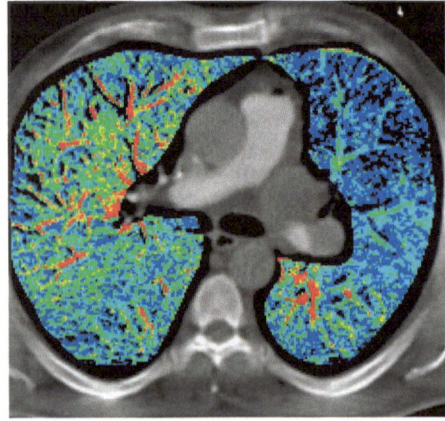

b

Fig. 9.1.2. a A 52-year-old male patient with chest pain and initial suspicion of pulmonary embolism. Volume scanning reveals a mediastinal soft-tissue mass encasing the left upper lobe artery. Biopsy later revealed small-cell lung cancer. b Dynamic EBCT scanning in the same patient. A perfusion deficit is seen in the left upper lobe due to obstruction of the feeding vessel by the soft tissue mass. Blood flow in the left lower lobe and the right lung is maintained. A scintigraphic test had revealed a ventilation-perfusion mismatch in the left upper lobe, compatible with lobar PE

pitfalls can be avoided by the combined use of both CT angiography and CT perfusion measurement for a comprehensive diagnosis.

9.1.2.4
Functional EBCT Perfusion Imaging of the Lung as a Method for Therapy Control

Primary pulmonary hypertension is associated with a worse mortality than many malignant diseases. Other than double lung transplantation there are very few therapeutic options. One therapeutic regimen comprises administration of prostaglandin analogues which can be administered as a continuous IV infusion or by inhalation. Inhalation treatment with Iloprost has certain advantages such as enabling the patient to self-administer the agent in an ambulatory setting.

A limitation of prostaglandin treatment is that there is a fraction of about 30% of patients who do not respond to this particular therapy. Usually, these have to be identified by right heart catheterization before and after administration of the drug, which is an invasive test and thus is associated with occasional complications.

Functional EBCT perfusion measurement of the lung offers a minimal invasive option for monitoring prostaglandin therapy and for identifying non-responders.

A dynamic perfusion scan of the lungs is performed before and after inhalation of Iloprost in the CT-scanner room. Scanning is performed using the ECG-triggered dynamic acquisition mode of the EBCT system as discussed earlier. The scan field with a z-axis extension of 7.6 cm is centered over the hilar region of the lung including the main pulmonary artery. A contrast bolus of 60 cc at a flow rate of 8–10 cc/s should be administered into an antecubital vein. After scan acquisition, ROIs are traced over the lung parenchyma for each z-axis position, excluding the central pulmonary vessels, and time-density curves are calculated. Typically, the area under the time-density curve of the scan prior to administration of the drug is smaller compared to the second scan acquired after Iloprost administration because of residual contrast enhancement from the first scan. Good responders to the drug typically show a higher and earlier peak in the attenuation curve of the second, post-therapeutic scan (Fig. 9.1.3) while non-responders show no change in the curves before and after administration of the drug (Fig. 9.1.4). The higher and earlier peak is the correlate of increased pulmonary blood flow and decreased pulmonary-

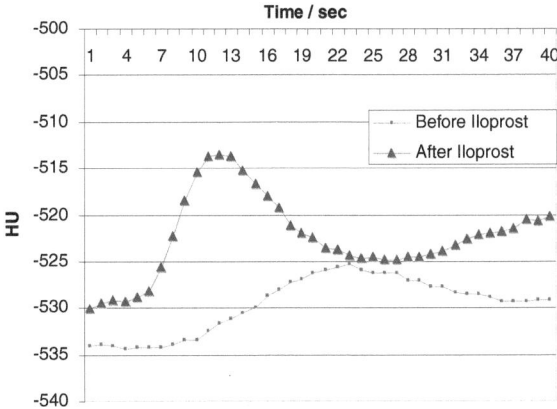

Fig. 9.1.3. Good responder to Iloprost

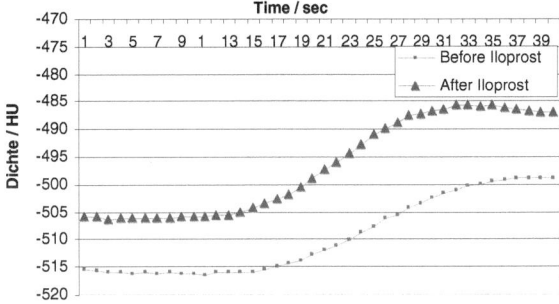

Fig. 9.1.4. Non-responder to Iloprost

arterial pressure. Using the attenuation curve of the main pulmonary artery as an input function, pulmonary blood flow per volume of lung tissue can be calculated using the perfusion models described before.

9.1.3
Multislice-CT

9.1.3.1
Structural and Functional Multislice CT Protocol

The introduction of multislice CT (MSCT) technology is the single most important development in the field of CT imaging after the advent of spiral CT (Kalender 1994). The single detector bank of conventional spiral scanners has been replaced by multiple detector banks (currently 2, 4, 8, or 16) which can be combined during read-out to acquire multiple slices simultaneously. The most prominent feature of this technology is increased speed. Compared with conventional 1 s rotation single-slice CT, the same volume can be covered up to 40 times faster with submillimeter resolution. Another option is to use narrower collimation in order to increase spatial resolution and reduce partial volume averaging. In a recent study (WILDBERGER et al. 2001) a commercially available 4-slice MSCT (SOMATOM Volume Zoom; Siemens, Forchheim, Germany) scanner was used for structural and functional assessment of the lung in pulmonary embolism. MSCT imaging was performed with 120/140 kV and 100 mA, using a rotation time of 0.5 s, a thin collimated spiral scan (4 1 mm) and a table speed of 7 mm/rot. (pitch: 1.75). With this protocol the entire chest can be examined within approximately 21 s. No additional late phase scanning is necessary. Non-ionic contrast media (Ultravist 370, Schering, Berlin, Germany) was applied intravenously using a double power injector (CT 9000 Digital Injection System; Liebel-Flarsheim, Cincinnati, OH). Flow parameters were 3 cc/s, with a total amount of 120 cc, followed by a saline chaser bolus (3 cc/s, 30 cc). Start delay was 27 s in caudo-cranial direction.

Radiological diagnoses regarding PE were established on thin collimated axial slices, slice thickness $_{eff.}$ 1.25 mm, reconstruction increment 0.8 mm, standard soft tissue kernel, in cine mode view on a workstation. Complete or incomplete occlusion of pulmonary arteries as well as direct visualization of emboli due to filling defects were considered as primary signs of PE.

9.1.3.2
Functional MSCT Analysis

The image post-processing algorithm for visualization of parenchymal density distribution is structured into 5 steps: contour definition/segmentation, vessel cutting, adaptive filtering, color-coding and overlay with the original images.

9.1.3.2.1
Contour Definition by Identifying the Lung by Threshold-Based Contour Finding/Segmentation Algorithm

In a first step, a binary mask of the image is derived by identifying lung areas and non-lung areas. A threshold-based contour finder is used for segmentation, as lung tissue (low HU) is generally enclosed by high HU areas (pleura/chest wall). The threshold HU value separated both HU ranges and defined the contour of the lung. A typical value for the segmen-

tation threshold is –300 HU. To ensure maximum robustness, the algorithm is modified for the special conditions within the lung parenchyma: A primary initial starting point is placed by the user in both lungs. From this point, six secondary starting points are calculated automatically; three points above and three points below the primary initial point are determined. The y-coordinate distance between the initial points is five pixels, the x-coordinates are equal to the primary initial point x-coordinate. Seven potential starting points are determined in strictly horizontal direction in each lung, from the initial points to the outer image border. The starting point with the maximum x-distance from its initial point is taken as the contour tracing starting point (Fig. 9.1.5).To trace the lung contour, a standard algorithm has been adapted. The search proceeds counterclockwise, starting from the detected starting point. The algorithm always considers the neighboring three pixels in the search direction and determines the first pixel with a value below the specified threshold as the next contour point. If the first of the three neighboring pixels is detected as a contour point, the search direction changes for –90°. If none of the three pixels fulfills the criterion, the search direction is changed for +90°. In any other case the search direction remains unaltered. The algorithm is allowed to perform U-turns and to walk back in its own tracks. If the number of iterations exceeds a specified value, tracing is aborted. Based on the extracted contour a binary segmentation mask is generated (Fig. 9.1.6).

To exclude pleural walls, seven layers of pixels are removed at the border of segmented lung areas by applying the morphological operation of erosion five times to the binary segmentation mask, using the four connected neighbors as structuring elements (Fig. 9.1.7).

The center positions of each lung are derived from the segmentation mask and used as initial points for the segmentation of the next image. Segmentation proceeds automatically until the whole dataset is processed.

All following steps are performed on the extracted lung areas, exclusively.

9.1.3.2.2
Vessel Cutting

Major vascular structures and airways are removed by HU range selection to prepare images for subsequent filtering. A lower threshold HU_B (B=Bronchi) and an upper threshold HU_V (V=vessel) are specified, pixels below HU_B are identified as airways, pixels

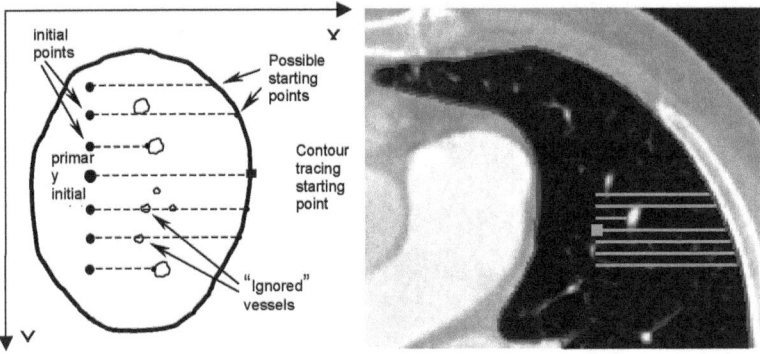

Fig. 9.1.5. Scheme for the detection of contour tracing starting point; clinical example

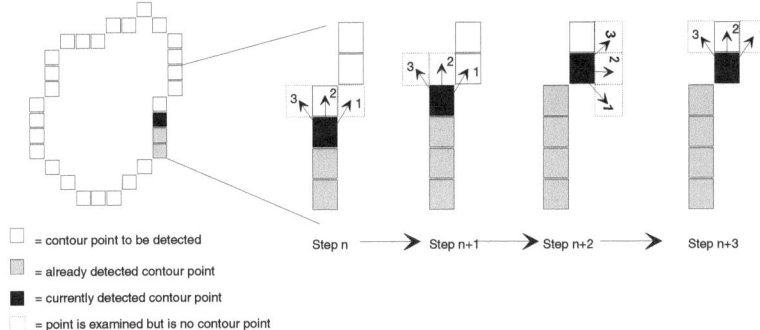

Fig. 9.1.6. Contour tracing algorithm

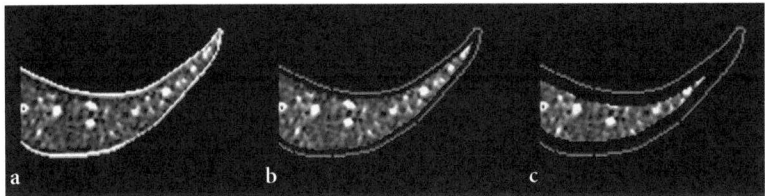

Fig. 9.1.7. Erosion of contour pixels (window: 300 HU, center: –880 HU). **a** No erosion, **b** three layers of pixels eroded, **c** seven layers of pixels eroded

above HU_V as vessels. For optimal image visualization, vessel cutting has to be balanced between cutting all vessels and retaining as many lung pixels as possible. As there is an interpatient and even an intrapatient variability of the optimum HU_V, a percentage-based definition of HU_V has proved to be more general. Therefore, a combination of threshold and percentage based removal scheme is applied. The algorithm removes all pixels below HU_B which is fixed at a value of –990 HU. HU_V is individually adjusted to reach a maximum share of 28% of removed pixels (Fig. 9.1.8), which was found to be the best compromise for achieving diagnostic results in the majority of patients.

9.1.3.2.3
Adaptive Filtering

The segmented dataset is reformatted by linear interpolation to obtain a dataset with isotropic voxel spacing. An adaptive sliding mean value filter is applied, using an isotropic (spherical) 3D kernel with a diameter of 5 mm. In most datasets, this corresponds to seven pixels (Fig. 9.1.9). In this process, up to seven adjacent slices are combined, pixels lost due to vessel cutting are replaced to a specified extent by the average pixel value from their 3D environment. When the filter is applied, the central pixel of each (kernel) volume is replaced by the mean value of all pixels in the volume.

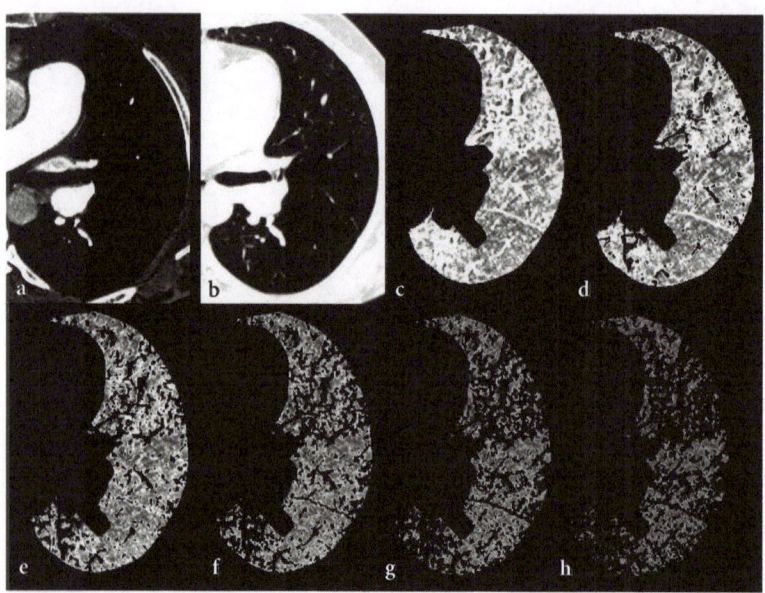

Fig. 9.1.8. Influence of vessel cutting. **a** Original image, soft tissue setting (window: 400 HU, center: 80 HU). **b** Original image, lung tissue setting (window: 1200 HU, center: –600 HU). **c** Settings for vessel cutting **d–h** (window: 300 HU, center: -880 HU). **d** HU_B=–990 HU, HU_V=–655 HU, 8% removed; **e** HU_B=–990 HU, HU_V=–783 HU, 18% removed; **f** HU_B=–990 HU, HU_V=–827 HU, 28% removed; **g** HU_B=–990 HU, HU_V=–849 HU, 38% removed; **h** HU_B=–990 HU, HU_V=–864 HU, 48% removed

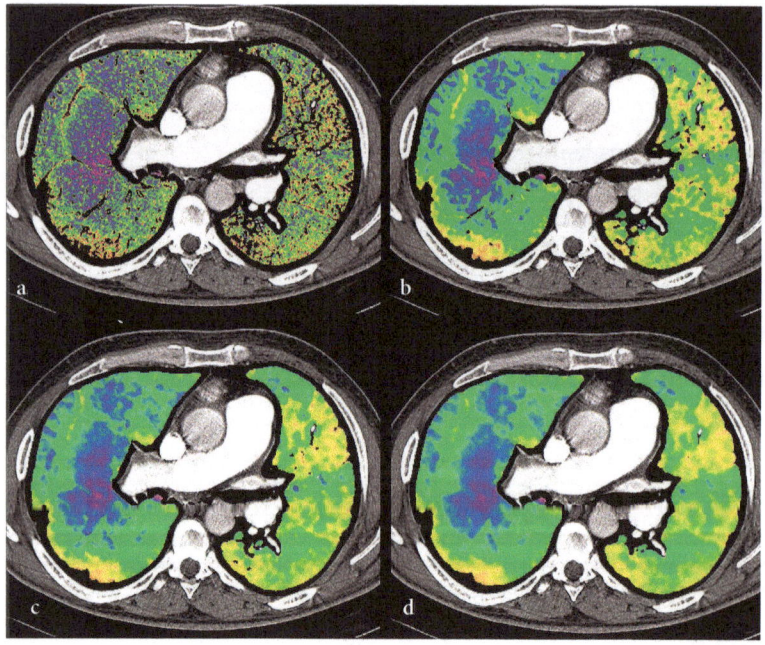

Fig. 9.1.9. Influence of kernel diameter. **a** No filter, **b** 7 pixels, 4.9 mm, **c** 9 pixels 6.4 mm, **d** 11 pixels, 7.8 mm

Pixels removed in preceding operations (segmentation, erosion, vessel cutting) are defined as invalid and do not contribute to the average density values.

The user defines the minimum share of valid pixels in the volume, required for a valid averaging result. The limit is specified as a percentile value; in the following it will be referred to as "inclusion threshold". If the share is below the limit, the central pixel is set as invalid. As in the last processing step, all invalid pixels are replaced by the corresponding pixels of the original image, the inclusion threshold specifies how many vascular structures and airways will appear in the processed parenchymal area. We used a heuristic approach with an inclusion threshold of 20% (Fig. 9.1.10).

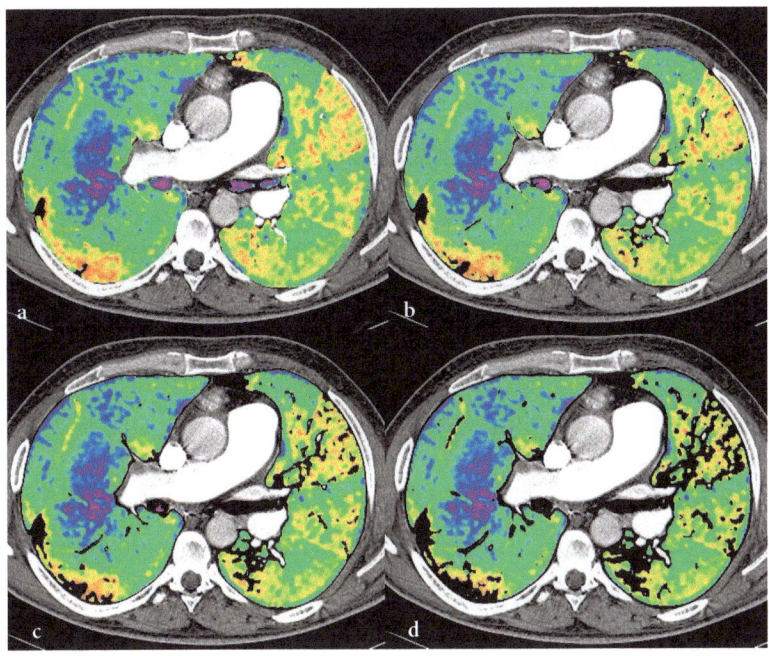

Fig. 9.1.10. Variation of the inclusion threshold: As all pixels that are invalid are set to zero and do not contribute to the averaging density values, the image impression also changes according to the chosen inclusion threshold. **a** 5%, **b** 20%, **c** 35%, **d** 50%

The implementation of the adaptive filtering takes advantage of a fast numerical convolution algorithm. In the following, it is described for a 2D filter, which can be easily generalized to 3D. In the image, all pixels that are invalid or outside the contour are set to zero. The image matrix and its binary mask are convoluted separately. The convolution of the binary mask yields the number of valid pixels in the volume, corresponding to the position of the value in the matrix. The inclusion threshold is implemented by thresholding the matrix and setting all values below the computed limit to zero. To produce the filtered image, the convoluted image is divided by the convoluted mask, the division is carried out element-by-element. If an element of the convoluted mask array is zero, the result is set to invalid.

9.1.3.2.4
Color Coding

To facilitate visualization of parenchymal enhancement, the resulting image is mapped onto a spectral color scale. Mapping is controlled by center and width, analogous to gray scale image mapping. As the subtraction technique from native data is not possible, no baseline is defined. A heuristic approach analyzes the parenchymal histogram and determines the window parameters automatically. The algorithm uses the median of the processed parenchymal pixels as the center value, the width range is fixed to 100 HU.

If requested, additional manual interactive windowing by the user is feasible.

9.1.3.2.5
Image Fusion with the Original Image

The resulting color-encoded parenchymal images are overlaid onto the original CT images, as these CT source images are crucial for spatial orientation in the data set. For overlay, all non parenchymal pixels are replaced by the original pixels of the respective slice position and displayed in the usual CT gray scale presentation. In addition, manual interactive windowing of the gray and the colored parts of the image by the user is possible.

A pictorial overview of the image processing algorithm is given in Fig. 9.1.11. These algorithms are implemented in a MatLab-based development environment (MatLab 5.3; The MathWorks Inc., Natick, MA) on a PC (Pentium III, 600 MHz). Modular software architecture allows easy modifications and tests of new algorithms. The user interface consists of an image processing module and a viewing module for axial, sagittal and coronal display. All relevant parameters are accessible by the user interface (Fig. 9.1.12).

These resulting color-coded images are evaluated for distribution of density values within the whole data set.

9.1.3.3
Clinical Examples

In patients with normal CT scans in axial scanning, color-coded display of lung parenchyma showed quite a homogeneous appearance of density values, displayed

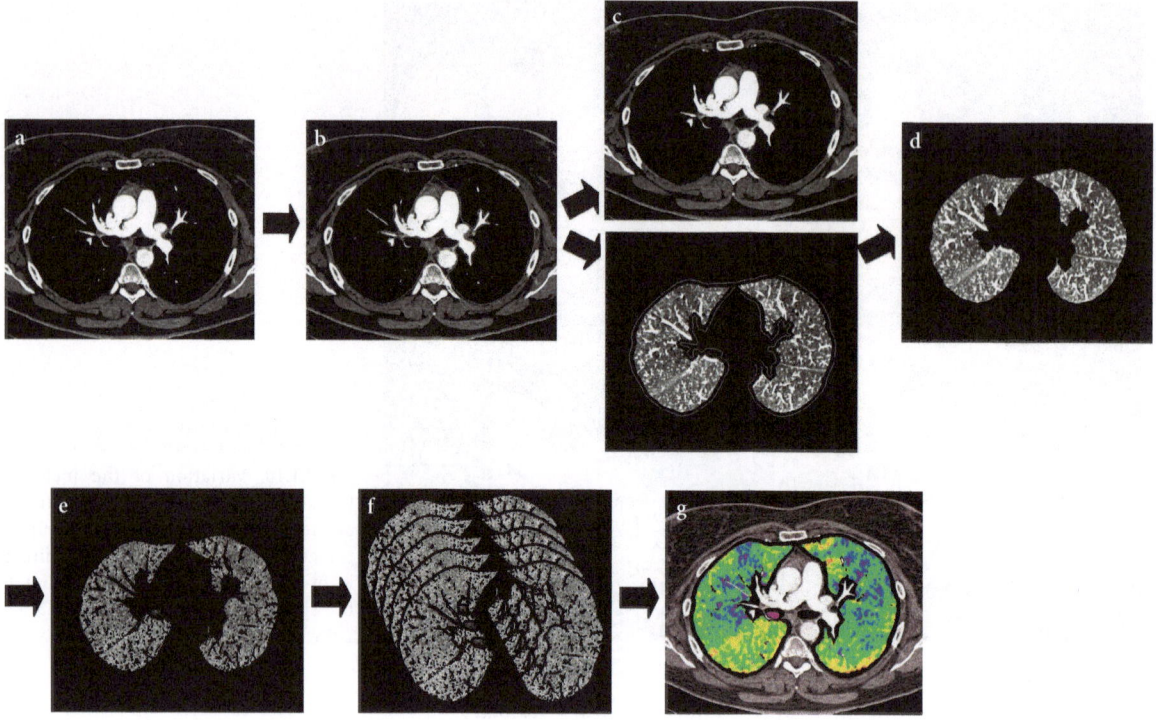

Fig. 9.1.11. Image processing algorithm: **a** original image, **b** contour detection, **c** split-image, lung/non-lung, **d** removal of pleura walls, **e** removal of vessels and bronchi., **f** 3D adaptive filtering on image-stack, **g** color coding and overlay to original image

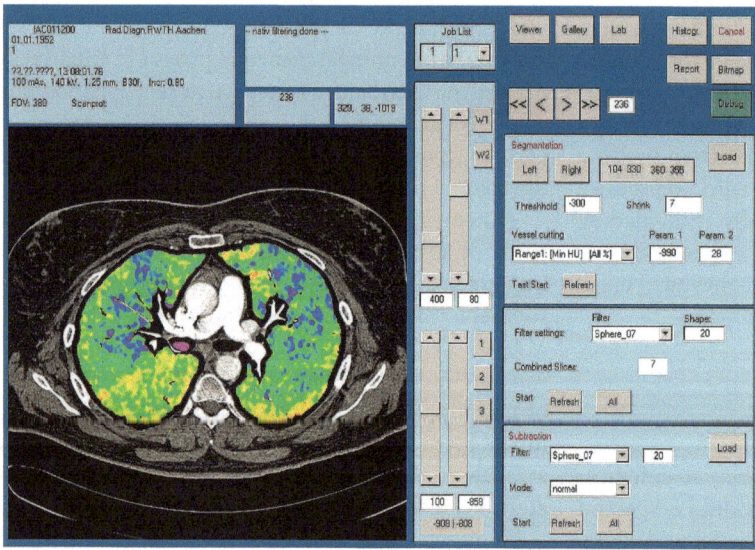

Fig. 9.1.12. User interface of the image-processing module for structural and functional assessment of the lung parenchyma

in bright green and partly yellow colors (Fig. 9.1.13a–c). Anatomical details, such as lung fissures, were sharply delineated and allowed anatomical orientation even on sagittal and coronal images. On axial and sagittal planes, a gravity-dependent gradient was visible in the ventro-dorsal direction in nearly every patient (green-blue encoded in the ventral lung areas).

In patients with acute vessel-occluding PE, filling defects on CTA corresponded to areas of decreased densities, as arterial inflow into the lung parenchyma was impaired significantly in these lung segments. These areas were predominantly displayed in violet and (dark) blue colors. Areas of increased density were displayed in red colors (Fig. 9.1.14a–c). Areas

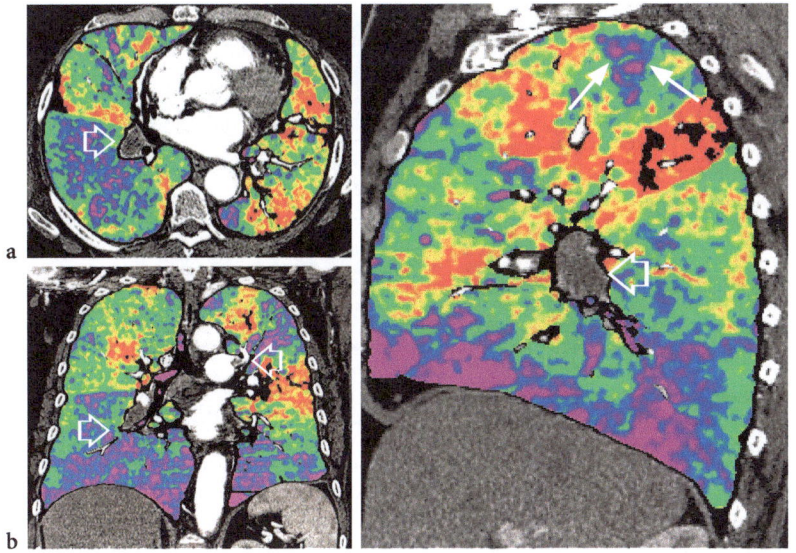

Fig. 9.1.13. a Axial view. Homogeneous distribution of lung densities in green-yellow colors, despite a gravity dependent gradient in ventro-dorsal direction. No circumscribed areas of decreased or increased densities. **b** In coronary view, also a homogenous color coding is displayed. Note streaking artifacts due to contrast media application via the right arm veins. **c** In sagittal view, both major and minor fissure are clearly delineated in the right lung (*arrows*). Again, the gravity-dependent gradient in ventro-dorsal direction is clearly delineated

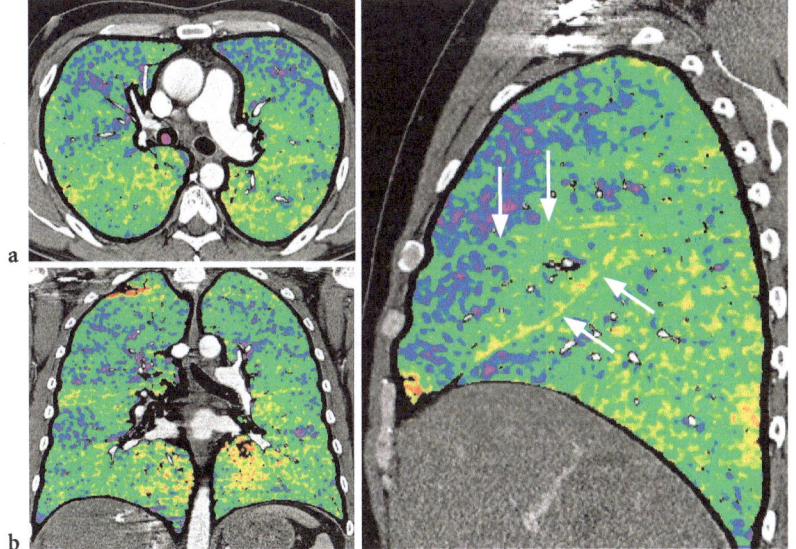

Fig. 9.1.14. a Axial view. Occluding embolus in the right descending artery (*arrow*). The adjacent lung parenchyma shows decreased density distribution. Additionally, also non-occluding wall-adherent emboli are depicted in segmental arteries on the left side. **b** In the coronal view at the level of the left atrium additional (sub)segmental areas with decreased densities are seen in the left upper (apicoposterior segment) and both lower lobes corresponding to the occluded vessels on CTA of the pulmonary arteries. Note the thrombus in the apicoposterior left lung artery (*arrow*). **c** In sagittal view, peripheral deficits are delineated in the right lower lobe due to extensive central thrombus (*open arrow*) as well as in the right upper lobe (apical segment). The latter shows triangular shaped perfusion deficits characteristic of peripheral pulmonary embolism (*solid arrows*)

above the selected threshold will not be color-coded at all.

In addition, also non-occluding emboli can easily be delineated, as the information of the source images regarding the direct visualization of PE is still displayed (Fig. 9.1.15). Also diagnoses on non-embolic extrapulmonary diseases can be made, as gray-scale windowing can be performed interactively by the user.

9.1.4
MSCT Assessment of Lung Perfusion: Future Directions

Recently, a new generation of 16-slice CT scanners was introduced. It has been shown that this technology has finally made routine isotropic volume data acquisition possible. Using 0.5 s rotation time, the table speed in spiral mode can be freely selected between 12 mm/s and 36 mm/s using 0.75 mm collimation, and between 24 mm/s and 72 mm/s using 1.5 mm collimation. As an example of a clinical protocol, a complete thorax examination over a scan range of, e.g., 30 cm can be completed within 8 s, even when using 0.75 mm collimation. This tremendous increase in scan speed allows multiphasic data acquisition from which perfusion data can be derived. Figure 9.1.16 illustrates the timing employed for a biphasic scan of the thorax, where the first scan is non-contrast enhanced and serves as a baseline image. A second, contrasted scan of the same range can be obtained during the same breath-hold. Using suitable volume matching algorithms, it will now be possible to obtain a subtraction image, providing information on the perfusion level of the lung parenchyma, which may be able to serve as an indicator for the hemodynamic effects of chronic or acute pulmonary embolism.

Considering the dramatic increase in performance CT has experienced during the last 5 years, it is questionable whether innovation will keep progressing at this pace. A number of new developments have recently been discussed as potential future pathways.

It remains questionable whether the developments towards more and more slices will continue for much longer. With 16 slices we can today achieve scan speeds of 36 mm/s and 72 mm/s. With a higher number of slices users would be able to cover a thorax in e.g. 2 s, rather than in 8 s, but the clinical benefit is doubtful. Several groups have started investigating flat panel detector CT technology. At the current time, however, available flat panel detector technology suffers from a limited dynamic range, generally not allowing

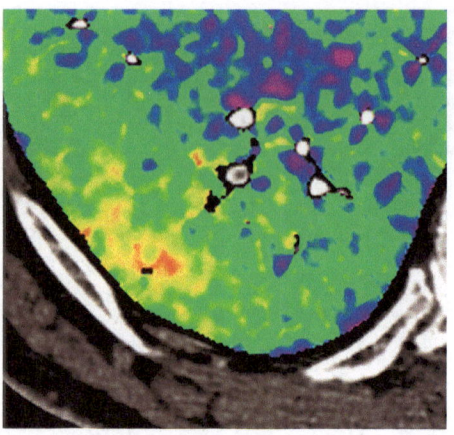

Fig. 9.1.15. Right lower lobe of the lung of a patient with PE. Note a small, non-occluding embolus in the right lower lobe artery on subsegmental level (*arrow*, soft tissue window settings; width: 400 HU; center: 80 HU). No perfusion defect is visualized on color-coded attenuation map since the embolus is non-occluding and near normal perfusion maintained despite the presence of the embolus

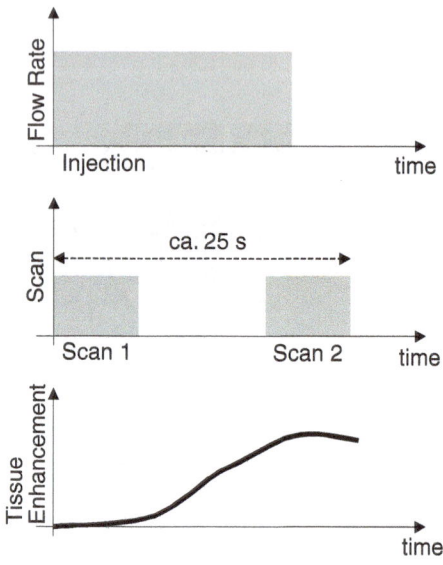

Fig. 9.1.16. Injection and scan timing for a biphasic thorax examination using a fast 16-slice CT scanner

uncompromised CT imaging capabilities. Also, with frame rates being limited to around 50 frames/s, a sufficient number of projections cannot be acquired during subsecond rotations. These problems might be overcome as technology progresses, but the application spectrum will be limited. The major field that could benefit from flat panel detector CT technology is those clinical applications that require following of an entire volume, e.g. an organ, over a certain period of time. Examples could be dynamic perfusion studies, i.e. in kidneys, liver, brain or lungs. This would allow tumor characterization or classification of hemodynamic effects from vascular disease. Another interesting example regarding lung scanning is ventilation assessment. Likewise, this 4D imaging could have a value in interventional applications. One of the most serious problems hindering this kind of application to be routinely useful will certainly be excessive patient dose. If dose values from today's protocols are extrapolated to 4D imaging of entire organs using flat panel detector CT technology, one easily arrives at effective patient dose values in the order of 50 mSv. Hence, for this type of application to be successful, it is essential to find technology and image processing algorithms to reduce doses to the lowest values possible.

References

Blachere H, Latrabe V, Montaudon M et al. (2000) Pulmonary embolism revealed on helical CT angiography: comparison with ventilation-perfusion radionuclide lung scanning. AJR Am J Roentgenol 174:1041–1047

Brundage B (1995) Beyond perfusion with ultrafast computed tomography. Am J Cardiol 75:D69–D73

Diffin D, Leyendecker JR, Johnson SP, Zucker RJ, Grebe PJ (1998) Effect of anatomic distribution of pulmonary emboli on interobserver agreement in the interpretation of pulmonary angiography. AJR Am J Roentgenol 171:1085–1089

Gobbel G, Cann CE, Iwamoto HS, Fike JR (1991) Measurement of regional cerebral blood flow in the dog using ultrafast computed tomography, experimental validation. Stroke 22: 772–779

Goldhaber S (1997) Pulmonary embolism thrombolysis. Broadening the paradigm for its application. Circulation 96:716–718

Groell R, Peichel KH, Uggowitzer MM, Schmid F, Hartwagner K (1999) Computed tomography densitometry of the lung: a method to assess perfusion defects in acute pulmonary embolism. Eur J Radiol 32:192–196

Hoffman E, McLennan G (1997) Assessment of the pulmonary structure-function relationship and clinical outcome measures: quantitative volumetric CT of the lung. Acad Radiol 4:758–776

Hoffman E, Tajik JK, Kugelmass SD (1995a) Matching pulmonary structure and perfusion via combined dynamic multislice CT and thin-slice high-resolution CT. Comput Med Imaging Graph 19:101–112

Hoffman E, Tajik JK, Petersen G, Reiners TJ, Thompson BH, Stanford W (1995b) Perfusion deficit versus anatomic visualization in detection of pulmonary emboli via electron-beam CT: validation in swine, Medical Imaging 1995: Physiology and Function from Multidimensional Images, Proc SPIE, p 2433

Hu H, He HD, Foley WD, Fox SH (2000) Four multidetector-row helical CT: image quality and volume coverage speed. Radiology 215:55–62

Kalender WA (1994) Technical foundations of spiral CT. Semin Ultrasound CT MR 15:81–89

Kauczor HU, Heussel CP, Thelen M (1999) Update on diagnostic strategies of pulmonary embolism. Eur Radiol 9: 262–275

Klingenbeck-Regn K, Schaller S, Flohr T, Ohnesorge B, Kopp AF, Baum U (1999) Subsecond multi-slice computed tomography: basics and applications. Eur J Radiol 31: 110–124

Konstantinides SGA, Olschewski M, Heinrich F et al. (1997) Association between thrombolytic treatment and the prognosis of hemodynamically stable patients with major pulmonary embolism. Results of a multicenter registry. Circulation 96:882–888

Lerman L, Taler SJ, Textor SC, Sheedy PF II, Stanson AW, Romero JC (1995) Computed tomography-derived intrarenal blood flow in renovascular and essential hypertension. Kidney Int 49:846–854

McCollough C, Morin RL (1994) The technical design and performance of ultrafast computed tomography. Radiol Clin North Am 32:521–536

McCollough CH, Zink FE (1999) Performance evaluation of a multi-slice CT system. Med Phys 26:2223–2230

Meaney J, Weg JG, Chenevert TL, Stafford-Johnson D, Hamilton BH, Prince MR (1997) Diagnosis of pulmonary embolism with magnetic resonance angiography. N Engl J Med 336:1422–1427

Oudkerk M, Beek EJ van, Wielopolski P, Ooijen PM van, Brouwers-Kuyper EM, Bongaerts AH, Berghout A (2002) Comparison of contrast-enhanced magnetic resonance angiography and conventional pulmonary angiography for the diagnosis of pulmonary embolism: a prospective study. Lancet 359:1643–1647

PIOPED-Investigators (1990) Value of the ventilation/perfusion scan in acute pulmonary embolism. JAMA 95: 498–502

Remy-Jardin M, Remy J (1999) Spiral CT angiography of the pulmonary circulation. Radiology 212:615–636

Rhodes C, Valind SO, Brudin LH, Wollmer PE, Jones T, Buckingham PD, Hughes JMB (1989) Quantification of regional V/Q ratios in humans by use of PET. II Procedure and normal values. J Appl Physiol 66:1905–1913

Roberts DA, Gefter WB, Hirsch JA et al. (1999) Pulmonary perfusion: respiratory-triggered three-dimensional MR imaging with arterial spin tagging – preliminary results in healthy volunteers. Radiology 212:890–895

Rumberger J (1992) Ultrafast computed tomography scanning modes, scanning planes and practical aspects of contrast administration. In: Stanford WRJ (ed) Ultrafast computed tomography in cardiac imaging: principles and practice. Futura, Mount Kisco NY, pp 17–24

Rumberger J, Feiring AJ, Lipton MJ, Higgins CB, Ell SR, Marcus ML (1987) Use of ultrafast computed tomography to quantitate regional myocardial perfusion: a preliminary report. J Am Coll Cardiol 9:59–69

Rumberger J, Bell MR, Feiring AJ, Behrenbeck T, Marcus ML, Ritman EL (1991) Measurement of myocardial perfusion using fast computed tomography. In: Marcus M, Schelbert HL, Skorton DJ, Wolf GL (eds) Cardiac imaging. A companion to Braunwald's heart disease. Saunders, Philadelphia, pp 688–702

Schoepf UJ, Bruning R, Becker C et al. (1999) Imaging of the thorax with multislice spiral CT. Radiologe 39:943–951

Schoepf U, Bruening R, Konschitzky H et al. (2000a) Pulmonary embolism: comprehensive diagnosis using electron-beam computed tomography for detection of emboli and assessment of pulmonary blood flow. Radiology 217: 693–700

Schoepf UJ, Helmberger T, Holzknecht N et al. (2000b) Segmental and subsegmental pulmonary arteries: evaluation with electron-beam versus spiral CT. Radiology 214:433–439

Stanford W, Rooholamini SA, Galvin JR (1992) Ultrafast computed tomography in the detection of intracardiac masses and pulmonary artery thrombembolism. In: Stanford WRJ (ed) Ultrafast computed tomography in cardiac imaging: principles and practice. Futura, Mount Kisco NY, pp 235–249

Stein PD, Henry JW, Gottschalk A (1999) Reassessment of pulmonary angiography for the diagnosis of pulmonary embolism: relation of interpreter agreement to the order of the involved pulmonary arterial branch. Radiology 210: 689–691

Thompson H, Starmer CF, Whalen RE, McIntosh HD (1964) Indicator transit time considered as a gamma variate. Circ Res 14:502–512

Van Erkel AR, Rossum AB van, Bloem JL, Kievit J, Pattynama PNT (1996) Spiral CT angiography for suspected pulmonary embolism: a cost-effectiveness analysis. Radiology 201:29–36

Wildberger JE, Niethammer MU, Klotz E, Schaller S, Wein BB, Gunther RW (2001) Multi-slice CT for visualization of pulmonary embolism using perfusion weighted color maps. Rofo Fortschr Geb Rontgenstr Neuen Bildgeb Verfahr 173:289–294

Wintersperger BJ, Stabler A, Seemann M, Holzknecht N, Helmberger T, Fink U, Reiser MF (1999). Evaluation of right heart load with spiral CT in patients with acute lung embolism. Rofo Fortschr Geb Rontgenstr Neuen Bildgeb Verfahr 170:542–549

Wolfkiel C, Rich S (1992) Analysis of regional pulmonary enhancement in dogs by ultrafast computed tomography. Invest Radiol 27:211–216

Wolfkiel C, Ferguson JL, Chomka EV, Law WR, Labin IN, Tenzer ML, Booker M, Brundage BH (1987) Measurement of myocardial blood flow by ultrafast computed tomography. Circulation 76:1262–1273

9.2 MR Pulmonary Perfusion

Hidematsu Uematsu, Hiroto Hatabu, Christian Fink, Hans-Ulrich Kauczor

CONTENTS

9.2.1 Introduction

Pulmonary perfusion is a fundamental parameter of lung function, since matched distribution of the regional pulmonary blood flow (perfusion) and ventilation is a prerequisite for gas exchange to occur efficiently. The balance between pulmonary perfusion and ventilation is altered in a variety of diseases of the pulmonary circulation and the lungs, such as pulmonary embolism, chronic obstructive pulmonary disease (COPD), and lung cancer. There are several motivations for a better knowledge of the regional perfusion of the lung in these disease entities: this includes a better understanding of the physiological and pathophysiological processes involved in gas

H. Uematsu, MD, PhD
Department of Radiology, Fukui Medical University, Fukui, Japan
H. Hatabu, MD, PhD
Department of Radiology, Beth Israel Deaconess Medical Center, Harvard Medical School, Boston, Massachusetts, USA
C. Fink, MD
Department of Radiology, Innovative Krebsdiagnostik und Therapie, Deutsches Krebsforschungszentrum (DKFZ), Im Neuenheimer Feld 280, Heidelberg, Germany
H.-U. Kauczor, MD
Professor, Department of Radiology, Innovative Krebsdiagnostik und Therapie, Deutsches Krebsforschungszentrum (DKFZ), Im Neuenheimer Feld 280, Heidelberg, Germany

exchange as well as additional functional information for the diagnosis, treatment planning and monitoring of lung diseases.

Radionuclide techniques using intravenous administration of radioactive macroaggregates have been used for the clinical assessment of regional lung perfusion (Wagner 1976). The scintigraphic method, an established clinical tool, is limited by poor spatial resolution, artifacts from the diaphragm and breast tissue, and overlap due to projection. Furthermore no temporal information is available from this method. Although the absolute pulmonary blood flow can be measured using positron emission tomography (PET) with $H_2^{15}O$ (Mintun et al. 1986; Schuster et al. 1995), it requires a cyclotron for production of a tracer with an extremely short half-life and still offers limited spatial resolution.

Magnetic resonance imaging (MRI) facilitates higher spatial and temporal resolutions than radionuclide methods; however, MRI of lung perfusion has been neglected for some time for several reasons. The air-soft tissue interfaces that facilitate gas exchange (Hatabu et al. 1999a) produce the extremely heterogeneous magnetic susceptibility of the lungs, resulting in a reduction of the MR signal. Moreover, other factors that decrease the signal from the lung parenchyma are intrinsic low proton density, respiratory and cardiac motion, pulmonary blood flow, and molecular diffusion (Bergin et al. 1993).

MRI of the pulmonary parenchyma is possible despite the short $T2^*$ of lung tissue (ranging from 0.9 to 2.2 ms) using ultra-short echo time (TE) spin-echo (Mayo et al. 1992), single-shot fast spin-echo (Hatabu et al. 1999c), and gradient-echo (Alsop et al. 1995; Hatabu et al. 1999b) pulse sequences. MR perfusion imaging has been accomplished using two techniques. Contrast-enhanced perfusion MRI rapidly scans the lung parenchyma following an intravenous bolus injection of a paramagnetic contrast agent. An alternative MR perfusion technique is known as arterial spin labeling (ASL) or spin tagging, which offers the advantage of requiring no exogenous contrast material.

9.2.2
Contrast-Enhanced Perfusion MRI

Contrast-enhanced perfusion MRI uses dynamic imaging of the first-pass of a contrast agent bolus through the pulmonary macrocirculation and microcirculation after peripheral contrast bolus injection (Fig. 9.2.1). In these images, in addition to a time-resolved MR angiography of the pulmonary vasculature, the perfusion of the lungs can be observed as a short blush of contrast enhancement of the lung parenchyma. Since the lung, like no other organ, is characterized by a very short transit time in the range of 3–5 s, rapid imaging techniques are required in order to enable the visualization of the peak enhancement of the lung parenchyma. Therefore contrast-enhanced perfusion MRI uses T1-weighted ultra-short repetition time (TR) and TE gradient echo pulse sequences (HATABU et al. 1996a). Both two-dimensional (2D) and three-dimensional (3D) acquisition techniques can be used for contrast-enhanced perfusion MRI (HATABU et al. 1996b).

The advantage of 2D perfusion MRI is the excellent temporal resolution of up to 0.3 s per image (LEVIN et al. 2001). This not only reduces artifacts from patient movement and cardiac and respiratory motion, but also allows a quantitative or at least semiquantitative approach to an estimation of lung perfusion by using deconvolution methods (WEISSKOFF et al. 1993; OSTERGAARD et al. 1996). Although 2D perfusion MRI often has sufficient in-plane resolution, limited anatomic coverage and

insufficient spatial resolution in the z-axis are major disadvantages which restrict its clinical value, e.g. for the assessment of pulmonary embolism. Consequently, 3D MRI which offers an improved anatomic coverage compared to 2D MRI, has been proposed for the assessment of regional lung perfusion in most recent studies (IWASAWA et al. 2002; MATSUOKA et al. 2002; NIKOLAOU et al. 2002; FINK et al. 2003). With the implementation of high gradient MR systems and alternative k-space sampling strategies such as zero interpolation, 3D imaging of the entire pulmonary vascular tree in less than 4 s has been enabled (GOYEN et al. 2001). Nevertheless, the temporal resolution of 3D MRI remains significantly lower than for 2D perfusion MRI, which reduces the chance to visualize the peak enhancement of the lungs during the first-pass of the contrast agent bolus.

A further improvement of contrast-enhanced perfusion MRI can be expected from the recent introduction of parallel MRI techniques, such as the simultaneous acquisitions of spatial harmonics (SMASH) or the sensitivity encoding for fast MRI (SENSE) (SODICKSON and MANNING 1997; PRUESSMANN et al. 1999; GRISWOLD et al. 2002; FINK et al. 2003). In contrast to conventional sequential MRI, parallel MRI uses the spatial information inherent in the geometry of surface coil arrays to reduce the number of phase-encoding steps, thus leading to a faster acquisition time. In practice, parallel techniques allow for a substantial improvement of the temporal and/or spatial resolution. Compared to perfusion MRI protocols without parallel imaging techniques, parallel MRI achieves an

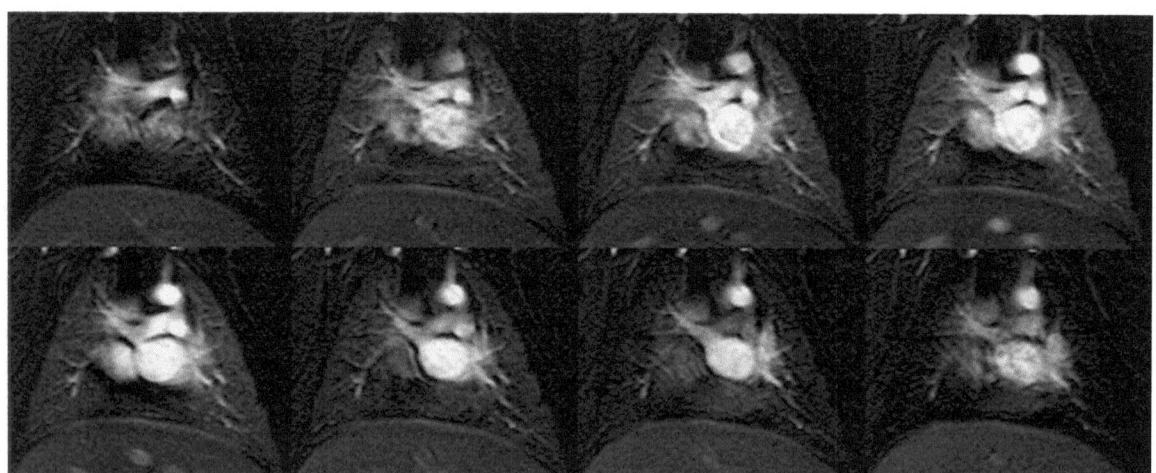

Fig. 9.2.1. Lung of a pig model visualized by perfusion MR imaging using a 2D gradient echo sequence (TR/TE/flip angle=6.3/1.3 ms/45, 10 mm section thickness, 128 128 matrix, 24 cm field of view). Gd-DTPA (5 ml) is administered, followed by a 20 ml saline flush. From the total of 40 coronal images obtained every 0.8 s in the plane of the pulmonary hila, 8 were selected . The images allowed visualization beyond the segmental branches of the pulmonary arterial tree followed by a faint diffuse blush of lung parenchyma. The lung parenchyma experienced diffuse gradual increase in signal intensity

improvement of the spatial and temporal resolution of perfusion MRI by a factor of 1.4–6 and 3–5, respectively (FINK et al. 2003). A major drawback of parallel MRI is that the time saving achieved over sequential MRI is accompanied by an increase of image noise, which is crucial with regards to the low signal-to-noise ratio (SNR) of the lungs. In detail, it can be assumed that the SNR of parallel MRI is reduced by approximately 30–40% compared to conventional non-parallel MRI when using conventional coil arrays (HUNOLD et al. 2002). Since the ultimate achievable SNR for parallel MR images is closely tied to the geometry and sensitivity patterns of the coil arrays, dedicated design of coil arrays for parallel MRI might overcome this limitation in the future (SODICKSON et al. 2002; MADORE and PELC 2001).

Various injection protocols have been described for contrast-enhanced perfusion MRI. However, only few studies have assessed the influence of the contrast agent dose and injection rate on the degree and duration of pulmonary enhancement (HATABU et al. 1996a; MATSUOKA et al. 2002). From the data of these studies it can be concluded that higher injection rates (i.e. faster than 3ml/s) and lower injection volumes will improve the separation of pulmonary and systemic circulation, while higher contrast agent doses will increase the peak enhancement of the lungs. In a recent study of contrast-enhanced 3D perfusion MRI the injection rate did not influence the degree of peak enhancement of the lungs (MATSUOKA et al. 2002). However, the temporal resolution of the pulse sequence used in this study was poor and exceeded the transit time, which might limit the conclusions for perfusion studies with higher temporal resolution.

There are several ways of processing contrast-enhanced perfusion MRI data. For a better visualization of the perfusion signal, contrast-enhanced 3D perfusion MRI is usually processed by subtraction of mask image data acquired before contrast bolus arrival. A rather simple approach for a semiquantitative analysis of contrast-enhanced perfusion MRI data consists of the calculation of signal time curves, SNR, and contrast-to-noise ratios (CNR) using region-of-interest (ROI) analysis of the signal of the lung tissue. In addition, MR perfusion can be quantified by using the indicator dilution principle (HATABU et al. 1999d; UEMATSU et al. 2001). However, a linear relationship between the signal intensity of perfusion MR images and the concentration of the contrast agent is assumed when obtaining these measurements. The quantitative indexes, such as relative regional transit time, blood volume and blood flow are derived from the time intensity curve, defined by the dynamic series of perfusion MR images. Recent attempts to visualize lung perfusion using parametric maps (LEVIN et al. 2001) are shown in Fig. 9.2.2. The correlation between the quantitative indexes and absolute perfusion measurements derived from microsphere studies in a pig model were promising (HATABU et al. 1999d).

Deconvolution analysis has already been applied to quantify tissue perfusion using MR techniques (WEISSKOFF et al. 1993; OSTERGAARD et al. 1996) and have also been proposed to generate a pixel-by-pixel map of perfusion (MURASE et al. 2002). It has been demonstrated that the MR perfusion technique is suitable for the quantitative analysis of regional pulmonary blood flow. The indicator dilution theory for intravascular contrast agent can determine the pulmonary blood flow (PBF) when the arterial input function (AIF) of the contrast agent entering the volume of interest (VOI) is known using the following equation:

$$C_{VOI}(t) = PBF \int_0^t C_{AIF}(\tau) \cdot R(t-\tau) \, d\tau$$

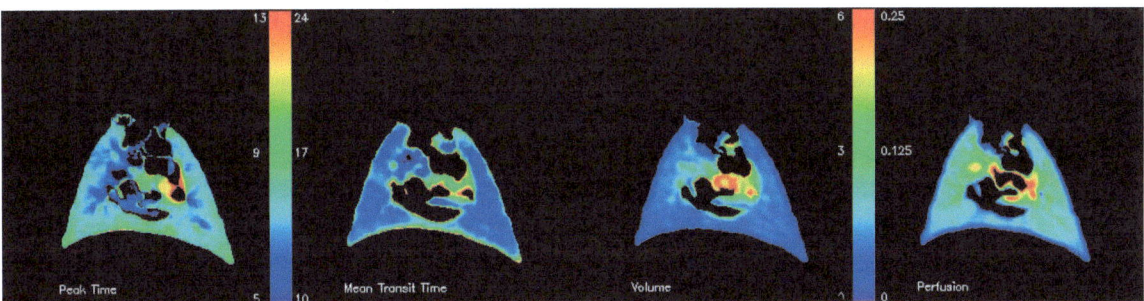

Fig. 9.2.2. Parametric maps of pulmonary perfusion in a pig. These parametric maps were produced from the data of Fig. 1. The peak time, mean transit time, blood volume, and blood flow maps are visualized (from left to right)

where $C_{VOI}(t)$ and $C_{AIF}(t)$ are the time-dependent concentrations of the contrast agent in the VOI and the AIF, respectively. $R(t)$ is known as the residue function, which is the relative amount of contrast agent in the VOI in an idealized perfusion experiment. In this experiment, a unit area bolus is instantaneously injected and subsequently washed out by the perfusion. The initial height of the deconvolved time-concentration curve determined by the tissue impulse response function $h(t)$ $[=PBF*R(t)]$, equals the PBF. Several methods can be implemented to calculate $h(t)$ from Eq. [1], which is based on the deconvolution technique. MURASE et al. (2002) adopted an algebraic approach, based on singular value decomposition (OSTERGAARD et al. 1996), which is robust against statistical noise. These PBF and absolute perfusion values as measured by positron emission tomography correlated well.

Despite the ability to obtain absolute quantitative parameters of lung perfusion the practicality of obtaining absolute measurements by MR imaging and the clinical value of these measurements for the management of disease is a matter of controversy (MAEDA et al. 1999). In the clinical management of acute lung disease the post-processing time that allows us to generate useful and meaningful parameters is limited. Therefore a routine processing of MR perfusion data to obtain absolute perfusion values may be unrealistic in the management of patients' diseases, such as acute pulmonary embolism. Moreover, even if an absolute value of pulmonary blood flow could be obtained with deconvolution analysis, exact clinical values for various lung diseases have not been well documented.

On the other hand, quantitative perfusion MRI gives important insights in the physiology and pathophysiology of the lungs. This includes the demonstration of physiological differences of regional lung perfusion with a well-known gradient of perfusion towards gravity-dependent lung regions (STOCK et al. 1999; KEILHOLZ et al. 2001) (Fig. 9.2.3). Another example is the demonstration of the effect of lung inflation on lung perfusion with a higher perfusion at expiration reflecting the pressure-flow relationship of the pulmonary vasculature (MAI et al. 2001). Other studies have visualized the dependency of ventilation and perfusion using perfusion MRI. In an animal model of airway obstruction, SUGA et al. (2002a) demonstrated a decreased perfusion of hypoventilated lung using perfusion MRI. In combination with MRI of ventilation, this will provide a valuable tool for V/Q imaging (RIZI et al. 2003).

In a clinical setting perfusion MRI may add relevant information in the diagnosis and management of various lung diseases. In patients with suspected pulmonary embolism, perfusion MRI offers information of perfusion defects in addition to the visualization of thrombotic material in the pulmonary arteries by contrast-enhanced MRA (Fig. 9.2.4) (MEANEY et al. 1997). When compared to the nuclear medicine gold standard (V/Q scintigraphy) perfusion MRI shows a comparably high sensitivity and specificity for the detection of perfusion defects (AMUNDSEN et al. 1997). Several studies have demonstrated a good correlation of perfusion MRI and conventional radionuclide scintigraphy (Fig. 9.2.5). In patients with suspected pulmonary embolism or severe pulmonary emphysema BERTHEZENE et al. (1999) found a kappa value of =0.63 for the agreement of perfusion scintigraphy and perfusion MRI. In another study of perfusion MRI in patients with pulmonary embolism, pneumonia and COPD, a kappa value of =0.51–0.56 was reported (AMUNDSEN et al. 2002).

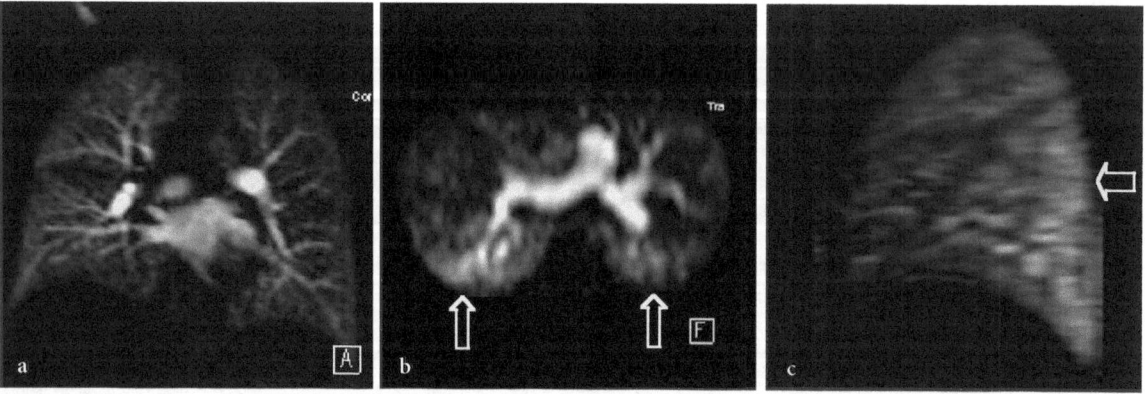

Fig. 9.2.3. Source image (**a**) and transverse (**b**) and sagittal (**c**) reconstructions of a coronal contrast-enhanced 3D perfusion MRI (FLASH 3D; TE/TR/ : 0.8/1.9 ms/400) obtained in a healthy volunteer. A higher signal intensity can be observed in gravity dependent lung (*open arrows*)

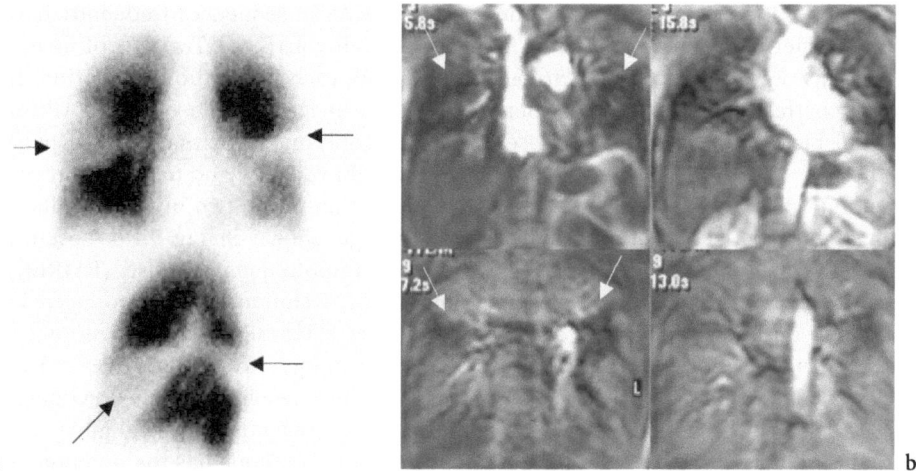

Fig. 9.2.4a, b. A 29-year-old patient with pulmonary embolism. (a) 99mTc-MAA perfusion scintigraphy shows perfusion defects in middle lobe, bilateral posterior superior segments, and lingula (*black arrows*). (b) Corresponding perfusion defects (*white arrows*) are demonstrated by 3D-dynamic contrast-enhanced MR imaging (TR/TE/flip angle=2.7 ms/ 0.6 ms/20°, 100 mm slab thickness, 10 partitions, 128 96 matrix, 450 mm field of view) during the first pass (pulmonary circulation phase). Late enhancement in these areas is demonstrated in systemic circulation phase (paradoxical enhancement)

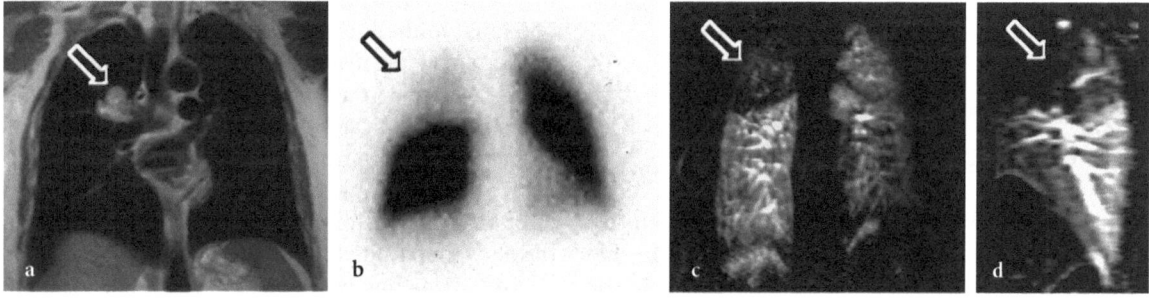

Fig. 9.2.5a–d. Comparison of contrast-enhanced 3D perfusion MRI with conventional radionuclide perfusion scintigraphy in a patient with lung cancer. **a** Coronal HASTE showing central lung cancer with infiltration of upper lobe vessels (arrow). **b** Conventional radionuclide perfusion scintigraphy shows corresponding perfusion defect in the right upper lobe (*arrow*). **c, d** Coronal source image (**c**) and sagittal reconstruction (**d**) demonstrate perfusion defect within segments 1 and 3 (*arrows*)

In addition to pulmonary embolism, MR perfusion imaging can be useful in the clinical assessment of lung function using perfusion as a surrogate, especially for postoperative lung function. This again is an approach which is derived from the routine applications of perfusion scintigraphy. In a study by IWASAWA et al. (2002) in 20 patients with lung cancer, perfusion MRI correlated well with perfusion scintigraphy (r=0.92) and allowed an accurate prediction of the postoperative FEV1 (r=0.68). Another application of perfusion MRI in patients with lung cancer is the evaluation of tumor perfusion and vascular supply or the assessment of vascular infiltration. Typically lung tumors show a delayed perfusion in comparison to the lung parenchyma reflecting the

vascular supply from the systemic circulation (JING-TAO et al. 2001). In patients with central lung tumors the combination of the angiographic and perfusion information of contrast-enhanced perfusion MRI allows an accurate classification of vascular involvement when compared to DSA or perfusion scintigraphy (LEHNHARDT et al. 2002).

In patients with pneumonia and chronic obstructive lung disease perfusion MRI might provide important information in the differential diagnosis to pulmonary embolism (AMUNDSEN et al. 2000).

With the possibility of semiquantitative or quantitative measurements of lung perfusion, perfusion MRI might be a valuable diagnostic tool for the follow-up and monitoring of therapeutic effects in various lung diseases. This includes the assessment of effectiveness and side effects of interventional radiology and radiation therapy (OGASAWARA et al. 2002).

A further improvement of contrast-enhanced perfusion MRI might arise from blood pool contrasts. The advantage of blood pool agents is the higher relaxivity and longer retention within the intravascular compartment allowing higher spatial resolution of pulmonary angiography with equal or even higher vascular contrast (AHLSTROM et al. 1999; NOLTE-ERNSTING et al. 1999). The single injection of a blood pool contrast agent yielded sufficient SNR and negligible artifact, thus demonstrating its potential for the acquisition of pulmonary perfusion and angiographic imaging (ZHENG et al. 2001). Although the routine clinical application of blood pool contrast agents is still limited, they may soon play an important role in perfusion MRI in general and in particular of the lung. Recently, blood-dissolved hyperpolarized ^{3}helium gas (microbubbles) was found to be effective as an intravascular contrast agent in animal models (VIALLON et al. 2000; CALLOT et al. 2001).

9.2.3
Non-Contrast Agent Technique

Arterial spin labeling (ASL) techniques, which use blood water as an endogenous, freely diffusible tracer, detect signal changes due to perfusion in the lung. The two types of ASL are pulsed and steady state. Signal targeting alternating radiofrequency (STAR) is the pulsed ASL method of perfusion imaging and is based on the pulsed magnetic labeling of the arterial water supplying an organ (EDELMAN et al. 1994). The contrast is created by the preinversion of the inflowing spins by selective inversion recovery radio frequency (RF) pulses. Subtraction of images with and without the selective inversion recovery RF pulse make near-complete background suppression a reality. The magnetic labeling of the proton spins prevents the use of a Gd-chelate as a contrast agent. Magnetic labeling also maintains the high contrast to noise ratio of the vasculature and separation of arteries from veins. Quantitative information about the blood flow is obtained through the repeated imaging of the vessels with various inversion times. A fast gradient-echo (HATABU et al. 1999e) or single shot half-Fourier turbo spin-echo (HASTE) pulse sequence (HATABU et al. 2000) were utilized for image acquisition of lung tissue instead of the STAR sequence using echo planar imaging. Fast

gradient-echo or HASTE sequences frequently have magnetic susceptibility artifacts. Two sets of images are acquired during each breath-holding period. In order to invert the magnetization of blood within those structures in only one set of images, an RF pulse is applied to the right ventricle and main pulmonary artery. The subtraction of the two images results in the perfusion image, shown in Fig. 9.2.6. Recently, FAIR with extra radiofrequency pulse (FAIRER), another pulsed ASL technique, was introduced by MAI and BERR (1999). This technique is a modification of the FAIR technique (KWONG et al. 1992; KIM 1995) and provides high-resolution perfusion images of the pulmonary parenchyma with negligible artifacts. The pulsed ASL technique has the potential to provide a non-invasive means of obtaining absolute measurements of pulmonary perfusion through the application of a mathematical model (BUXTON et al. 1998; HATABU et al. 2000). The measured signal is a qualitative index of local perfusion; the signal is proportional to the amount of blood that enters the tissue during the delay interval, which is proportional to the local perfusion. This model yielded a pulmonary perfusion value of 2.0 ml/min ml^{-1} in tissue. A cardiac output of 5 l/min passing through lungs with a typical 2–3 l/min volume, would create an average local perfusion of about 1.7–2.5 ml/min ml^{-1} of tissue. Thus, new MR approaches to image pulmonary perfusion are promising when combining the abovementioned recent MR technologies.

An ASL method based on the steady state magnetic labeling of the arterial water supplying an organ is another approach to the evaluation of perfusion (DETRE et al. 1992). The measurable changes in the steady-state magnetization can be used to calculate tissue-specific perfusion. The respiratory-triggered 3D implementation of this technique has been used to depict regional pulmonary perfusion in healthy subjects without the need for breath holding (ROBERTS et al. 1999). The perfusion deficit in pulmonary arterial occlusion in a pig model has also been demonstrated using this technique (ROBERTS et al. 2001). Although steady-state ASL techniques theoretically produce larger signal changes, magnetization-transfer and transit-time delays cause signal loss. Furthermore, the specific absorption rate is a concern.

Although the effectiveness of the ASL technique has been reported from a limited number of institutions, this technique has only been used in animal experiments or normal subjects, not in experiments using human subjects with lung disease. Therefore, the significance of this technique in clinical settings must still be determined.

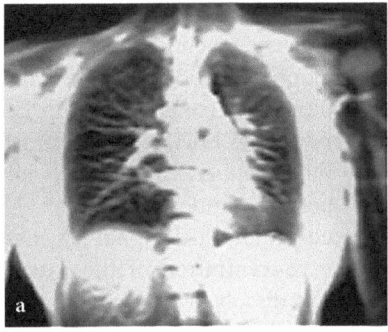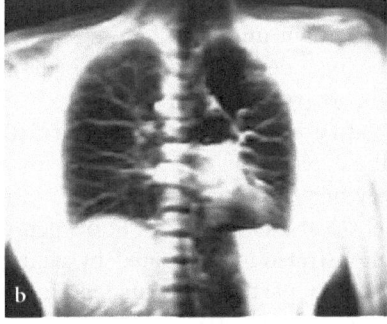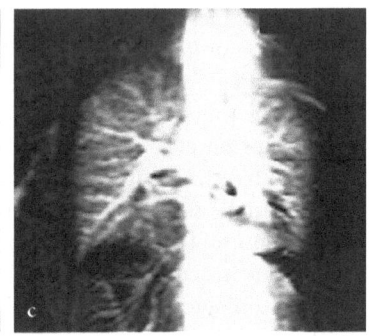

Fig. 9.2.6. Coronal HASTE images of the chest in a healthy volunteer without (**a**) and with (**b**) an RF inversion pulse applied to the right ventricle and pulmonary artery with a TI time of 600 ms. Pulmonary perfusion (**c**) is a subtraction image between (**a**) and (**b**). (Reprinted with permission from HATABU et al. 2000)

9.2.4
Subtraction Technique

The signal intensity of the HASTE sequence increases gradually in the cardiac diastolic phase and decreases in the systolic phase. The MR signal intensity change in the pulmonary parenchyma during the cardiac cycle can be utilized to visualize pulmonary perfusion (TADAMURA and HATABU 2001). Thus, pulmonary perfusion could potentially be depicted through the subtraction of diastolic and systolic HASTE images (Fig. 9.2.7). In a pig model with pulmonary embolism, the subtracted images were able to depict perfusion abnormalities with accuracy (TADAMURA and HATABU 2001). This simple method has potential in evaluating pulmonary perfusion.

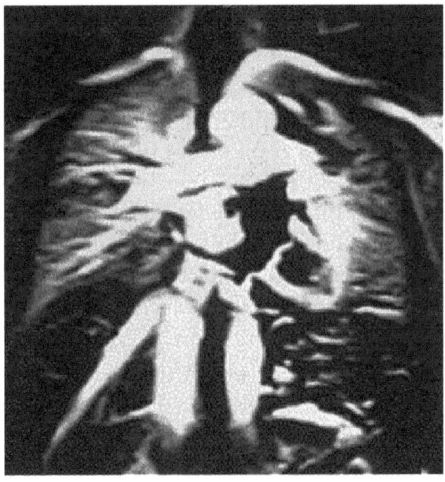

Fig. 9.2.7. Coronal HASTE images that are the result of a subtraction between diastole and systole in a healthy volunteer (TR/effective TE/echo SPACING=infinite/43 ms/4.2 ms, 12 mm section thickness, 128×256 acquisition matrix, 40×40 cm field of view). (Reprinted with permission from TADAMURA and HATABU 2001)

9.2.5
Ventilation/Perfusion MRI

Gas exchange is probably the most important function of the lungs. In order for this process to occur efficiently, there must be a balance between oxygen provided by ventilation and oxygen removed by perfusion to supply the body as well as between carbon dioxide produced by the body which is delivered to the lungs by perfusion and its elimination by expiration. At a steady state of inspired gas mixture and body metabolism, alveolar gas composition is assumed to be homogeneously distributed throughout the functional units of the lung. This is achieved by a local balance of pulmonary ventilation and perfusion, and we say the distribution of regional perfusion and ventilation is matched (for more details see Chap 8). This balance between pulmonary perfusion

and ventilation is affected in various disease states. However, there are compensatory mechanisms which are activated to counteract impairments of the balance of ventilation and perfusion. Most efficient are two mechanisms called hypoxic pulmonary vasoconstriction and hypocapnic bronchoconstriction (see Chap 8). It is important to realize that in disease significant regional mismatches of ventilation and perfusion can be present at rest even if global measurements of gas exchange, such as blood gases, inspiratory and expiratory gas analysis or lung function tests, are still within normal limits. They may be become apparent in stress situations, such as exercise or anesthesia. In such cases of decompensation it is difficult to take appropriate therapeutic measures since the underlying causes of the ventilation/perfusion mismatch remain unclear because they are still masked. Thus, dedicated tools depicting regional mismatches of ventilation and perfusion at a compensated state are not only highly interesting with regard to the explanation of physiopathological interactions, they may also serve as an important step in early detection of airway or pulmonary vascular disease. As demonstrated in this book radiological imaging methods, in particular MRI, provide a wide array of new tools to investigate pulmonary perfusion and ventilation regionally. Obviously, they provide highly complementary data about gas exchange within the lungs. However, at the moment the adequate way to display the information of ventilation and perfusion studies is unclear. Is it mere side-by-side presentation of projection images or coronal slices as we know from nuclear medicine? Is simple image fusion adequate? Some authors have created maps of ventilation/perfusion ratios (MAI et al. 2002; RIZI et al. 2003). Although they are rather impressive we still have to learn how they have to be interpreted. In addition, ventilation and perfusion can be measured quantitatively, and we can provide a numerical output for both which then can be translated into a color-coded map.

As mentioned above, there are two major methods of imaging pulmonary perfusion by MRI. The first uses contrast-enhancement and different contrast agents have already been applied (HATABU et al. 1996a; BERTHEZENE et al. 1997; AHLSTROM et al. 1999; NOLTE-ERNSTING et al. 1999), the second works without a contrast agent and uses endogenous water instead (MAI and BERR 1999; MAI et al. 1999). Both methods have already been applied for the perfusion side in joint imaging of ventilation/perfusion ratios.

Currently, there are several methods of imaging pulmonary ventilation by MRI. In summary, they can be divided in proton and non-proton MRI techniques. When using proton MRI aerosolized Gd-compounds (MONTGOMERY et al. 1987; MISSELWITZ et al. 1997; SUGA et al. 2002b; HAAGE et al. 2003) or oxygen (EDELMAN et al. 1996) can be used. Gd-compounds will only visualize the distribution of an aerosol which depends on the size and weight of the droplets as well as the underlying disease. Their signal intensity neither coincides with perfusion or gas exchange nor does it provide any further information about them. SUGA et al. (2002b) applied aerosolized Gd-DTPA together with a contrast-enhanced perfusion technique in animal models of acute airway obstruction and pulmonary embolism. They consistently observed regionally matched ventilation-perfusion deficits in

acute airway obstruction, whereas they consistently found regionally mismatched ventilation-perfusion deficits in the acute pulmonary embolism model. Thus, aerosolized Gd-compounds seem to be an appropriate complementation of perfusion MRI, either contrast-enhanced or non contrast-enhanced, to regionally impaired ventilation-perfusion ratios.

In oxygen-enhanced proton MRI, signal is generated by changing the concentration of the oxygen inhaled. Since it is mainly dissolved molecular oxygen which contributes to the signal change, it represents a very complex mixture of ventilation, oxygen uptake and perfusion. Thus, it might not be an ideal partner for straightforward perfusion MRI as described above. However, it may serve as a comprehensive V/Q imaging method by itself (see Chap 8). MAI et al. (2002) used oxygen-enhanced and non-contrast-enhanced perfusion imaging with arterial spin labeling in 10 healthy volunteers to investigate the distribution of ventilation-perfusion (V/Q) signal intensity ratios. The plots of the V/Q ratio were similar to the logarithmic normal distribution obtained by multiple inert gas elimination techniques which serves as the invasive gold standard for measuring V/Q ratios. Thus, they proved the concept of combined semiquantitative V/Q imaging using MR methods without the use of a contrast agent (MAI et al. 2002). NAKAGAWA et al. (2001) proved the feasibility of oxygen-enhanced ventilation imaging together with first-pass contrast enhanced perfusion imaging in patients suffering from different diseases, such as pulmonary embolism and emphysema with an overall success rate of 80% for ventilation and 94% for perfusion imaging. In pulmonary embolism the typical ventilation-perfusion mismatches were observed.

In non-proton MRI, hyperpolarized noble gases, such as ^{3}He or ^{129}Xe (ALBERT et al. 1994; MIDDLETON et al. 1995; KAUCZOR et al. 1996), or sulfur hexafluoride in ^{19}F MRI are used. All of these are well suited for V/Q imaging since they not only visualize regional ventilation but they also allow for dedicated measurements of gas exchange, oxygen uptake etc. (see Chap 8). Hyperpolarized ^{3}He-MRI was used in several studies together with contrast-enhanced perfusion imaging to investigate ventilation-perfusion matches and mismatches in animals with and without airway or pulmonary artery obstruction (CHEN et al. 1999; CREMILLIEUX et al. 1999; ZHENG et al. 2002). Matched ventilation and perfusion abnormalities were identified in all animals with airway obstruction, whereas MR perfusion defects without ventilation abnormalities were seen in all animals with pulmonary emboli (CHEN et al. 1999). Additional high-resolution MR angiography can unambiguously reveal the location and size of the pulmonary emboli (ZHENG et al. 2002). However, if ³He-MRI is combined with any method for perfusion imaging not only mismatches of ventilation and perfusion can be depicted but deep insights into respiratory physiology and its regulation will be obtained. We are optimistic that they will allow for the determination of the underlying disease, whether it is bronchial or pulmonary arterial in origin, as well as the reversibility of the compensatory mechanisms. Additionally, further aspects, such as the contribution of the bronchial (=systemic) circulation of the lungs to gas exchange, will be addressed. Since the signal of hyperpolarized ³He gas is also dependent on the ambient partial oxygen pressure, it can also be used alone for the measurement of partial oxygen pressure within the lung and its development over time, such as during an inspiratory breath-hold. The decrease rate of the partial oxygen pressure is then an indicator of oxygen uptake which is determined by regional perfusion (EBERLE et al. 1999; DENINGER et al. 2000). Thus, ³He-MRI alone has the potential for a comprehensive regional evaluation of ventilation and perfusion of the lung (see Chap 8). Similar observations have been made when using ¹²⁹Xe-MRI (RUPPERT et al. 2000) or even sulfur hexafluoride in ¹⁹F-MRI (KUETHE et al. 2000).

9.2.6
Conclusions

Contrast-enhanced perfusion MRI and non-contrast arterial spin labeling (ASL) perfusion MRI are feasible methods for the non-invasive assessment of pulmonary perfusion. Contrast-enhanced perfusion MRI contrast is a simple and robust technique providing 2D and 3D images of lung perfusion. The ASL technique is noninvasive and eliminates the risks and expense associated with contrast material administration. However, the clinical utility of this method may be limited by several technical factors, especially difficulty in implementation as well as lower SNR and spatial resolution.

In addition to the analysis of physiological and pathophysiological processes perfusion MRI offers valuable information for the diagnosis, follow-up and treatment monitoring of various lung diseases. In addition to the qualitative assessment of regional lung perfusion, quantitative measurements can be obtained from perfusion MRI. These measurements can be of clinical relevance for the diagnostic assessment of lung function. The combination of MR ventilation imaging and MR perfusion imaging is opening the door to multi-functional MR imaging of the lung, so-called V/Q MRI, which might substantially improve the diagnostic evaluation of lung disease.

References

Ahlstrom KH, Johansson LO, Rodenburg JB et al. (1999) Pulmonary MR angiography with ultrasmall superparamagnetic iron oxide particles as a blood pool agent and a navigator echo for respiratory gating: pilot study. Radiology 211:865–869

Albert MS, Cates GD, Driehuys B et al. (1994) Biological magnetic resonance imaging using laser-polarized 129Xe. Nature 370:199–201

Alsop DC, Hatabu H, Bonnet M et al. (1995) Multi-slice, breath-hold imaging of the lung with submillisecond echo times. Magn Reson Med 33:678–682

Amundsen T, Kvaerness J, Jones RA et al. (1997) Pulmonary embolism: detection with MR perfusion imaging of lung – a feasibility study. Radiology 203:181–185

Amundsen T, Torheim G, Waage A et al. (2000) Perfusion magnetic resonance imaging of the lung: characterization of pneumonia and chronic obstructive pulmonary disease. A feasibility study. J Magn Reson Imaging 12:224–231

Amundsen T, Torheim G, Kvistad KA et al. (2002) Perfusion abnormalities in pulmonary embolism studied with perfusion MRI and ventilation-perfusion scintigraphy: an intra-modality and inter-modality agreement study. J Magn Reson Imaging 15:386–394

Bergin CJ, Glover GM, Pauly J (1993) Magnetic resonance imaging of lung parenchyma. J Thorac Imaging 8:12–17

Berthezene Y, Croisille P, Bertocchi M et al. (1997) Lung perfusion demonstrated by contrast-enhanced dynamic magnetic resonance imaging. Application to unilateral lung transplantation. Invest Radiol 32:351–356

Berthezene Y, Croisille P, Wiart M et al. (1999) Prospective comparison of MR lung perfusion and lung scintigraphy. J Magn Reson Imaging 9:61–68

Buxton RB, Frank LR, Wong EC et al. (1998) A general kinetic model for quantitative perfusion imaging with arterial spin labeling. Magn Reson Med 40:383–396

Callot V, Canet E, Brochot J et al. (2001) MR perfusion imaging using encapsulated laser-polarized 3He. Magn Reson Med 46:535–540

Chen Q, Levin D, Kim D et al. (1999) Pulmonary disorders: ventilation-perfusion MR imaging with animal models. Radiology 213:871–879

Cremillieux Y, Berthezene Y, Humblot H et al. (1999) A combined 1H perfusion/3He ventilation NMR study in rat lungs. Magn Reson Med 41:645–648

Deninger AJ, Eberle B, Ebert M et al. (2000) 3He-MRI-based measurements of intrapulmonary p(O2) and its time course during apnea in healthy volunteers: first results, reproducibility, and technical limitations. NMR Biomed 13:194–201

Detre JA, Leigh JS, Williams DS et al. (1992) Perfusion imaging. Magn Reson Med 23:37–45

Eberle B, Weiler N, Markstaller K et al. (1999) Analysis of intrapulmonary O(2) concentration by MR imaging of inhaled hyperpolarized helium-3. J Appl Physiol 87:2043–2052

Edelman RR, Siewert B, Adamis M et al. (1994) Signal targeting with alternating radiofrequency (STAR) sequences: application to MR angiography. Magn Reson Med 31:233–238

Edelman R, Hatabu H, Tadamura E et al. (1996) Noninvasive assessment of regional ventilation in the human lung using oxygen-enhanced magnetic resonance imaging. Nature Med 2:1236–1239

Fink C, Bock M, Puderbach M et al. (2003) Partially parallel three-dimensional magnetic resonance imaging for the assessment of lung perfusion – initial results. Invest Radiol 38:482–488

Goyen M, Laub G, Ladd ME et al. (2001) Dynamic 3D MR angiography of the pulmonary arteries in under four seconds. J Magn Reson Imaging 13:372–377

Griswold MA, Jakob PM, Heidemann RM et al. (2002) Generalized autocalibrating partially parallel acquisitions (GRAPPA). Magn Reson Med 47:1202–1210

Haage P, Karaagac S, Spuntrup E, Adam G, Gunther RW (2003) MR imaging of lung ventilation with aerosolized gadolinium-chelates. Fortschr Geb Rontgenstr Neuen Bildgeb Verfahr 175:187–193

Hatabu H, Gaa J, Kim D et al. (1996a) Pulmonary perfusion: qualitative assessment with dynamic contrast-enhanced MRI using ultra-short TE and inversion recovery turbo FLASH. Magn Reson Med 36:503–508

Hatabu H, Gaa J, Kim D et al. (1996b) Pulmonary perfusion and angiography: evaluation with breath-hold enhanced three-dimensional fast imaging steady-state precession MR imaging with short TR and TE. AJR Am J Roentgenol 167:653–655

Hatabu H, Alsop DC, Listerud J et al. (1999a) T2* and proton density measurement of normal human lung parenchyma using submillisecond echo time gradient echo magnetic resonance imaging. Eur J Radiol 29:245–252

Hatabu H, Chen Q, Stock KW et al. (1999b) Fast magnetic resonance imaging of the lung. Eur J Radiol 29:114–132

Hatabu H, Gaa J, Tadamura E et al. (1999c) MR imaging of pulmonary parenchyma with a half-Fourier single-shot turbo spin-echo (HASTE) sequence. Eur J Radiol 29:152–159

Hatabu H, Tadamura E, Levin DL et al. (1999d) Quantitative assessment of pulmonary perfusion with dynamic contrast- enhanced MRI. Magn Reson Med 42:1033–1038

Hatabu H, Wielopolski PA, Tadamura E et al. (1999e) An attempt of pulmonary perfusion imaging utilizing ultra-short echo time turbo FLASH sequence with signal targeting and alternating radio-frequency (STAR). Eur J Radiol 29:160–163

Hatabu H, Tadamura E, Prasad PV et al. (2000) Noninvasive pulmonary perfusion imaging by STAR-HASTE sequence. Magn Reson Med 44:808–812

Hunold P, Maderwald S, Ladd ME et al. (2002) Parallel acquisition techniques for cardiac cine MRA: comparison of image quality. Proc ISMRM 1664

Iwasawa T, Saito K, Ogawa N et al. (2002) Prediction of postoperative pulmonary function using perfusion magnetic resonance imaging of the lung. J Magn Reson Imaging 15:685–692

Jingtao M, Yuesong Y, Zhengyu L et al. (2001) Neovascular perfusion in lung cancer: qualitative and quantitative study using MR perfusion imaging. Imaging Dec MRI 5:2–6

Kauczor H-U, Hofmann D, Kreitner K-F et al. (1996) Normal and abnormal pulmonary ventilation: visualization at hyperpolarized He-3 MR imaging. Radiology 201:564–568

Keilholz SD, Knight-Scott J, Christopher JM et al. (2001) Gravity-dependent perfusion of the lung demonstrated with the FAIRER arterial spin tagging method. Magn Reson Imaging 19:929–935

Kim SG (1995) Quantification of relative cerebral blood flow change by flow-sensitive alternating inversion recovery (FAIR) technique: application to functional mapping. Magn Reson Med 34:293–301

Kuethe DO, Caprihan A, Gach HM et al. (2000) Imaging obstructed ventilation with NMR using inert fluorinated gases. J Appl Physiol 88:2279–2286

Kwong KK, Belliveau JW, Chesler DA et al. (1992) Dynamic magnetic resonance imaging of human brain activity during primary sensory stimulation. Proc Natl Acad Sci USA 89:5675–5679

Lehnhardt S, Winterer TJ, Strecker R et al. (2002) Assessment of pulmonary perfusion with ultrafast projection magnetic resonance angiography in comparison with lung perfusion scintigraphy in patients with malignant stenosis. Invest Radiol 37:594–599

Levin DL, Chen Q, Zhang M et al. (2001) Evaluation of regional pulmonary perfusion using ultrafast magnetic resonance imaging. Magn Reson Med 46:166–171

Madore B, Pelc NJ (2001) SMASH and SENSE: experimental and numerical comparisons. Magn Reson Med 45:1103–1111

Maeda M, Yuh WT, Ueda T et al. (1999) Severe occlusive carotid artery disease: hemodynamic assessment by MR perfusion imaging in symptomatic patients. AJNR Am J Neuroradiol 20:43–51

Mai VM, Berr SS (1999) MR perfusion imaging of pulmonary parenchyma using pulsed arterial spin labeling techniques: FAIRER and FAIR. J Magn Reson Imaging 9:483–487

Mai VM, Hagspiel KD, Christopher JM et al. (1999) Perfusion imaging of the human lung using flow-sensitive alternating inversion recovery with an extra radiofrequency pulse (FAIRER). Magn Reson Imaging 17:355–361

Mai VM, Chen Q, Bankier AA et al. (2001) Effect of lung inflation on arterial spin labeling signal in MR perfusion imaging of human lung. J Magn Reson Imaging 13:954–959

Mai VM, Liu B, Polzin JA et al. (2002) Ventilation-perfusion ratio of signal intensity in human lung using oxygen-enhanced and arterial spin labeling techniques. Magn Reson Med 48:341–350

Matsuoka S, Uchiyama K, Shima H et al. (2002) Effect of the rate of gadolinium injection on magnetic resonance pulmonary perfusion imaging. J Magn Reson Imaging 15:108–113

Mayo JR, MacKay A, Muller NL (1992) MR imaging of the lungs: value of short TE spin-echo pulse sequences. AJR Am J Roentgenol 159:951–956

Meaney JF, Weg JG, Chenevert TL et al. (1997) Diagnosis of pulmonary embolism with magnetic resonance angiography. N Engl J Med 336:1422–1427

Middleton H, Black RD, Saam B et al. (1995) MR imaging with hyperpolarized 3He Gas. Magn Reson Med 33:271–275

Mintun MA, Ter-Pogossian MM, Green MA et al. (1986) Quantitative measurement of regional pulmonary blood flow with positron emission tomography. J Appl Physiol 60:317–326

Misselwitz B, Muhler A, Heinzelmann I et al. (1997) Magnetic resonance imaging of pulmonary ventilation. Initial experiences with a gadolinium-DTPA-based aerosol. Invest Radiol 32:797–801

Montgomery A, Paajanen H, Brasch R, Murray J (1987) Aerosolized gadolinium-DTPA enhances the magnetic resonance signal of extravascular lung water. Invest Radiol 22:377–381

Murase K, Matsuda T, Yasuhara Y et al. (2002) Mapping of regional pulmonary blood flow using contrast-enhanced MRI and deconvolution analysis. Proceedings of the 1st international workshop on pulmonary functional imaging. Philadelphia, PA, vol 1, p 12

Nakagawa T, Sakuma H, Murashima S et al. (2001) Pulmonary ventilation-perfusion MR imaging in clinical patients. J Magn Reson Imaging 14:419–424

Nikolaou K, Schoenberg SO, Nittka M et al. (2002) Magnetic resonance imaging in the diagnosis of pulmonary arterial hypertension: high resolution angiography and fast perfusion imaging using intelligent parallel acquisition techniques (IPAT) (abstract). Radiology 225:473

Nolte-Ernsting CC, Krombach G, Staatz G, Kilbinger M, Adam GB, Gunther RW (1999) Virtual endoscopy of the upper urinary tract based on contrast-enhanced MR urography data sets. Fortschr Geb Rontgenstr Neuen Bildgeb Verfahr 170:550–556

Ogasawara N, Suga K, Karino Y et al. (2002) Perfusion characteristics of radiation-injured lung on Gd-DTPA-enhanced dynamic magnetic resonance imaging. Invest Radiol 37:448–457

Ostergaard L, Weisskoff RM, Chesler DA et al. (1996) High resolution measurement of cerebral blood flow using intravascular tracer bolus passages, part I. Mathematical approach and statistical analysis. Magn Reson Med 36:715–725

Pruessmann KP, Weiger M, Scheidegger MB et al. (1999) SENSE: sensitivity encoding for fast MRI. Magn Reson Med 42:952–962

Rizi RR, Saha PK, Wang B et al. (2003) Co-registration of acquired MR ventilation and perfusion images – validation in a porcine model. Magn Reson Med 49:13–18

Roberts DA, Gefter WB, Hirsch JA et al. (1999) Pulmonary perfusion: respiratory-triggered three-dimensional MR imaging with arterial spin tagging – preliminary results in healthy volunteers. Radiology 212:890–895

Roberts DA, Rizi RR, Lipson DA et al. (2001) Dynamic observation of pulmonary perfusion using continuous arterial spin-labeling in a pig model. J Magn Reson Imaging 14:175–180

Ruppert K, Brookeman JR, Hagspiel KD, Mugler JP III (2000) Probing lung physiology with xenon polarization transfer contrast (XTC). Magn Reson Med 44:349–357

Schuster DP, Kaplan JD, Gauvain K et al. (1995) Measurement of regional pulmonary blood flow with PET. J Nucl Med 36:371–377

Sodickson DK, Manning WJ (1997) Simultaneous acquisition of spatial harmonics (SMASH): fast imaging with radiofrequency coil arrays. Magn Reson Med 38:591–603

Sodickson DK, McKenzie CA, Ohliger MA et al. (2002) Recent advances in image reconstruction, coil sensitivity calibration, and coil array design for SMASH and generalized parallel MRI. MAGMA 13:158–163

Stock KW, Chen Q, Levin D et al. (1999) Demonstration of gravity-dependent lung perfusion with contrast-enhanced magnetic resonance imaging. J Magn Reson Imaging 9:557–561

Suga K, Ogasawara N, Okada M et al. (2002a) Potential of non-contrast electrocardiogram-gated half-fourier fast-spin-echo magnetic resonance imaging to monitor dynamically altered perfusion in regional lung. Invest Radiol 37:615–625

Suga K, Ogasawara N, Okada M et al. (2002b) Regional lung functional impairment in acute airway obstruction and pulmonary embolic dog models assessed with gadolinium-based aerosol ventilation and perfusion magnetic resonance imaging. Invest Radiol 37:281–291

Tadamura E, Hatabu H (2001) Assessment of pulmonary perfusion using a subtracted HASTE image between diastole and systole. Eur J Radiol 37:179–183

Uematsu H, Levin DL, Hatabu H (2001) Quantification of pulmonary perfusion with MR imaging: recent advances. Eur J Radiol 37:155–163

Viallon M, Berthezene Y, Decorps M et al. (2000) Laser-polarized (3)He as a probe for dynamic regional measurements of lung perfusion and ventilation using magnetic resonance imaging. Magn Reson Med 44:1–4

Wagner HN Jr (1976) The use of radioisotope techniques for the evaluation of patients with pulmonary disease. Am Rev Respir Dis 113:203–218

Weisskoff RM, Chesler D, Boxerman JL et al. (1993) Pitfalls in MR measurement of tissue blood flow with intravascular tracers: which mean transit time? Magn Reson Med 29:553–558

Zheng J, Carr J, Harris K et al. (2001) Three-dimensional MR pulmonary perfusion imaging and angiography with an injection of a new blood pool contrast agent B-22956/1. J Magn Reson Imaging 14:425–432

Zheng J, Leawoods J, Nolte M et al. (2002) Combined MR proton lung perfusion/angiography and helium ventilation: potential for detecting pulmonary emboli and ventilation defects. Magn Reson Med 47:433–438

10 Respiratory Mechanics: CT and MRI

PHILIPPE CLUZEL

CONTENTS

10.1
Introduction

Imaging of the lung and its diseases has traditionally been the main focus of the chest radiology literature, whereas imaging of the respiratory mechanics has received little attention. However, the inspiratory pump (consisting of the various inspiratory muscle groups and of the various components of the rib cage) is the effector that translates automatic and voluntary inspiratory commands into alveolar ventilation by applying expanding forces to the lungs. As such, it is a major determinant of the physiology of the act of breathing and its functioning is of paramount importance to the pathophysiology of many pathological situations, including various respiratory diseases or respiratory repercussions of neurological diseases.

Major difficulties to study the function of the inspiratory pump in relationship to its structure arise from the extremely complex geometrical disposition of its active and passive elements. However, imaging techniques are a major tool to visualize the geometrical changes corresponding to the actions of inspiratory muscles (WHITELAW 1987; GAUTHIER

P. CLUZEL, PhD
Service de Radiologie Polyvalente Diagnostique et Interventionelle. Groupe Hospitalier Pitié-Salpetrière, 47-83 Bd de l'Hopital, 75651 Paris Cedex 13, France

et al. 1994; CLUZEL et al. 2000), and can be useful to understand various mechanical aspects of the respiratory system in health and diseases and to establish structure-function relationships.

The present chapter will discuss three-dimensional (3D) geometrical parameters of the chest wall for assessing the respiratory mechanics in healthy subjects, and their modification in chronic obstructive pulmonary disease patients, before and after lung volume reduction surgery. In the latter part, the way geometrical data can be incorporated into mechanical models will be illustrated. These models are able to gather information about how the respiratory pump works that no amount of experimentation can provide.

10.2
Imaging the Inspiratory Pump
in Healthy Subjects

In the past, the inspiratory pump was investigated using a number of radiologic and non-radiologic techniques (BERGOFSKY 1964; KONNO and MEAD 1967; WANG and JOSENHANS 1971; SHARP et al. 1975; ROCHESTER and FARKAS 1987; KRAYER et al. 1987; CASSART et al. 1996). These only provided an indirect assessment of diaphragm shortening or volume displacement because of the complex 3D shape of the diaphragm. The complexity of the chest wall in terms of its 3D geometry and the mechanical interdependence of its constituent components makes the analysis of only one of these elements in one or two dimensions very difficult, thus emphasizing the necessity for a 3D analysis. Such 3D analyses have to take into account (a) the shape of the rib cage and the diaphragm, (b) the respective volumes displaced, and (c) the relationship between these volumes and the surface areas of the relevant diaphragmatic zones over the respiratory cycle. To understand the parameters used in such analyses, the anatomy of the diaphragm has been reviewed (CLUZEL et al. 2000). The diaphragm is attached to the inferior thoracic aperture with the sternal part to the back of the xiphoid process, the costal part to the internal surface of the lower six costal cartilages and their adjoining ribs, and the lumbar part is fixed to the aponeurotic medial and lateral arcuate ligaments and to some lumbar vertebrae by means of the crura (Fig. 10.1). The diaphragmatic silhouette can be divided into two zones (CLUZEL et al. 2000). The first zone, where the diaphragm is roughly horizontal and apposed to the lungs above and the abdominal contents below, is referred to as the "diaphragmatic dome." The second zone, where the diaphragm assumes a roughly vertical direction and through which the rib cage is apposed to the abdominal contents and thus exposed to abdominal pressure, is referred to as the "apposition zone" (Figs. 10.1, 10.2). The silhouettes of the rib cage can be drawn similarly (Fig. 10.2). Three-dimensional reconstructions were based on the Delaunay triangulation algorithm of object contours (CLUZEL et al. 2000). All these small triangles provide a 3D polygonal representation. From these 3D objects diaphragmatic lengths can be calculated in three different planes, one coronal and two sagittal (right and left, respectively). These planes correspond roughly to the orientations of diaphragmatic fibers, which radiate from the central tendon (Fig. 10.1). Along these selected planes, total diaphragmatic length, length of the apposition zone and the diaphragmatic dome can be calculated. The length of the apposition zone includes the anterior and posterior parts of the apposition zone in the sagittal sections and both the right and left parts of the apposition zone for coronal sections. Diaphragmatic surface areas, such as total diaphragmatic surface area, area of the diaphragmatic dome, and area of apposition, are the sum of all the appropriate small triangular surfaces.

WHITELAW (1987) was the first to provide a 3D rendition of the diaphragm, which he constructed from serial transverse sections obtained with a CT scanner. A total of 12 sections were made at 0.5-cm intervals during 12 separate breath-holds. Sections were scanned at relaxed functional residual capacity (FRC) and FRC plus 1 l. The configurations of the rib cage and abdomen were monitored using magnetometers attached to the thorax and abdomen and were kept constant. Because the CT scan sections did not include the bottom of the rib cage, the muscle length between the lowest section and the insertion of the diaphragm was estimated from plane X-rays of the same subject, assuming that the line of insertion ran from the transverse process of the first lumbar vertebra horizontally to the 12th rib and along the lower border of the rib cage 1 cm above the ends of the ribs. The calculated area of apposition in that study was 385 cm^2 at FRC and 161 cm^2 at FRC plus 1 l. The contracted diaphragm

occupied a thoracic volume 680 ml smaller than the relaxed one. No information was provided regarding total diaphragmatic surface areas, length of the diaphragm, absolute volume under the diaphragm or the rib cage. PAIVA et al. studied four normal males at FRC using MRI (PAIVA et al. 1992). In the spin-echo mode, 11 apneas of 54 s (permitted by the administration of supplemental oxygen) allowed the acquisition of 36 sagittal and 30 coronal images. A flexible tube filled with a $CuSO_4$ solution was fixed around the lower costal border of each subject to identify the loci of origin of the costal fibers of the diaphragm. The subsequent study by GAUTHIER et al. (1994) reported data on the diaphragm at three lung volumes. Four normal males were studied with the spin-echo mode and three sets of six slices (30 s data acquisition period each), one in the coronal plane and one in the sagittal plane for each hemidiaphragm were obtained. The loci of diaphragm insertions onto the rib cage were approximated in the same manner as in the previous study by PAIVA et al. (1992). Finally, PETTIAUX et al. (1994), from the same laboratory, developed a technique of diaphragm imaging by using spiral CT and studied the same 4 normal subjects previously investigated with MRI and another group of 10 subjects in a subsequent study.

The shape of the diaphragm changed markedly along the sagittal plane with lung inflation in the three previous studies (GAUTHIER et al. 1994; PETTIAUX et al. 1994; CASSART et al. 1996). However, there were no major changes in the coronal plane. This finding was not confirmed in the study from CLUZEL et al. (2000) where changes in the coronal plane were marked. This difference is mainly explained by the choice of loci of diaphragm insertions. In those studies (GAUTHIER et al. 1994; PETTIAUX et al. 1994; CASSART et al. 1996) the choice of an external costal marker could have resulted in a systematic overestimation of the length of the diaphragm. Using anatomic landmarks (see Fig. 10.1), CLUZEL et al. (2000) found that the apposition zone was entirely eliminated at TLC (see Fig. 10.2). This finding is not unexpected (FARKAS and ROCHESTER 1988), and agrees well with previous studies (WHITELAW et al. 1983; FARKAS and ROCHESTER 1988; PETROLL et al. 1990).

Lung inflation to total lung capacity (TLC) is associated with an important length reduction of the human diaphragm. Lengths estimated with MRI are in reasonable agreement with those reported in the literature (WHITELAW 1987; FARKAS and ROCHESTER 1988; PETROLL et al. 1990; PAIVA et al. 1992; GAUTHIER et al. 1994; CASSART et al. 1997). Slight discrepancies could come from the different methods used to measure length.

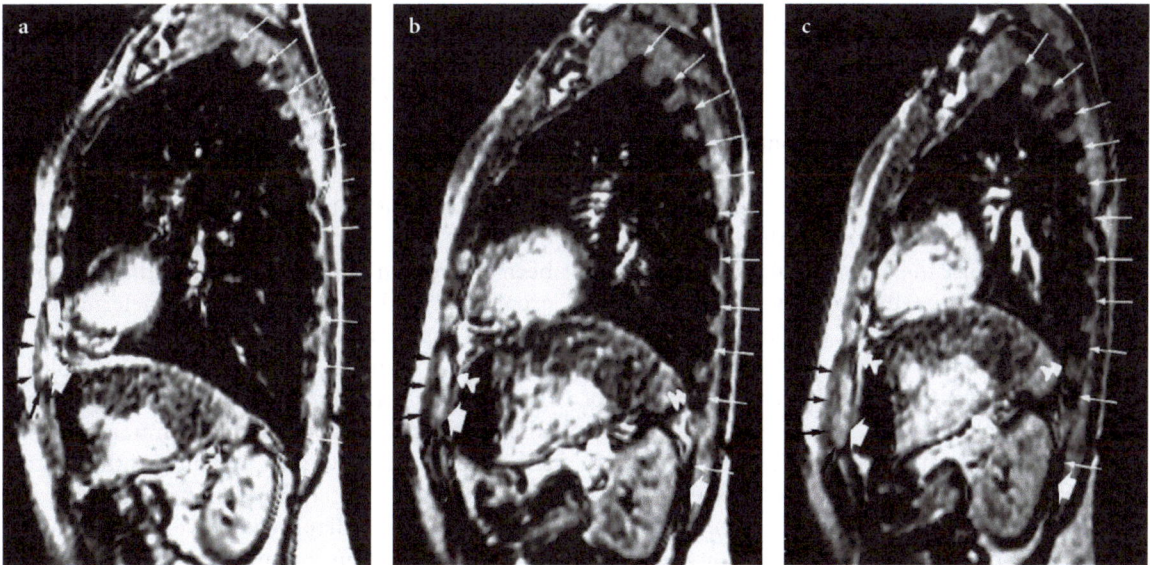

Fig. 10.1a–c. Fast GRE images of the diaphragm and rib cage (TR 6.7 ms, TE 2.2 ms, flip angle 30°, 10 mm thickness, FOV 48 cm, matrix size 128 256). Sagittal images were used to determine diaphragm attachments to the thoracic outlet. (**a**) Right sagittal image obtained at TLC, 7 cm away from the midline. The first rib is no longer visible on this sagittal plane, but the subsequent ribs are identified by *thin white arrows*. A *large white arrow* points to the 12th rib onto which the diaphragm is inserted. Lower part of the anterior chest wall is identified by *small black arrows* and the anterior insertion of the diaphragm onto the internal surfaces of the cartilage and adjoining ribs is identified by another *large white arrow*. Note that the apposition zone is "close to zero" at TLC. Slices corresponding to (**b**) FRC and (**c**) RV, respectively. *Rounded arrows* indicate the upper limit of the apposition zone. Anatomic landmarks can be identified by paying close attention to the modifications caused by lung-volume variations

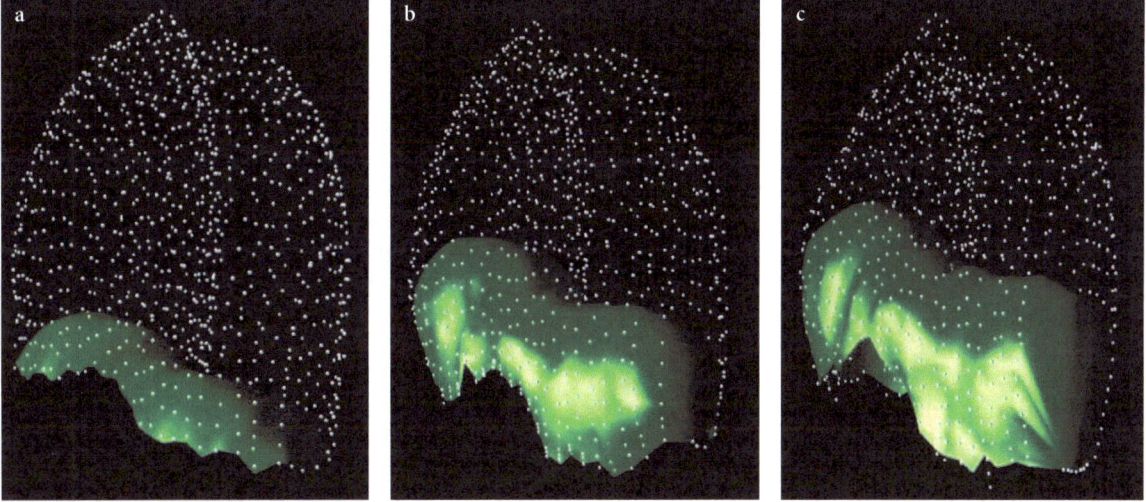

Fig. 10.2a–c. Representative 3D reconstructions of the diaphragm in green and rib cage as dots at TLC (**a**), FRC (**b**) and RV (**c**) in a left anterolateral view. Note the marked changes of diaphragmatic shape that occur in: both sagittal and coronal planes and that the apposition zone is totally absent at TLC. The apposition zone is defined as the roughly vertical portion of the diaphragm that separates the lower rib cage from the abdominal contents

Area variations estimated with MRI confirm the remarkable shortening of relaxed human diaphragm fibers (55±11%) over vital capacity (VC), consistent with the length-tension characteristics of diaphragm muscle fibers in vitro (AGOSTONI and HERMANN 1960). The average surface area of the muscle value at FRC in the study of CLUZEL et al. was 997±93 cm², which is close to 898±146 cm², a necropsy estimate reported by ARORA and ROCHESTER (1982) for normal humans. Slight discrepancies could come from the different methods used to measure length. The explanation for these differences is probably methodologic and probably mainly related to the means for identifying diaphragm attachments.

Investigators who used various indirect techniques to determine diaphragm displacement obtained conflicting results. MRI produces a more accurate diaphragm volume displacement than indirect measurements because the volume under the diaphragm can be measured directly. According to Cluzel et al. (2000) the volume displaced by the diaphragm as it descends within the thorax, or diaphragm volume displacement, over the entire VC, represents 59.7% of the inspired volume.

Few reports exist describing the use of MRI to calculate lung volumes (O'Callaghan et al. 1987; Chapman et al. 1990; Clausen 1997). Cluzel et al. (2000) estimates of lung volumes fit well with the spirometric values but are not identical. Similar observations have been made when using hyperpolarized ^{3}He MRI (Kauczor et al. 2001). Lung volume measurements obtained by MRI take into account the pulmonary blood volume and the lung tissue volume which could be estimated at 500 ml and 600 ml, respectively (Cander and Forster 1959; Dock et al. 1961; Yu 1969). Using these values, calculated values for residual volume (RV) and FRC were 1,745±458 cm^3 and 2,718±964 cm^3, which are slightly above (300–400 cm^3) the expected values of 1,410±310 cm^3 and 2,290±660 cm^3, respectively (Cluzel et al. 2000). This discrepancy could be explained by the inclusion of a part of the dead space and pulmonary veins or arteries in the calculated volume. Conversely, calculated values for TLC, 6,036±1,118 cm^3, are slightly below the expected value of 6,780±1,360 cm^3. Potential explanations for this difference include consumption of oxygen during the period of apnea or the decrease of pulmonary blood volume between FRC and TLC. Also important could be compression of intrathoracic gas during relaxation against the buccal valve during the acquisition, leading to a slight decrease of intrathoracic volume.

A complete analysis of the mechanics of the diaphragm needs to give the relationships among tension, transdiaphragmatic pressure, muscle fiber length, thickness, and shape for various conditions of load and of activation of muscle fibers. In keeping with this, Smith and Bellemare (1987) demonstrated the loss in effectiveness of the diaphragm as a pressure generator as lung volume increases in a sitting position. At a volume close to TLC, the diaphragm was still able to generate a transdiaphragmatic pressure but without converting it into

a useful inspiratory pressure. However, simultaneous measurement of all the key factors has not yet been achieved and correlations between geometric and functional data are still lacking.

In order to gain further insights in the mechanics of the diaphragm, diaphragm areas, total area and apposition zone, obtained from 3D reconstructions of MRI of the thorax at different lung volume, have been correlated with twitch transdiaphragmatic, esophageal and gastric pressures, determined by bilateral supramaximal electrical stimulation of the phrenic nerve in the neck (personal unpublished data). An inspiratory action at TLC still persists, whereas the zone of apposition, inspiratory actor as a site of action of abdominal pressure on the rib cage, disappears (see Fig. 10.3). However, imaging and stimulation of the diaphragm have been performed in a relaxed position, and the dome of a contracting diaphragm is still able to flatten and tilt, carrying volume displacement and pleural pressure falling. As Smith and Bellemare (1987), Gauthier et al. (1994) and Farkas and Roussos (1988) reported, the observed decline in transdiaphragmatic pressure twitch amplitude with increasing lung volume can be accounted for on the basis of the length-tension properties of the diaphragm alone (see Fig. 10.4). This in turn would suggest that the geometric advantage of the diaphragmatic muscle fibers does not vary with lung volume. This is consistent with other studies and with the piston analogy.

The abdominal muscles are primarily expiratory muscles through their action on the diaphragm and the lung, and they play important roles in activities such as coughing and speaking. The biomechanical effect of abdominal belts have been described as having a hydraulic effect on the diaphragm putting

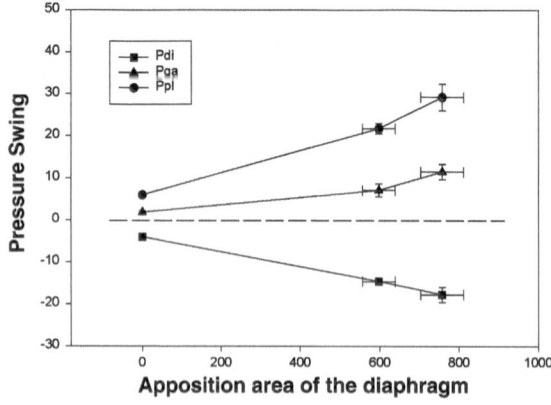

Fig. 10.3. Twitch transdiaphragmatic (Pdi), gastric (Pga), and pleural (Ppl) pressures swings are plotted against apposition area of the diaphragm. *Dashed line* indicates zero pressure change

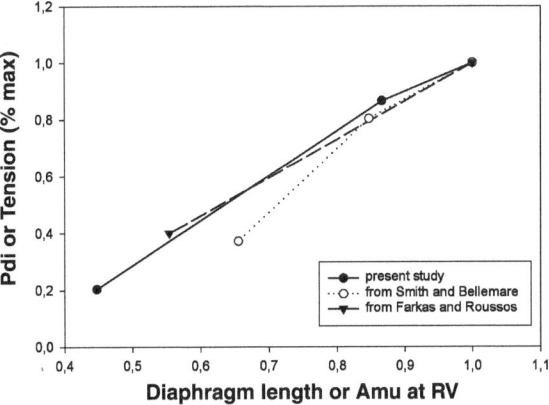

Fig. 10.4. Comparison between in vitro length-twitch force and in vivo muscular area-transdiaphragmatic twitch pressure (Pdi,T) relationships. *Inverted solid triangles* in vitro data adapted from FARKAS and ROUSSOS (1988), *open circles* in vivo data adapted from SMITH and BELLEMARE (1987), *solid circles* in vivo data from personal study (unpublished data)

an extension force on the spine. This effect has been assessed with fast magnetic resonance imaging quantifying the abdominal geometric changes during the Valsalva maneuvers (MIYAMOTO et al. 2002).

10.3
Changes in Chronic Obstructive Pulmonary Disease (COPD) Patients Before and After Lung Volume Reduction Surgery

In COPD patients, hyperinflation of the lungs decreases the operating length of the diaphragm, impairing the inspiratory function of this muscle. However, very few studies have investigated the effect of chronic hyperinflation on diaphragm dimensions and the role of the inspiratory muscles of the rib cage seems controversial in COPD patients (SIMILOWSKI et al. 1991). Using chest X-ray, SHARP et al. (1974) reported that the diaphragm was 40% shorter at FRC in COPD patients than in normal subjects, and ROCHESTER and BRAUN (1985) showed that diaphragm length was reduced by 28% at RV. This difference disappears at similar absolute lung volumes. Using spiral CT and 3D reconstruction, CASSART et al. (1997) studied the effect of chronic hyperinflation on diaphragm length and surface area in COPD patients. They showed that at FRC, patients with COPD have marked reduction in total diaphragm surface area to 73% of the normal value whereas surface area of the zone of apposition decrease to 54%. At the same time diaphragm dimensions were similar to those of

normal subjects when compared at similar absolute lung volumes. Few studies have specifically assessed rib cage dimensions in patients with COPD. Anthropometric techniques (PIERCE and EBERT 1958; KILBURN and ASMUDSSON 1969; GILMARTIN and GIBSON 1984) and radiographic measurements (SHARP et al. 1986; WALSH et al. 1992) have yielded conflicting results. Using CT, CASSART et al. (1996) demonstrated an increase in the anteroposterior but not in transverse diameters. However, at a given absolute lung volume, anteroposterior diameters are smaller in patients with COPD than in normal subjects in the upper but not in the lower portion of the rib cage.

In a study based on an MR analysis of lung volumes and thoracic dimensions in patients with emphysema before and after lung volume-reduction surgery, GIERADA et al. (1998) found a significant correlation between MR and plethysmographic measurements of inspiratory lung volume, although the former measurements were consistently lower than the latter. However, no spirometric monitoring was performed during the acquisition of the images and the lung volumes were determined by plethysmography in the seated position, different from the position of MR image acquisition. Impaired respiratory mechanics in pulmonary emphysema was evaluated with dynamic breathing MRI by SUGA et al. (1999) in 28 patients and respiratory motions of the diaphragm and chest wall were assessed by a cine-loop view. The maximal inspiratory and expiratory amplitude of diaphragm/chest wall and the length of apposition of the diaphragm were significantly reduced in patients compared to controls and the patients frequently showed reduced, irregular or asynchronous motions. These paradoxical diaphragmatic motions have been quantitatively evaluated by IWASAWA et al. (2002) in nine patients but the exact mechanism of the paradoxical motion of the diaphragm is not fully understood. It could be related in part to reduced efficiency of diaphragmatic contraction in emphysema, or abdominal muscle recruitment, or accessory muscles such as parasternal inspiratory intercostal, scalenus and sternocleidomastoid muscles which are less affected by hyperinflation.

Since its introduction by COOPER et al. (1995) and WAKABAYASHI et al. (1994), lung volume reduction surgery has emerged as a useful therapeutic option for patients with severe, non-bullous emphysema. Lung volume reduction surgery improves dyspnea, exercise tolerance and quality of life although the pathophysiologic mechanisms responsible for these improvements are still not fully understood. The major early effects of lung volume reduction surgery are a reduction in static lung volume in particular FRC and RV, and an increase in lung elastic recoil (which leads to a reduction in the degree of airflow obstruction and dynamic pulmonary hyperinflation. In addition to its effect on lung and airway mechanics, lung volume reduction surgery improves the function of the respiratory muscles in particular the diaphragm.

Measuring the most vertically oriented portion of the right hemidiaphragm muscle on the AP chest X-ray, LANDO et al. (1999) found that lung volume reduction surgery leads to a significant increase in diaphragm length, especially in the area of apposition. This lengthening is most likely the result of a reduction in lung volume and is well correlated with postoperative improvements in diaphragm strength, exercise capacity, and maximum voluntary ventilation. Using spiral CT and 3D reconstruction, CASSART et al. (2001) showed that at FRC, the total diaphragm surface area increased by 17±4% and the surface area of the zone of apposition increase 43±8%, but the surface of the dome did not change. The curvature of the dome was unaffected by surgery. The effect of single-lung transplantation has also been studied by CASSART et al. (1999) in nine patients using CT and 3D reconstructions. After single lung transplantation, diaphragm configuration comes back to normal but dome area and with it the average surface area of the muscle value, remains smaller than in normal subjects because the mediastinum is displaced toward the graft.

Few studies using MRI have been published. Anatomic changes in the thorax after a lung resection have been studied by NONAKA et al. (2000). For GIERADA et al. (1998), changes in thoracic dimensions were consistent with improved respiratory mechanics. All thoracic dimensions decreased, and such decreases were greatest at expiration. In 10 patients, surgery produced significant reductions in TLC, FRC and RV, between 20 and 25%, and all patients reported improved function during their activities and daily living after lung volume reduction surgery. The total surface of the diaphragm and the zone of apposition at FRC increased by 18% and 34%, respectively (see Fig. 10.5), but the area of the dome did not change (personal unpublished data).

10.4
Mechanical Model of the Inspiratory Pump

Beside imaging techniques describing the actions of inspiratory muscles through the corresponding geometrical changes (GAUTHIER et al. 1994; PETTIAUX et al. 1994; CASSART et al. 1996; CLUZEL et al. 2000), different attempts have been used to explain the mechanics of inspiration through models of various degrees of complexity (PRIMIANO 1982; BEN-HAIM and SAIDEL 1990; WARD et al. 1992). However, it appears from this literature that an important gap between the models and the data provided by the images remains to be filled. Since KONNO and MEAD (1967) a large number of models have been proposed to describe the respiratory system, two-compartment models having repre-

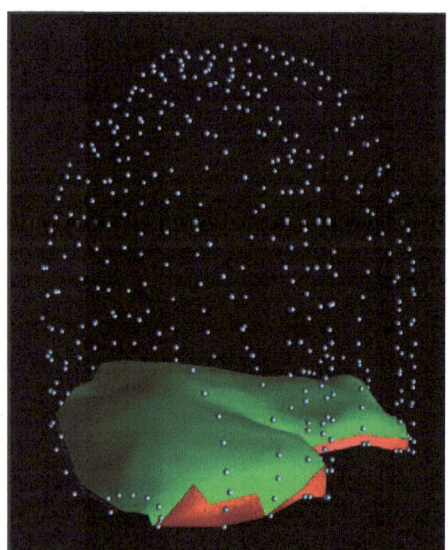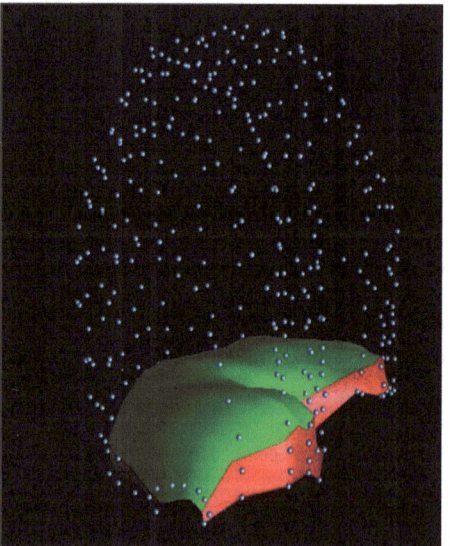

Fig. 10.5a,b. Effect of lung volume reduction surgery on rib cage and diaphragm dimensions before (**a**) and 6 months after (**b**) at FRC, the surface area of the zone of apposition increase and the curvature of the dome was changed by surgery

sented a major advance. To our knowledge, PRIMIANO (1982) was the first to use a mathematical approach to describe the behavior of the two-compartment model he proposed. The architecture of his model is very simple and assumes a series of restricting hypotheses. For example, the thorax is considered a rigid cylindrical body and the compartments are delimited by rigid membranes moving vertically. It is very difficult, if at all possible, to establish a one-to-one correspondence between the parameters of the model and their anatomical counterparts. However it is important to stress the introduction by PRIMIANO (1982) of a complete muscle model with active and passive components. BEN-HAIM and SAIDEL (1990) proposed a more anatomically realistic model, based on a spatial disposition of fixed and moving segments. The spine and shoulders are considered fixed, the chest and abdominal walls can translate horizontally and the apposition zone of the diaphragm is featured. As a further improvement, BEN-HAIM and SAIDEL (1990) simulated different respiratory maneuvers to test and validate their model. The results of these simulations were consistent with physiology. However, the equation used to describe the muscles used in the model

was a global one, thus not accounting for the active, elastic and viscous components of force generation. WARD et al. (1992) developed a model that was the first to account for the chest wall distortion phenomenon. This model is adequate from the geometrical point of view but no numerical simulations have been generated with it. To our knowledge, BASSO-RICCI et al. (2002) were the first to directly relate human anatomical data obtained from MRI of the thorax with the parameters of a mathematical model permitting numerical simulations. Their model is a two-compartment one, with the addition of an elastic membrane representing the diaphragm (see Fig. 10.6). Splitting the equations of the system in kinematic, constitutive, and force balance equations provides an improved description of the system. It is shown that the behavior of the inspiratory pump can be simulated in a straightforward manner using modern computational techniques. Of note, the muscle modeling used in their approach is physiologically realistic, because it includes and distinguishes an active, an elastic, and a viscous component. Functioning validation of the model has been obtained through a series of numerical simulations (see Fig. 10.7). This opens the

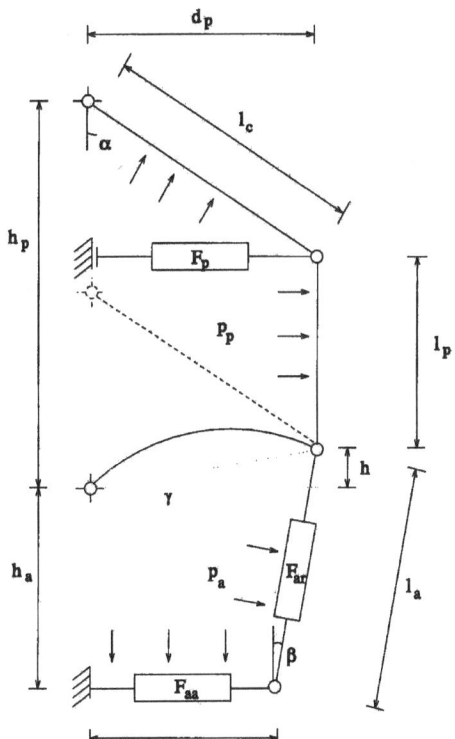

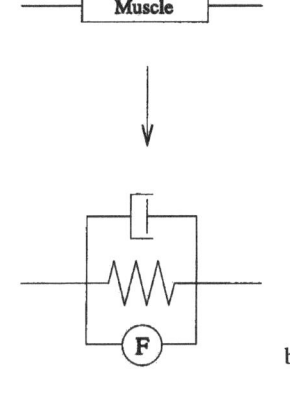

Fig. 10.6a,b. Lateral projection of a mechanical model of the inspiratory pump. (a) The proposed model is a two-compartment one (rib cage-abdominal), composed of a set of rigid and deformable elements fixed on a rigid vertebral column. The moving segments of the rib cage are represented by two rigid elements l_c and l_p. The abdominal wall is represented by two deformable elements l_a and d_a. Between the thoracic and the abdominal compartments, the diaphragm is described by an elastic membrane with an added active muscular component. The chest wall is considered anchored to a fixed point over the vertebral column around which it can rotate, the angle α between the rib cage and the chest wall defining a first degree of freedom. The inferior abdominal wall can translate horizontally, its position being described by d_a. The anterior abdominal wall l_a can accomplish a rototranslation, its position being described by the length d_a and the angle . The forces acting on the respiratory system are given by the rib cage muscles Fp, the oblique and the right abdominal muscles Faa and Far, the diaphragmatic tension $Tdia$ and the abdominal and pleural pressures, p_a and p_p. (b) Active, elastic and viscous components of force generation used in the model

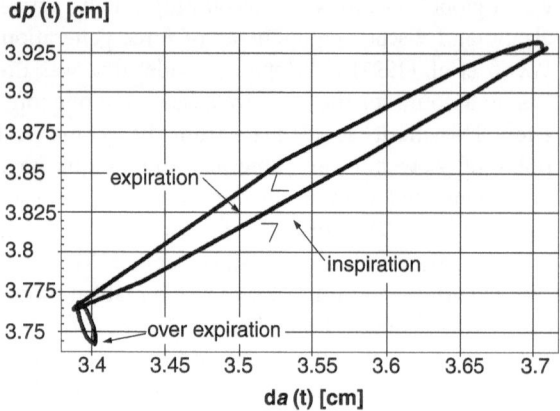

Fig. 10.7. Evolution with time of the chest wall motion (dp) in relationship with abdominal wall motion (da) in response to a combined diaphragm and extradiaphragmatic inspiratory muscle contraction. Rib cage and abdominal dimensions vary in phase during normal respiration, and become out of phase during an active expiration below FRC. This corresponds to physiological observations. It also shows that the model adequately exhibits hysteresis in the inspiratory-expiratory loop, in line with the viscous properties of the respiratory system that it includes

possibility of predicting the behavior of the respiratory system during diseases involving changes in its mechanical or geometrical characteristics. Among possible applications could be the prediction of the results of functional surgery of the thorax, such as lung volume reduction surgery in emphysema. Adding a mathematical description of ventilation taking into account viscoelastic properties of the lung is the next logical step but certainly not the last one toward a perfect model.

10.5
Conclusion and Perspectives

MRI and CT are excellent tools to explore the respiratory mechanics, and have promising applications in clinical and fundamental research. However, supine posture and passive conditions of examinations are the main limitations for physiologic studies. More relevant, but as yet unexplored, is the configuration of the active chest wall. This could be achieved using open MRI, dynamic acquisition using EPI and permanent spirometric control of lung volume.

References

Agostoni E, Hermann R (1960) Abdominal and thoracic pressures at different lung volumes. J Appl Physiol 15:1087–1092

Arora NS, Rochester DF (1982) Effect of body weight and muscularity on human diaphragm muscle mass, thickness and area. J Appl Physiol 52:64–70

Basso Ricci S, Cluzel P, Constantinescu A, Similowski T (2002) Mechanical model of the inspiratory pump. J Biomech 35: 139–145

Ben-Haim SA, Saidel GM (1990) Mathematical model of chest wall mechanics: a phenomenological approach. Ann Biomed Eng 18:37–56

Bergofsky EH (1964) Relative contributions of the rib cage and the diaphragm to ventilation in man. J Appl Physiol 19:698–706

Cander L, Forster RE (1959) Determination of pulmonary parenchymal tissue volume and pulmonary capillary blood flow in man. J Appl Physiol 14:541–551

Cassart M, Genevois PA, Estenne M (1996) Rib cage dimensions in hyperinflated patients with severe chronic obstructive pulmonary disease. Am J Respir Crit Care Med 154: 800–805

Cassart M, Pettiaux N, Genevois PA, Paiva M, Estenne M (1997) Effect of chronic hyperinflation on diaphragm length and surface area. Am J Respir Crit Care Med 156:504–508

Cassart M, Verbandt Y, Francquen P de, Gevenois PA, Estenne M (1999) Diaphragm dimensions after single-lung transplantation for emphysema. Am J Respir Crit Care Med 159: 1992–1997

Cassart M, Hamacher J, Verbandt Y et al. (2001) Effects of lung volume reduction surgery for emphysema on diaphragm dimensions and configuration. Am J Respir Crit Care Med 163:1171–1175

Chapman B, O'Callaghan C, Coxon R et al. (1990) Estimation of lung volume in infants by echo planar imaging and total body plethysmography. Arch Dis Child 65:168–170

Clausen J (1997) Measurement of absolute lung volumes by imaging techniques. Eur Respir J 10:2427–2431

Cluzel P, Similowski T, Chartrand-Lefebvre C, Zelter M, Derenne J-P, Grenier PA (2000) Diaphragm and chest wall: assessment of the inspiratory pump with MR imaging – preliminary observations. Radiology 215:574–583

Cooper JD, Trulock EP, Triantafillou AN et al. (1995) Bilateral pneumectomy (volume reduction) for chronic obstructive pulmonary disease. J Thorac Cardiovasc Surg 109:106–116

Dock DS, Kraus WL, McGuire LB, Hyland JW, Haynes FW, Dexter L (1961) The pulmonary blood volume in man. J Clin Invest 40:317–328

Farkas GA, Rochester DF (1988) Functional characteristics of canine costal and crural diaphragm. J Appl Physiol 65: 2253–2260

Gauthier AP, Verbanck S, Estenne M, Segebarth C, Macklem PT, Paiva M (1994) Three-dimensional reconstruction of the in vivo human diaphragm shape at different lung volumes. J Appl Physiol 72:495–506

Gierada DS, Hakimian S, Slone RC, Yusen RD (1998) MR analysis of lung volume and thoracic dimensions in patients with emphysema before and after lung volume reduction surgery. AJR Am J Roentgenol 170:707–714

Gilmartin JJ, Gibson GJ (1984) Abnormalities of chest wall motion in patients with chronic airflow obstruction. Thorax 39:264–271

Iwasawa T, Kagei S, Gotoh T et al. (2002) Magnetic resonance analysis of abnormal diaphragmatic motion in patients with emphysema. Eur Respir J 19:225–231

Kauczor H-U, Markstaller K, Puderbach M et al. (2001) Volumetry of ventilated airspaces using 3He MRI: preliminary results. Invest Radiol 36:110–114

Kilburn KH, Asmudsson T (1969) Anteroposterior chest diameter in emphysema. Arch Intern Med 123:379–382

Konno K, Mead J (1967) Measurement of the separate volume changes in rib cage and abdomen during breathing. J Appl Physiol 22:407–422

Krayer S, Rehder K, Beck KC, Cameron PD, Didier Edward P, Hoffman EA (1987) Quantification of thoracic volumes by three-dimensional imaging. J Appl Physiol 62:591–598

Lando Y, Boiselle PM, Shade D et al. (1999) Effect of lung volume reduction surgery on diaphragm length in severe chronic obstructive pulmonary disease. Am J Respir Crit Care Med 159:796–805

Miyamoto K, Shimizu K, Masuda K (2002) Fast magnetic resonance imaging used to evaluate the effect of abdominal belts during contraction of trunk muscles. Spine 27:1749–1755

Nonaka M, Kadokura M, Yamamoto S et al. (2000) Analysis of the anatomic changes in the thoracic cage after a lung resection using magnetic resonance imaging. Surg Today 30:879–885

O'Callaghan C, Small P, Chapman B et al. (1987) Determination of individual and total lung volumes using nuclear magnetic resonance echo-planar imaging. Ann Radiol (Paris) 30:470–472

Paiva M, Verbanck S, Estenne M, Poncelet B, Segebarth C, Macklem PT (1992) Mechanical implications of in vivo human diaphragm shape. J Appl Physiol 72:1407–1412

Petroll WM, Knight H, Rochester DF (1990) Effect of lower rib cage expansion and diaphragm shortening on the zone of apposition. J Appl Physiol 68:484–488

Pettiaux N, Cassart M, Paiva M, Estenne M (1994) Three-dimensional reconstruction of human diaphragm with the use of spiral computed tomography. J Appl Physiol 76:495–506

Pierce JA, Ebert RV (1958) The barrel deformity of the chest, the senile lung and obstructive pulmonary emphysema. Am J Med 25:13–22

Primiano FP (1982) Theoretical analysis of chest wall mechanics. J Biomech 15:919–931

Rochester DF, Braun NMT (1985) Determinants of maximal inspiratory pressure in chronic obstructive pulmonary disease. Am Rev Respir Dis 132:42–47

Rochester DF, Farkas GA (1987) Airway pressure responses to phrenic stimulation: dependance on lung volume as well as diaphragm length. Am Rev Respir Dis 135:A330

Sharp JT, Danon J, Druz WS, Goldberg NB, Fishùman H, Machnach W (1974) Respiratory muscle function in patients with chronic obstructive pulmonary disease: its relationship to disability and to respiratory therapy. Am Rev Respir Dis 110:154–167

Sharp JT, Goldberg NB, Druz WS, Danon J (1975) Relative contributions of rib cage and abdomen to breathing in normal subjects. J Appl Physiol 39:608–618

Sharp JT, Beard GAT, Sunga M et al. (1986) The rib cage in normal and emphysematous subjects: a roentgenographic approach. J Appl Physiol 61:2050–2059

Similowski T, Yan S, Gauthier AP, Macklem PT, Bellemare F (1991) Contractile properties of the human diaphragm during chronic hyperinflation. N Engl J Med 325:917–923

Smith J, Bellemare F (1987) Effect of lung volume on in vivo contraction characteristics of human diaphragm. J Appl Physiol 62:1893–1900

Suga K, Tsukuda T, Awaya H et al. (1999) Impaired respiratory mechanics in pulmonary emphysema: evaluation with dynamic breathing. J Magn Reson Imaging 10:510–520

Wakabayashi A (1994) Thoracoscopic partial lung resection in patients with severe chronic obstructive pulmonary disease. A preliminary report. Arch Surg 129:940–943

Walsh JM, Webber CL, Fahey PJ, Sharp JT (1992) Structural change of the thorax in chronic obstructive pulmonary disease. J Appl Physiol 72:1270–1278

Wang CS, Josenhans WT (1971) Contribution of diaphragmatic/abdominal displacement to ventilation in supine man. J Appl Physiol 31:576–580

Ward ME, Ward JW, Macklem PT (1992) Analysis of human chest wall motion using a two-compartment rib cage model. J Appl Physiol 72:1338–1347

Whitelaw WA, Hajdo LE, Wallace JA (1983) Relationship among pressure, tension, and shape of the diaphragm. American Physiological Society, pp 1899–1905

Whitelaw WA (1987) Shape and size of the human diaphragm in vivo. J Appl Physiol 62:180–186

Yu PN (1969) Pulmonary blood volume in health and disease. Lea and Febiger, Philadelphia

11 Respiration Therapy

Klaus Markstaller

11.1
Acute Lung Injury:
Challenges in Respiratory Therapy

Artificial ventilation is provided under circumstances in which a sufficient gas exchange cannot be secured by the patient's own respiratory function. Artificial ventilation might be supportive or completely controlled by the respirator. Within the last few years a large variety of different respiratory modes have been established in critical care medicine, clinical anesthesia and pneumonology to offer optimal ventilatory support under any circumstances. The challenge of artificial ventilation increases dramatically when the lung itself is affected of the patient's disease. In critical care medicine, the acute respiratory distress syndrome (ARDS) is one of the most important diseases which influence the outcome of these critically ill patients. ARDS represents a syndrome which is defined by an inhomogeneous distribution of ventilation and perfusion (V/P) followed by low oxygenation (oxygenation index, PaO_2/F_IO_2 200) without cardiac dysfunction (wedge pressure 18 mmHg). Lung imaging, especially CT, reveals the morphological aspects of this syndrome and helps to understand

K. MARKSTALLER, MD
Department of Anesthesiology, Johannes Gutenberg University Medical School, Langenbeckstrasse 1, 55131 Mainz, Germany

the underlying processes over time, especially the high inhomogeneity of V/P. Within the last decades pathophysiological mechanisms which contribute to ARDS have been intensively investigated.

Several mechanisms were detected which may lead to or aggravate lung damage during artificial ventilation. These mechanisms are thought to contribute to the "ventilator associated lung injury" (VALI):

1. A cyclical collapse and recruitment of atelectasis, mainly in the basal and dependent parts of the lung. This phenomenon leads to high shear forces between collapsed and aerated alveoli as well as to a varying shunt fraction with every respiratory cycle (DREYFUSS and SAUMON 1998; SLUTSKY and RANIERI 2000).

2. An overdistension of aerated lung parenchyma, the so-called baby lung. This yields an inflammatory stimulus triggering the process of lung injury and potentially the transition to multiple organ dysfunction syndrome (MODS) (International Consensus Conferences in Intensive Care Medicine 1999).

Based on this theory prospective randomized multicenter studies were designed to address the contribution of different ventilatory strategies on the outcome of artificially ventilated patients in ARDS. These studies showed that artificial ventilation with low tidal volumes, avoidance of high peak airway pressures and permissive hypercapnia, significantly increased the outcome in this syndrome (The Acute Respiratory Distress Syndrome Network 2000). However, these outcome-oriented studies are also controversially discussed, and mortality in ARDS remains high (SLUTSKY and RANIERI 2000; DE DURANTE et al. 2002). Recent studies report an outcome between 52% and 34% in this patient group that also yields high treatment and social costs (ESTEBAN et al. 2002; BERSTEN et al. 2002).

In clinical routine, the effectiveness of respiratory therapy is monitored primarily by clinical investigation, inspiratory and expiratory gas concentrations, airway pressures and the oxygenation index (PaO_2/F_IO_2). These parameters serve as easy measurements to quantify the effectiveness of the global system of lung ventilation and perfusion. Additional information is gained by the mixed venous oxygen concentration, cardiac output, dead space ventilation, oxygen carrying capacity, and the intrapulmonary shunt fraction. Lung imaging, e.g. conventional chest radiography, spiral and high resolution CT or lung scintigraphy, is used to depict regional pathologies and adverse events during artificial ventilation such as pneumonia or pneumothorax due to barotrauma.

The challenge for the intensive care physician is to evaluate the optimal respiratory parameters for each individual patient and to adapt these parameters during the clinical course, rapidly and prospectively in such a way to minimize VALI. Nowadays, the basic concept in respiratory therapy is focused on avoiding cyclical collapse and recruitment of atelectasis, as well as high F_IO_2, and facilitating spontaneous breathing as soon as possible (International Consensus Conferences in Intensive Care Medicine 1999; The Acute Respiratory Distress Syndrome Network 2000).

Experimental and clinical studies developing these concepts usually compared different airway pressures, inspiratory-to-expiratory time ratios, flow curves and tidal volumes. Recent studies also demonstrated the importance of respiratory rate on the dynamic behavior of atelectasis and their cyclical collapse (BAUMGARDNER et al. 2002).

Despite a huge amount of studies, which addressed an optimal positive end-expiratory pressure (PEEP) level, the best way to titrate PEEP is still discussed controversially (BRUNET et al. 1995). Traditionally PEEP is derived from the lower inflection point (LIP) of the static pressure volume (PV) curve (AMATO et al. 1998). In clinical practice, the assessment and interpretation of the PV curve is limited, as it is dangerous to perform complete derecruitment of the lung, and the information summarizes only the PV relationship of the total, but very inhomogeneously ventilated lung (HARRIS et al. 2000; BROCHARD 2001; DREYFUSS and SAUMON 2001; HICKLING 2001; HUBMAYR 2002).

Radiological imaging such as chest radiography (CXR) yields only limited information about the ventilated lung areas. However, it allows detection of adverse effects caused by artificial ventilation. It was shown that CT offers additional information in these patients in 66% compared with conventional CXR, which led to clinical consequences in 22% (TAGLIABUE et al. 1994). Spiral CT or HRCT is performed during breath-hold at end inspiration or at any predefined airway pressure. Thus, the total amount of atelectasis at a predefined airway pressure can be visualized and quantified by CT.

Clearly, this imaging strategy does not reflect the dynamic behavior of lung ventilation or perfusion during the respiratory cycle, which will both vary regionally and are influenced by the respiratory parameters during artificial ventilation. Ventilation is determined regionally by local lung resistance and lung compliance. New monitoring techniques aim to assess lung ventilation and perfusion with high temporal and regional resolution to address regional V/P ratios. In the ICU setting two novel techniques seem to be promising: dynamic computed tomography (dCT), which offers high temporal and spatial resolution and allows visualization of the effect of artificial ventilation on different functional lung compartments such as atelectasis, ventilated lung area and overinflated lung area (Fig. 11.1) (NEUMANN et al. 1998a,b; MARKSTALLER et al. 2001b). Also electrical impedance tomography (EIT) allows differences of lung aeration during artificial ventilation to be visualized and does so as a bedside technique (FRERICHS 2000).

Recent studies showed the potential of both imaging techniques, yielding regional and dynamic information so far unmatched by any other modality. These technical possibilities led a number of researchers to explore which additional information can be gained by functional lung imaging in the challenging process to optimize respiratory parameters in ARDS. In the following chapter an overview of this interdisciplinary effort is provided.

11.2
Functional Imaging Techniques in ARDS

11.2.1
Radiological Imaging

Previously, ARDS was thought to be a homogeneous process, and imaging of the aerated lung was limited to conventional chest radiography. Since the 1970s, CT offers a new radiological imaging technology. However, it took about 10 years until this technique was applied in ARDS patients. CT imaging revealed that ARDS represents an inhomogeneous pathology, which leads to atelectatic, hypoventilated and well-aerated lung regions next to each other, dependent on etiology, ventilatory parameters and positioning (ROMMELSHEIM et al. 1983; GATTINONI et al. 1986a,b; MAUNDER et al. 1986).

CT visualizes lung morphology with high spatial resolution. A slice thickness of approximately 1 mm

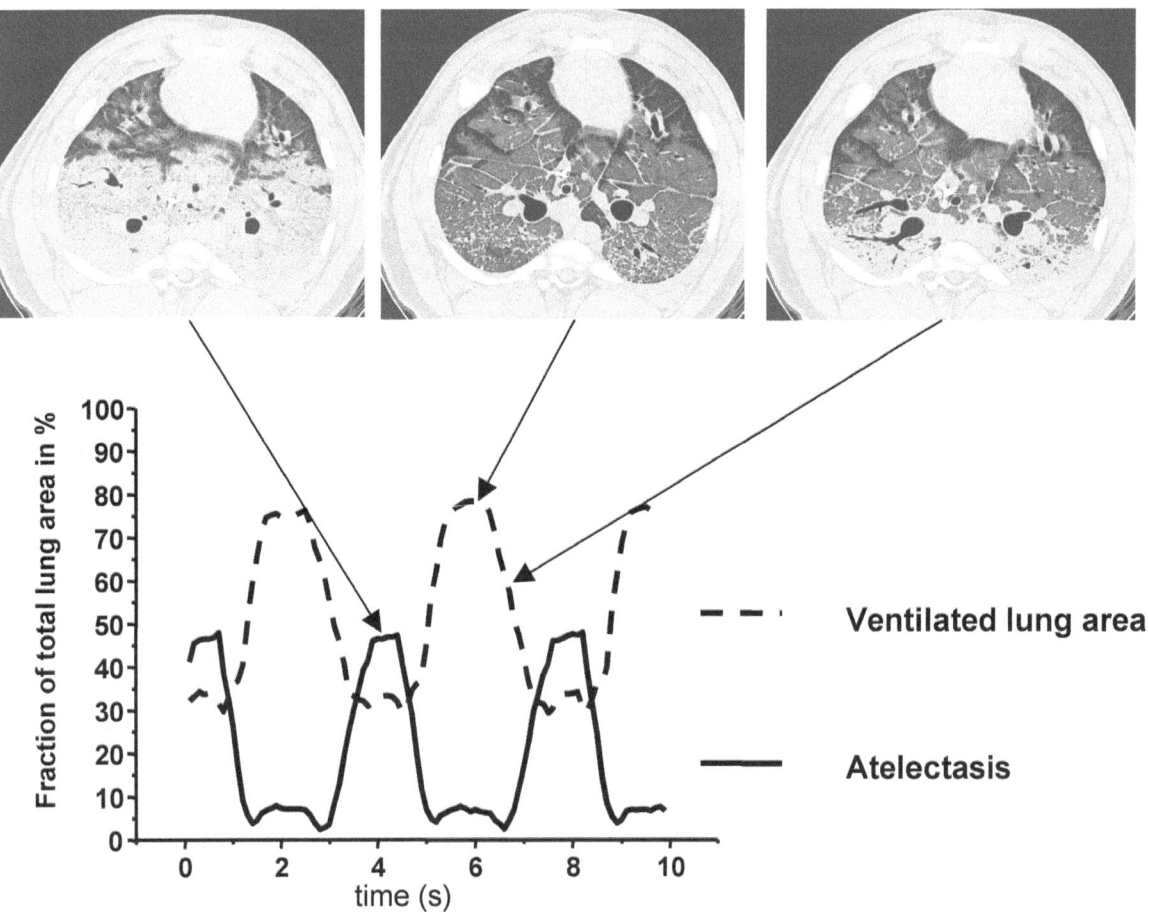

Fig. 11.1. Illustration of dynamic CT images (porcine lavage ARDS model) and the image-derived lung density ranges reflecting ventilated lung and atelectasis during pressure-constant ventilation (*PEEP*=15cmH$_2$O; *Paw*=23cmH$_2$O; *I:E*=1:1; F_lO_2=1.0) (from MARKSTALLER et al. 2003, with permission)

and high resolution reconstruction algorithms yield high spatial resolution. Functional information is obtained by quantitative density measurements within the imaged lung parenchyma.

11.2.1.1
Computed Tomography

Images in CT reflect the radiation absorption visualized by a gray scale, scaled in Hounsfield Units (HU) ranging from –1024 HU to 3071 HU. This technique allows an absolute calibration which is defined by water (0 HU) and air (–1000 HU), as the density absorption of CT is linearly correlated to the physical tissue density (MULL 1984). This makes an important difference with regard to other imaging techniques, such as e.g. EIT or MRI, that visualize changes in lung ventilation or perfusion, but do not allow absolute numbers of air, water or tissue contents within a voxel to be calculated.

Within the lung parenchyma, four different factors influence density in CT (at functional residual capacity, FRC): air (approx. 60%), blood (approx. 15%), water (interstitial, alveolar and intracellular, approx 20%) and solid tissue (approx. 5%) (BENUMOF 1995).

A lung acinus comprises about 2,000 alveoli at a volume of about 16–22 mm^3. The volume of a CT voxel is defined by the slice thickness and the CT matrix, e.g. 0.56 mm^3 with the following imaging parameters for spiral CT: 1 mm slice thickness, matrix 512 512, voxel dimensions of 0.75 0.75 1 mm, summarizing over about 50 alveoli. Whereas the CT voxel does not change in size and position during ventilation, the lung acinus does (Fig. 11.2) (GATTINONI et al. 2001). In addition, the content of air, edema, vessels, septa, collapsed alveoli, aerated alveoli etc. of the imaged lung voxel will vary during the respiratory cycle.

Recently, the dominating mechanism which is responsible for VALI and also for the beneficial effect of lung protective ventilation, has been controver-

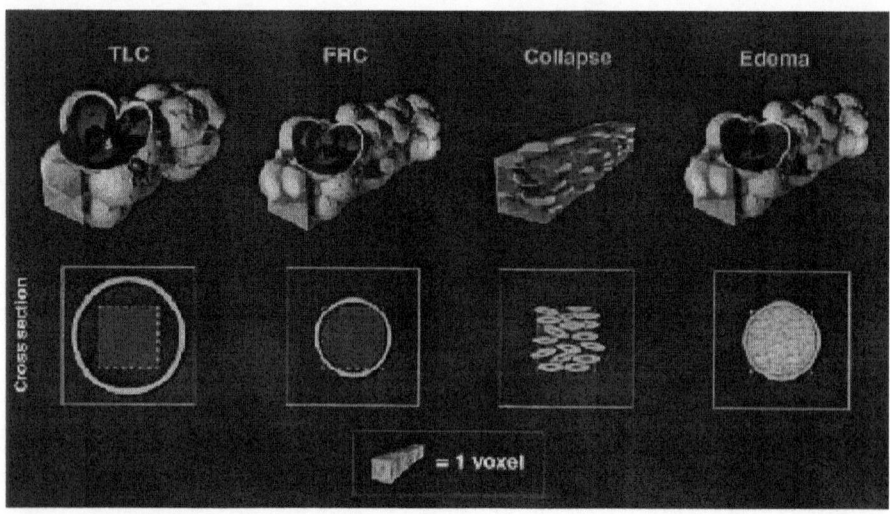

Fig. 11.2. Relationship between the "CT pulmonary unit" and alveolar acini in normal condition at functional residual capacity (FRC), at two-thirds total lung capacity (TLC), with collapsed alveoli otherwise normal, and with edema (from Gattinoni et al. 2001, with permission)

sially discussed. The traditional concept is based on the theory of atelectatic collapse and recruitment as discussed above, and CT studies were used to prove this hypothesis. However, this hypothesis was recently doubted since there was evidence that high density areas at CT may reflect water-flooded alveoli next to aerated alveoli (Fig. 11.3), which cannot be discriminated by CT due to its limited spatial resolution (Hubmayr 2002). Obviously, this theory also puts the contributing factors for VALI into a new light. Further studies have to prove whether CT shows atelectatic collapse and recruitment which might be the dominating pathomechanism of VALI.

Dynamic CT Imaging

Dynamic CT (dCT) represents an imaging technique originally performed for perfusion imaging. In this field, dCT did not enter clinical routine but recently, this imaging modality was applied to visualize respiration processes in acute lung injury.

From a technical point of view, dCT is performed by a continuous rotation of the x-ray tube at a fixed table position. Minimal slice thickness and a high resolution reconstruction algorithm at a matrix of 512 512 yield maximal spatial resolution. Modern CT scanners reach a complete rotation of the x-ray tube in less than 500 ms, and raw data acquired within a 270° rotation of the x-ray tube are already sufficient to reconstruct an image. Once the raw data for a complete image reconstruction have been acquired, additional raw data over 50° of the x-ray tube rotation are used to reconstruct

subsequent images (sliding-window reconstruction technique). Thus, this algorithm yields a temporal increment of 100 ms in dCT image series. Respiratory motion can now be visualized over several respiratory cycles within dCT series of several seconds duration. The total scan duration is limited by the RAM capacity of the scanner for raw data storage and the performance of the x-ray tube. Obviously, the scan duration should be held as short as possible due to the relatively high amount of radiation exposed during continuous imaging in one single slice.

Quantitative Image Analysis in CT

Different density parameters for quantitative CT studies have been developed and presented over the past few years.

The mean lung density (MLD) represents a single number descriptor summarizing all pixels of a transversal lung slice or a predefined region of interest (ROI). It is most often used in the radiological literature and is implemented in nearly every image postprocessing software. However, this parameter does not allow for any distinction of air, water or tissue within the observed lung structure (Gattinoni et al. 2001). The contribution of these factors to the MLD value can only be estimated by a density histogram of the analyzed ROI (Neumann et al. 1998a,b).

A density parameter is given by a predefined density value which is 0 HU, and air has a density of −1000 HU. Thus, a density of −700 HU in a pixel reflects 70% of air in that pixel. As the exact dimen-

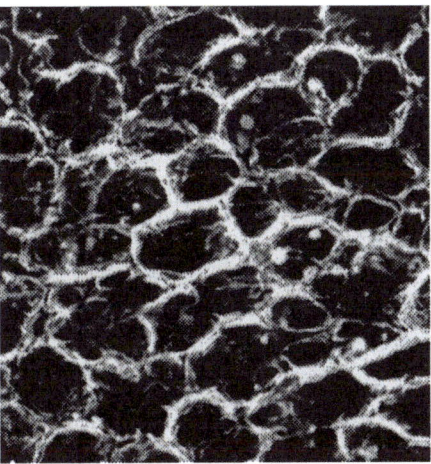 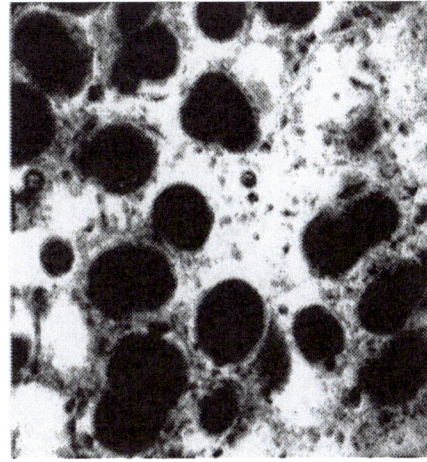

Fig. 11.3. Normal (*left*) versus edematous (*right*) rat lung imaged by laser confocal images, where the perfusate was labeled with fluorescein-labeled dextran (from HUBMAYR 2002, with permission)

sions of a voxel are well known in CT, the total amount of air in milliliters can be easily calculated (PUYBASSET et al. 2000). This parameter offers a very comprehensible value, i.e. milliliters of air. Compared to the MLD, the calculated amount of air will also vary during respiration (as the transversal lung slice, i.e. the pixel number, will vary), even if the MLD remains unchanged. However, the limitations of a single number descriptor still apply: comparable to the MLD this parameter is not able to distinguish the amount of atelectatic lung, poorly aerated lung, over-inflated lung or airways within a lung slice or a ROI. This discrimination can only be reached by combining the total amount of air information with a density histogram of the observed ROI, which is not feasible for a large number of CT images, e.g. in a dCT series.

A density parameter taking this implication into account was introduced by predefined density ranges, which are individually assigned to different functional lung compartments, e.g. atelectasis, ventilated or over-inflated lung. Several density ranges have been correlated to these functional lung compartments by different authors and in different species, pathologies and imaging parameters (DAMBROSIO et al. 1997; VIEIRA et al. 1998; MARKSTALLER et al. 1999). Furthermore, dedicated software tools are necessary, which planimetrically calculate the respective lung area (in cm^2) of each density range. To allow for inter-individual comparison of the data, the respective lung area of a density range is usually given as the fraction of the total lung area (MARKSTALLER et al. 2001a).

Whatever density parameter is used for quantitative analysis of lung aeration and atelectasis in ARDS, an automatic lung segmentation tool is required to postprocess the high number of images which are produced by dCT imaging (Fig. 11.4) (MARKSTALLER

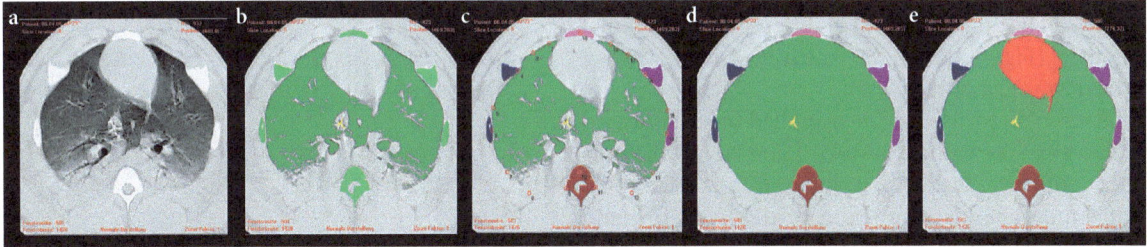

Fig. 11.4a–e. Automatic lung segmentation of a CT image showing a porcine lavage ARDS model (from MARKSTALLER et al. 2001a, with permission). (**a**) CT image from a dynamic CT series to demonstrate the software-based automatic lung segmentation. The image was acquired at end expiration after induction of lavage ARDS in a 24 kg pig. The depletion of surfactant in this model leads to high levels of atelectasis in the posterior, dependent parts of the lung. (**b**) Detection of bones (density-derived and by morphologic opening and closing). All pixels within the thorax between –1024 and +300 HU are marked. (**c**) Classification of bones using expert anatomic knowledge (color coded representation of sternum, left and right rib cage and vertebral column) and definition of interpolation points. (**d**) Segmentation of lung area within the interpolation curve. (**e**) Detection of the heart using morphological opening and closing algorithms and subtraction of this area from the total lung area

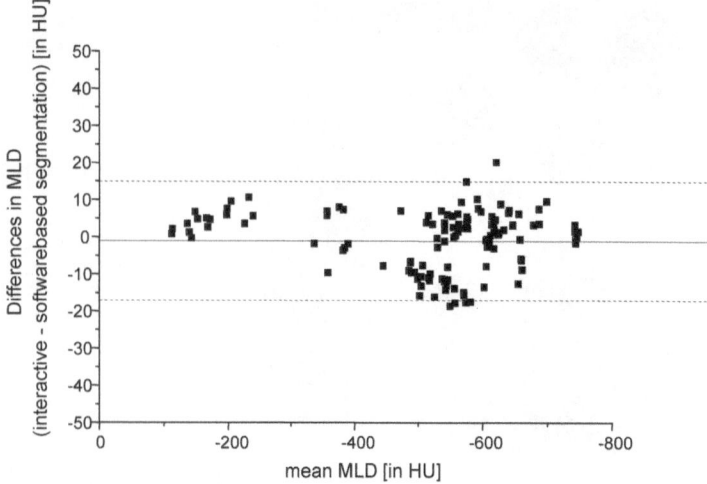

Fig. 11.5. Bland-Altman plot illustrating the agreement of the mean lung density (MLD) derived from an interactive versus automatic lung segmentation (healthy and lavage ARDS lungs, n=120) (data from MARKSTALLER et al. 2001a, with permission). An overestimation of the MLD of only 1 HU, and a slight variation of only 20 HU can be noted at a level of technical noise of the scanner of 10 HU

et al. 2001a). Software tools that use density ranges in combination with expert anatomic knowledge have been presented, and demonstrated an excellent correlation with interactive lung segmentation by experienced radiologists (Fig. 11.5).

Within the last few years, several experimental studies have been published, which focused on a quantitative analysis of lung aeration in ARDS with dCT imaging.

Time constants (TCs) in pulmonary physiology define the time necessary to reach 63.2% of a total inspiration or expiration. TCs are usually measured by spirometry and allow adaptation of the inspiration-expiration time and the respiratory rate in artificial ventilation. Functional lung imaging by dCT offers the possibility to determine TCs even on a regional basis. This will be of special interest in a diseased lung with an inhomogeneous distribution of ventilation. In recent studies, regional TCs of lung aeration derived from dCT measurements have been assessed, using the MLD or predefined density ranges in healthy lungs and in different ARDS animal models (Figs. 11.6, 11.7) (NEUMANN et al. 1998a; MARKSTALLER et al. 2001b).

This method also has been applied in anesthetized volunteers (ROTHEN et al. 1999) and ARDS patients (KARMRODT et al. 2002). Compared to conventional lung function, discrimination of different TCs and localization of the respective functional compartments, which respond differently to changes in airway pressure, are possible (Fig. 11.8). This application of dCT has not yet been analyzed in a prospective patient study. Thus, no data are available to prove its importance for the patient's benefit in the process of optimization of respiratory therapy. However, it has been shown that dCT can visualize and quantify

the kinetics of alveolar recruitment in a single respiratory cycle, and it does so with high spatial resolution. Especially in the inhomogeneously aerated lung in ARDS patients, this technique offers a control of the effect of respiratory parameters, such as PEEP, plateau pressure or respiratory rate in predefined lung regions, a feature which is not attained by any other clinical investigation.

The shunt fraction is an important determinant factor on lung oxygenation in ARDS. The iso-shunt diagram illustrates the influence of different shunt fractions of the cardiac output on the arterial partial pressure. In this model the shunt fraction represents

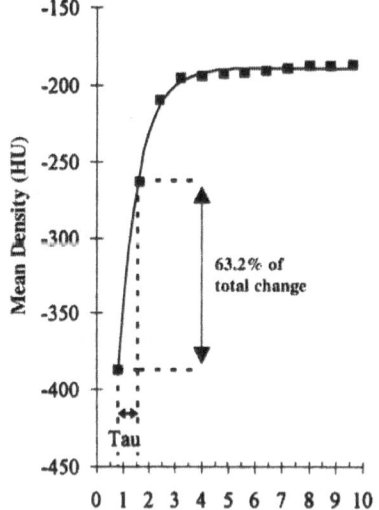

Fig. 11.6. Exponential curve fitting of the MLD values acquired from dynamic CT during an airway pressure step-down in a porcine oleic acid ARDS model, t is the time constant, defined as the time in which 63.2% of the total change in MLD occurs (from NEUMANN et al. 1998b, with permission)

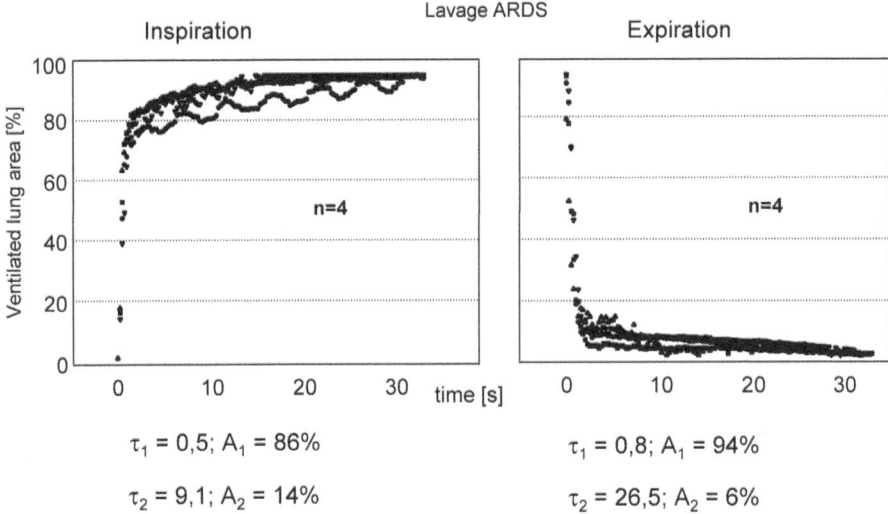

Fig. 11.7. Relative increase/decrease of aerated lung area after airway pressure steps of 50 cm H_2O in a lavage ARDS model. Aerated lung area (pixels with densities ranging from –910 to –300 HU) is expressed as the fraction of total lung area (in %), which becomes aerated/deaerated from the airway pressure steps. Two different time constants (τ_1 and τ_2) and their respective contribution to total lung area (A_1 and A_2) are calculated by bi-exponential fitting of the curve (from MARKSTALLER et al. 2001b, with permission)

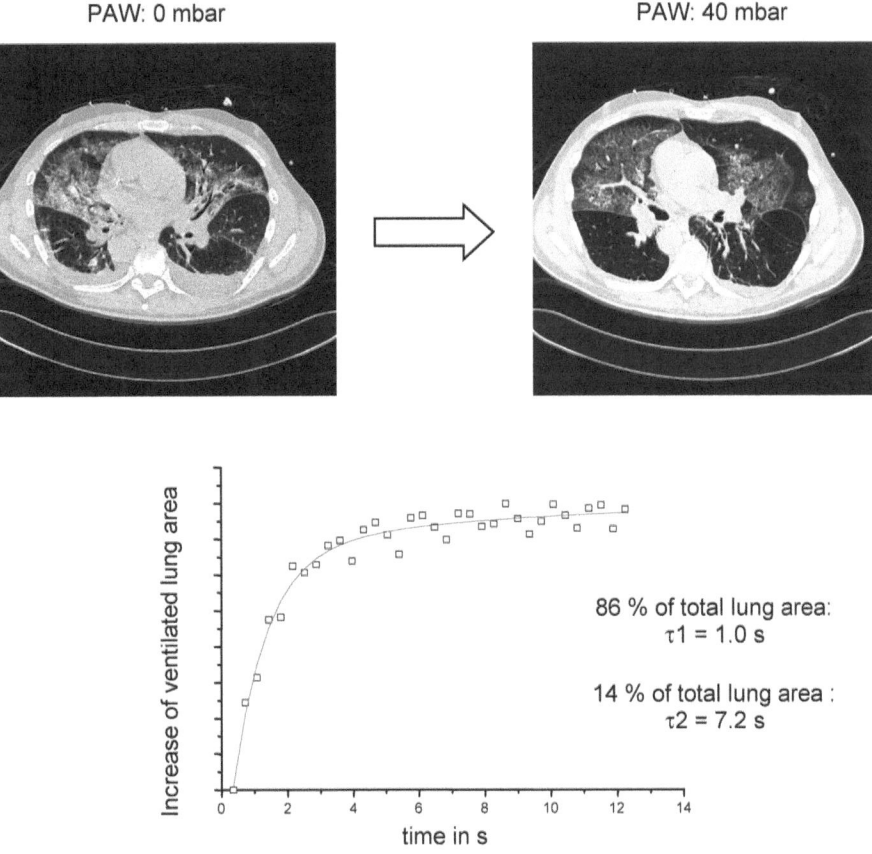

Fig. 11.8. Airway pressure step maneuver from 0–40 mbar in a patients with early ARDS due to aspiration and sepsis. Dynamic CT images are acquired during the airway pressure step-up maneuver and the increase of ventilated lung area is plotted versus time. Two different time constants (τ_1 and τ_2) and their individual contribution to lung aeration were determined by bi-exponential fitting of the curve. A very slow aeration process (τ_2) parallels, similar to the results in ARDS animal models, a predominant faster time constant (τ_1)

the mixed venous admixture to arterial blood, which in ARDS is mainly influenced by the amount of atelectasis (i.e. perfused but not ventilated lung compartments). In the iso-shunt diagram the influence of different inspiratory oxygen concentrations is also integrated (Fig. 11.9) (LUMB 2000).

Studies using the multiple inert gas technique (MIGET) suggested that the total amount of atelectasis should be the major contributing factor to the intrapulmonary shunt fraction in ARDS. Thus, several studies tried to correlate the intrapulmonary shunt fraction to the image-derived amount of atelectasis. However, these studies reported controversial results.

Whereas the atelectatic lung area correlated positively with the calculated shunt fraction by the MIGET during expiratory breath-hold (TOKICS et al. 1987, 1996), other authors showed a correlation at inspiratory, but not expiratory breath-holds (NEUMANN et al. 1998). It is clear that breath-hold maneuvers during imaging will not reflect the (patho-) physiological situation during continuous respiration. An experimental study using a lavage-ARDS model showed a positive correlation of the mean atelectatic lung area during continuous respiration gained by dCT imaging, and the calculated shunt fraction measured by arterial and mixed venous blood gases (Fig. 11.10) (MARKSTALLER et al. in press).

Recruitment of the atelectatic lung has been addressed by CT imaging (CROTTI et al. 2001; PELOSI et al. 2001) demonstrating that recruitment takes place over the whole inspiration process, i.e. independent of the lower and upper inflection point as determined by the static pressure-volume (PV) curve. Own experiments using dCT imaging during pressure constant ventilation demonstrated that higher PEEP levels as suggested by the LIP are beneficial to avoid end-expiratory lung collapse in a lavage ARDS model (KARMRODT et al. 2002).

First clinical applications demonstrated that dCT imaging and postprocessing are feasible in ARDS patients. Further studies are necessary to investigate whether this technique may contribute to adapt respiratory parameters individually and prospectively in ARDS patients and potentially offer a beneficial effect for respiratory therapy in this patient group.

Limitations are a relatively high radiation exposure, which has been estimated to about 50 mGy for a 10 s dCT scan in humans (HEUSSEL et al. 2001). Furthermore, transport of ICU patients to the CT scanner is demanding in sophisticated transport ventilators, continuous invasive monitoring and manpower, and so far increases therapy costs.

New volumetric CT scanners might offer dynamic 3D-imaging of the lung in the future. Up to now, this imaging technique has been limited to one or several slices very close to each other. Thus, the investigation has to be repeated at several lung levels (which dramatically increases radiation exposure) or the information will be limited to one single lung region, typically a slice next to the diaphragm. Whether a representative ARDS lung section compared to the disease process in the whole organ exists has also been controversially discussed in the literature (TAGLIABUE et al. 1998) vs. (GATTINONI et al. 1994; BRUNET et al. 1995; LU et al. 2001). Recent studies suggest that an increase of lung tissue (consistent with the theory of increased interstitial edema and alveolar flooding) is homogeneously distributed

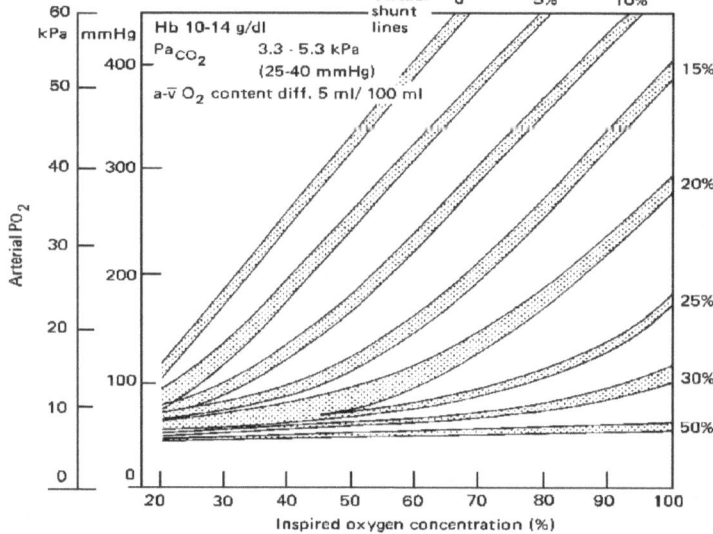

Fig. 11.9. Iso-shunt diagram. On co-ordinates of inspired oxygen concentration (abscissa) and arterial PO₂ (ordinate), iso-shunt bands have been drawn to include all values of Hb, PaCO₂, and arteriovenous content difference shown in the graph (from LUMB 2000, with permission)

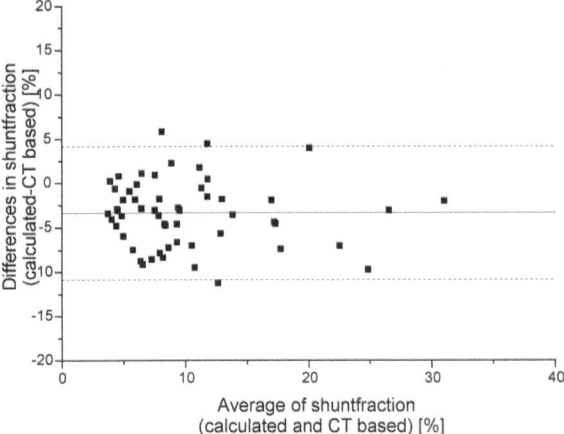

Fig. 11.10. Bland-Altman analysis of the relationship between atelectatic lung area determined by dynamic CT and shunt fraction determined by blood gas analyses (*solid line* bias, *dashed lines* limits of agreement, i.e. bias ±2 SD). Linear regression analysis showed a good correlation of both methods (healthy lungs: $r=+0.76$, lavaged lungs: $r=+0.89$) (from MARKSTALLER et al. 2003, with permission)

throughout the lung, whereas the loss of aeration (consistent with the alveolar collapse theory) is inhomogeneously distributed, i.e. predominantly found in caudal parts of the lung (ROUBY et al. 2003).

11.2.2
Electrical Impedance Tomography

Electrical impedance tomography (EIT) represents a new functional imaging technique, which allows transthoracic visualization of changes in lung aeration. This non-invasive and radiation-free technique offers new perspectives as a bed-side monitoring tool of artificial ventilation in ICU patients (FRERICHS 2000; FRERICHS et al. 2001).

EIT measures changes in electrical impedance at the thorax wall during uninterrupted ventilation. The changes in impedance are produced by continuous electrical impulses, whereby 16 or 32 skin surface electrodes (e.g. ECG electrodes) are placed around the thorax in a predefined transversal plane. The electrical impulse is generated between two electrodes next to each other and rotates through all the electrodes placed in the respective transversal plane. The transmitting electrode creates an electrical impulse (50 KHz, 5 mA peak-to-peak) and subsequently the impedance generated from this impulse is received by all the remaining electrodes (Fig. 11.11).

One measurement cycle is finished once each single electrode has served as transmitter of such an electrical impulse. This leads to 208 different data points for changes in impedance during one measurement cycle. Each measurement consists of a reference measurement and the data acquisition of the whole data set. A dedicated software tool compares the measured impedances with the impedances of the reference scan and normalizes the data (BARBER 1989). Finally, these normalized changes in impedance and their regional distribution are used to generate a two-dimensional image with a matrix of 32 32. This matrix is visualized in a gray or colored scale, which represents the intrathoracic gas volumes and the time-dependent changes during inspiration and expiration. It is important to note that the EIT images do not reflect absolute gas contents, but relative changes in aeration compared to the reference EIT scan. In addition, only processes which influence the impedances at the surface of the thoracic wall are visible by this technique.

A more advanced version of EIT, so-called functional EIT (fEIT), summarizes approximately 240 single EIT acquisitions in 1 image. For each pixel the relative change in impedance over time is given

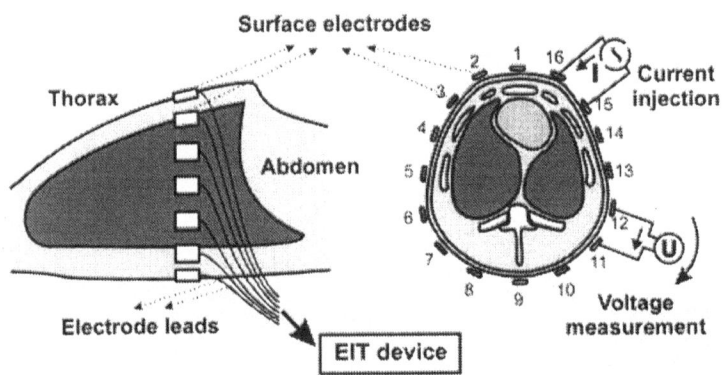

Fig. 11.11. Measuring principle of electrical impedance tomography: Electrical excitation currents (I) are consecutively applied between pairs of adjacent surface electrodes (1–16). After each current injection, resulting voltages (U) are measured between the remaining electrode pairs (from FRERICHS et al. 2002, with permission)

as standard deviation of several impedance measurements (HAHN et al. 1996). Tissues which have only minimal influence on the electrical impedance during ventilation, such as muscles or ribs, show only small changes of the regional impedance variations over time. Aeration is the major influence on impedance, and so far this technique is able to depict changes in aeration and lung volumes.

Functional EIT has been validated against various other imaging techniques in the last few years. These quantitative comparisons proved that fEIT is able to visualize local hypoventilation and to detect changes in aeration down to 20 ml (HAHN et al. 1995). It also allows localization of gravitational changes with high accuracy, e.g. enhanced ventilation in dependent lung areas which correspond to measurements performed by ventilation scintigraphy (FRERICHS et al. 1996; HAHN et al. 1996). Comparison of strain gauge plethysmography and changes in electrical impedance showed a significant correlation between both techniques, e.g. the inspired tidal volumes and changes of impedance showed a linear correlation (KUNST et al. 2000b). EIT also showed a good agreement with reference techniques to quantify extracellular lung water. KUNST et al. (1998) used EIT to estimate the distribution of ventilation and perfusion between the right and left lungs and compared these results to ventilation/perfusion scintigraphy. As EIT detects changes in lung aeration it allows estimation of alveolar collapse and recruitment in ARDS (KUNST et al. 2000b). Thus, EIT was used to perform an image-based static P/V curve. In healthy animals, no differences in the LIP were found compared to a conventional P/V curve, whereas in a lavage ARDS model the LIP determined by EIT differed regionally: in the dependent lung a significantly higher LIP was found than in the aerated non-dependent lung (KUNST et al. 2000a). Recently, EIT was compared to electron beam computed tomography (EBCT) and changes in lung aeration (by variation of PEEP and V_T) showed a good agreement between changes in electrical impedance and the MLD measured by CT in different lung regions (FRERICHS et al. 2002).

tilatory parameters in a patient with ARDS is based on physiological concepts of lung ventilation and perfusion and monitored by mostly static measurement techniques such as the PV curve, blood gas analyses or CT imaging during breath-hold. In the last few years new techniques have been developed that offer high temporal and/or spatial resolution of lung function. Imaging techniques such as dCT or EIT visualize regional aeration processes in the diseased lung and the individual responses to changes in the ventilatory pattern. As a bed-side technique EIT may become an ideal technique to follow lung aeration over time. Moreover, it does so without any radiation exposure of the patient and the personnel. In contrast, CT offers an intriguingly high spatial resolution and a real quantitative evaluation of the visual information. As the CT density is correlated linearly to the physical tissue density, the exact amount of air and atelectasis can be calculated by dedicated software. It therefore might be the optimal diagnostic technique for a fast and prospective optimization of respiratory parameters in early ARDS, e.g. at a first diagnostic evaluation of a patient with acute lung injury. Advanced CT techniques at reduced costs now question the role of conventional CXR in the diagnosis and follow up of ARDS patients.

Both dCT and EIT techniques require adequate image handling, image transfer, storage and postprocessing for optimal visual and quantitative presentation of the observed pathology.

Further studies may lead to a further development of both techniques and hopefully allow a three-dimensional representation of the lung parenchyma during the respiratory cycle with high temporal resolution. Preclinical studies are under way to implement diagnostic routines for optimization of respiratory parameters in this patient population. In the future, randomized, prospective trials will show whether this additional diagnostic information is helpful to prospectively optimize respiratory parameters and thus shorten the treatment period of artificial ventilation. This would not only have a positive effect on the individual outcome of patients but also reduce costs in the ICU setting.

11.3
Conclusions and Perspectives

Respiration therapy plays a major role in the treatment of ARDS patients. Protective lung ventilation specifically avoids further damage of the lung and progression of the disease. The titration of the individual ven-

References

Amato MB, Barbas CS et al. (1998) Effect of a protective-ventilation strategy on mortality in the acute respiratory distress syndrome. N Engl J Med 338:347–354
Barber DC (1989) A review of image reconstruction techniques for electrical impedance tomography. Med Phys 16:162–169

Baumgardner JE, Markstaller K et al. (2002) Effects of respiratory rate, plateau pressure, and positive end-expiratory pressure on PaO_2 oscillations after saline lavage. Am J Respir Crit Care Med 166:1556–1562

Benumof J (1995) Anesthesia for thoracic surgery. Saunders, Philadelphia

Bersten AD, Edibam C et al. (2002) Incidence and mortality of acute lung injury and the acute respiratory distress syndrome in three Australian States. Am J Respir Crit Care Med 165:443–448

Brochard L (2001) Watching what PEEP really does. Am J Respir Crit Care Med 163:1291–1292

Brunet F, Jeanbourquin D et al. (1995) Should mechanical ventilation be optimized to blood gases, lung mechanics, or thoracic CT scan? Am J Respir Crit Care Med 152:524–530

Crotti S, Mascheroni D et al. (2001) Recruitment and derecruitment during acute respiratory failure: a clinical study. Am J Respir Crit Care Med 164:131–140

Dambrosio M, Roupie E et al. (1997) Effects of positive end-expiratory pressure and different tidal volumes on alveolar recruitment and hyperinflation. Anesthesiology 87:495–503

De Durante G, Turco M del et al. (2002) ARDSNet lower tidal volume ventilatory strategy may generate intrinsic positive end-expiratory pressure in patients with acute respiratory distress syndrome. Am J Respir Crit Care Med 165:271–1274

Dreyfuss D, Saumon G (1998) Ventilator-induced lung injury: lessons from experimental studies. Am J Respir Crit Care Med 157:294–323

Dreyfuss D, Saumon G (2001) Pressure-volume curves: searching for the Grail or laying patients with adult respiratory distress syndrome on Procrustes' bed? Am J Respir Crit Care Med 163:2–3

Esteban A, Anzueto A et al. (2002) Characteristics and outcomes in adult patients receiving mechanical ventilation: a 28-day international study. JAMA 287:345–355

Frerichs I (2000) Electrical impedance tomography (EIT) in applications related to lung and ventilation: a review of experimental and clinical activities. Physiol Meas 21: R1–21

Frerichs I, Hahn G et al. (1996) Gravity-dependent phenomena in lung ventilation determined by functional EIT. Physiol Meas 17 [Suppl 4A]:A149–A157

Frerichs I, Schiffmann H et al. (2001) Non-invasive radiation-free monitoring of regional lung ventilation in critically ill infants. Intensive Care Med 27:1385–1394

Frerichs I, Hinz J et al. (2002) Detection of local lung air content by electrical impedance tomography compared with electron beam CT. J Appl Physiol 93:660–666

Gattinoni L, Mascheroni D et al. (1986a) Morphological response to positive end expiratory pressure in acute respiratory failure. Computerized tomography study. Intensive Care Med 12:137–142

Gattinoni L, Presenti A et al. (1986b) Adult respiratory distress syndrome profiles by computed tomography. J Thorac Imaging 1:25–30

Gattinoni L, Bombino M et al. (1994) Lung structure and function in different stages of severe adult respiratory distress syndrome. JAMA 271:1772–1779

Gattinoni L, Caironi P et al. (2001) What has computed tomography taught us about the acute respiratory distress syndrome? Am J Respir Crit Care Med 164:1701–1711

Hahn G, Sipinkova I et al. (1995) Changes in the thoracic impedance distribution under different ventilatory conditions. Physiol Meas 16 [Suppl A]:A161–A173

Hahn G, Frerichs I et al. (1996) Local mechanics of the lung tissue determined by functional EIT. Physiol Meas 17 [Suppl 4A]:A159–A166

Harris RS, Hess DR et al. (2000) An objective analysis of the pressure-volume curve in the acute respiratory distress syndrome. Am J Respir Crit Care Med 161:432–439

Heussel CP, Hafner B et al. (2001) Paired inspiratory/expiratory spiral CT and continuous respiration cine CT in the diagnosis of tracheal instability. Eur Radiol 11:982–989

Hickling KG (2001) Best compliance during a decremental, but not incremental, positive end-expiratory pressure trial is related to open-lung positive end-expiratory pressure: a mathematical model of acute respiratory distress syndrome lungs. Am J Respir Crit Care Med 163:69–78

Hubmayr RD (2002) Perspective on lung injury and recruitment. A skeptical look at the Opening and Collapse Story. Am J Respir Crit Care Med 165:1647–1653

International Consensus Conferences in Intensive Care Medicine (1999) Ventilator-associated lung injury in ARDS. American Thoracic Society, European Society of Intensive Care Medicine, Societe de Reanimation Langue Francaise. Intensive Care Med 25:1444–1452

Karmrodt J, Markstaller K et al. (2002) Determination of different coexisting pulmonary time constants in human ARDS by dynamic CT. Intensive Care Med 28:S141

Kunst PW, Vonk Noordegraaf A et al. (1998) Influences of lung parenchyma density and thoracic fluid on ventilatory EIT measurements. Physiol Meas 19:27–34

Kunst PW, Bohm SH et al. (2000a) Regional pressure volume curves by electrical impedance tomography in a model of acute lung injury. Crit Care Med 28:178–183

Kunst PW, Vazquez de Anda G et al. (2000b) Monitoring of recruitment and derecruitment by electrical impedance tomography in a model of acute lung injury. Crit Care Med 28:3891–3895

Lu Q, Malbouisson LM et al. (2001) Assessment of PEEP-induced reopening of collapsed lung regions in acute lung injury: are one or three CT sections representative of the entire lung? Intensive Care Med 27:1504–1510

Lumb A (2000) Nunn's applied respiratory physiology. Butterworth Heinemann, Oxford

Markstaller K, Kauczor HU et al. (1999) Multi-rotation CT during continuous ventilation: comparison of different density areas in healthy lungs and in the ARDS lavage model. Rofo Fortschr Geb Rontgenstr Neuen Bildgeb Verfahr 170:575–80

Markstaller K, Arnold M et al. (2001a) A software tool for automatic image-based ventilation analysis using dynamic chest CT-scanning in healthy and in ARDS lungs. Rofo Fortschr Geb Rontgenstr Neuen Bildgeb Verfahr 173: 830–835

Markstaller K, Eberle B et al. (2001b) Temporal dynamics of lung aeration determined by dynamic CT in a porcine model of ARDS. Br J Anaesth 87:459–468

Markstaller K, Kauczor HU et al. (2003) Lung density distribution in dynamic CT correlates with oxygenation in ventilated pigs with lavage ARDS. Br J Anaesth in press

Maunder RJ, Shuman WP et al. (1986) Preservation of normal lung regions in the adult respiratory distress syndrome. Analysis by computed tomography. JAMA 255:2463–2465

Mull RT (1984) Mass estimates by computed tomography: physical density from CT numbers. AJR Am J Roentgenol 143:1101–1104

Neumann P, Berglund JE et al. (1998a) Dynamics of lung collapse and recruitment during prolonged breathing in porcine lung injury. J Appl Physiol 85:1533–1543

Neumann P, Berglund JE et al. (1998b) Effect of different pressure levels on the dynamics of lung collapse and recruitment in oleic acid-induced lung injury. Am J Respir Crit Care Med 158:1636–1643

Pelosi P, Goldner M et al. (2001) Recruitment and derecruitment during acute respiratory failure: an experimental study. Am J Respir Crit Care Med 164:122–130

Puybasset L, Cluzel P et al. (2000) Regional distribution of gas and tissue in acute respiratory distress syndrome. I. Consequences for lung morphology. CT Scan ARDS Study Group. Intensive Care Med 26:857–869

Rommelsheim K, Lackner K et al. (1983) Respiratory distress syndrome of the adult in the computer tomograph. Anasthesiol Intensivther Notfallmed 18:59–64

Rothen HU, Neumann P et al. (1999) Dynamics of re-expansion of atelectasis during general anaesthesia. Br J Anaesth 82:551–556

Rouby JJ, Puybasset L et al. (2003) Acute respiratory distress syndrome: lessons from computed tomography of the whole lung. Crit Care Med 31 [Suppl]:S285–S295

Slutsky AS, Ranieri VM (2000) Mechanical ventilation: lessons from the ARDSNet trial. Respir Res 1:73–77

Tagliabue M, Casella TC et al. (1994) CT and chest radiography in the evaluation of adult respiratory distress syndrome. Acta Radiol 35:230–234

Tagliabue P, Giannatelli F et al. (1998) Lung CT scan in ARDS: are three sections representative of the entire lung? Intensive Care Med 24 [Suppl 1]:93

The Acute Respiratory Distress Syndrome Network (2000) Ventilation with lower tidal volumes as compared with traditional tidal volumes for acute lung injury and the acute respiratory distress syndrome. N Engl J Med 342:1301–1308

Tokics L, Hedenstierna G et al. (1987) Lung collapse and gas exchange during general anesthesia: effects of spontaneous breathing, muscle paralysis, and positive end-expiratory pressure. Anesthesiology 66:157–167

Tokics L, Hedenstierna G et al. (1996) V/Q distribution and correlation to atelectasis in anesthetized paralyzed humans. J Appl Physiol 81:1822–1833

Vieira SR, Puybasset L et al. (1998) A lung computed tomographic assessment of positive end-expiratory pressure-induced lung overdistension. Am J Respir Crit Care Med 158:1571–1577

Subject Index

List of Contributors

ALEXANDER A. BANKIER, MD
Department of Radiology
University of Vienna
Währinger Gürtel 18–20
1090 Wien
Austria

CATHERINE BEIGELMANN-AUBRY, MD
Service de Radiologie
Hôpital de la Pitié-Salpêtrière
47-83, boulevard de l'Hôpital
75651 Paris Cedex 13
France

ANDRÉ CAPDEROU, MD
Hôpital Marie Lannelongue
133 Avenue de la Résistance
92350 Plessis Robinson
France

PHILIPPE CLUZEL, PhD
Service de Radiologie Polyvalente Diagnostique
et Interventionelle
Groupe Hospitalier Pitié-Salpetrière
47-83 Bd de l'Hopital
75651 Paris Cedex 13
France

SUJAL R. DESAI, MD, MRCP, FRCR
Department of Radiology
King's College Hospital
Denmark Hill
London SE5 9RS
UK

BALTHASAR EBERLE, MD
Department of Anesthesiology
University of Bern and Inselspital Bern
3010 Bern
Switzerland

CATALIN FETITA, PhD
Department ARTEMIS
Institut National des Télécommunications
9, rue Charles Fourier
91011 Evry Cedex
France

CHRISTIAN FINK, MD
Department of Radiology
Innovative Krebsdiagnostik und Therapie
Deutsches Krebsforschungszentrum (DKFZ)
Im Neuenheimer Feld 280
69120 Heidelberg
Germany

PIÈRRE ALAIN GEVENOIS, MD
Université Libre des Bruxelles – Hopital Erasme
Route de Lennik 808
1070 Bruxelles
Belgium

PHILIPPE A. GRENIER, MD
Service de Radiologie
Hôpital de la Pitié-Salpêtrière
47-83, boulevard de l'Hôpital
75651 Paris Cedex 13
France

DAVID M. HANSELL, MD, FRCP, FRCR
Professor of Thoracic Imaging
Department of Radiology
Royal Brompton National
Sydney Street
London SW3 6NP
UK

HIROTO HATABU, MD, PhD
Department of Radiology
Beth Israel Deaconess Medical Center
Harvard Medical School
330 Brookline Avenue
Boston, MA 02215
USA

PETER HERZOG, MD
Department of Radiology
University Hospital Grosshadern
University Munich
Marchioninistr. 15
81377 Munich
Germany

HANS-ULRICH KAUCZOR, MD
Professor of Radiology
Innovative Krebsdiagnostik und Therapie
Deutsches Krebsforschungszentrum (DKFZ)
Im Neuenheimer Feld 280
69120 Heidelberg
Germany

Klaus Markstaller, MD
Department of Anesthesiology
Johannes Gutenberg University Medical School
Langenbeckstrasse 1
55131 Mainz
Germany

Matthias U. Niethammer, PhD
Siemens Medical Solutions
Computed Tomography
Siemensstrasse 1
91301 Forchheim
Germany

Françoise Preteux, PhD
Department ARTEMIS
Institut National des Télécommunications
9, rue Charles Fourier
91011 Evry Cedex
France

Stefan Schaller, PhD
Siemens Medical Solutions, CT Concepts
Siemensstr. 1
91301 Forchheim
Germany

U. Joseph Schoepf, MD
Department of Radiology
Brigham and Women's Hospital
Harvard Medical School
75 Francis Street
Boston, MA 02115
USA

Christian Straus, MD
Service Central d'Explorations Fonctionnelles Respiratoires
Groupe Hospitalier Pitié Salpêtrière
AP-HP
47-83 boulevard de l'Hôpital
75651 Paris Cedex 13
France

Andrew Swift, MD
Academic Unit of Radiology
Floor C, Royal Hallamshire Hospital
Glossop Road
Sheffield S10 2JF
UK

Hidemasa Uematsu, MD, PhD
Department of Radiology
Fukui Medical University
Fujui
Japan

Edwin J. R. van Beek, MD, PhD, FRCR
Senior Lecturer/Honorary Consultant Radiology
Academic Unit of Radiology
Floor C, Royal Hallamshire Hospital
Glossop Road
Sheffield S10 2JF
UK

Johny Verschakelen, MD
Department of Radiology
U. Z. Gasthuisberg
Herestraat 49
3000 Leuven
Belgium

Jim M. Wild, MD
Academic Unit of Radiology
Floor C, Royal Hallamshire Hospital
Glossop Road
Sheffield S10 2JF
UK

Joachim Ernst Wildberger, MD
Department of Diagnostic Radiology
University Hospital
RWTH Aachen
Pauwelsstrasse 30
52074 Aachen
Germany

Marc Zelter, MD
Service Central d'Explorations Fonctionnelles Respiratoires
Groupe Hospitalier Pitié Salpêtrière
AP-HP and Université Paris VI
47-83 boulevard de l'Hôpital
75651 Paris Cedex 13
France

MEDICAL RADIOLOGY Diagnostic Imaging and Radiation Oncology

Titles in the series already published

DIAGNOSTIC IMAGING

IInnovations in Diagnostic Imaging
Edited by J. H. Anderson

**Radiology of the
Upper Urinary Tract**
Edited by E. K. Lang

**The Thymus - Diagnostic Imaging,
Functions, and Pathologic Anatomy**
Edited by E. Walter, E. Willich,
and W. R. Webb

Interventional Neuroradiology
Edited by A. Valavanis

Radiology of the Pancreas
Edited by A. L. Baert,
co-edited by G. Delorme

**Radiology of the
Lower Urinary Tract**
Edited by E. K. Lang

Magnetic Resonance Angiography
Edited by I. P. Arlart,
G. M. Bongartz, and G. Marchal

**Contrast-Enhanced MRI
of the Breast**
S. Heywang-Köbrunner
and R. Beck

Spiral CT of the Chest
Edited by M. Rémy-Jardin
and J. Rémy

**Radiological Diagnosis
of Breast Diseases**
Edited by M. Friedrich
and E.A. Sickles

Radiology of the Trauma
Edited by M. Heller and A. Fink

Biliary Tract Radiology
Edited by P. Rossi

**Radiological Imaging
of Sports Injuries**
Edited by C. Masciocchi

**Modern Imaging
of the Alimentary Tube**
Edited by A. R. Margulis

**Diagnosis and Therapy
of Spinal Tumors**
Edited by P. R. Algra, J. Valk,
and J. J. Heimans

**Interventional Magnetic
Resonance Imaging**
Edited by J. F. Debatin and G. Adam

Abdominal and Pelvic MRI
Edited by A. Heuck and M. Reiser

**Orthopedic Imaging
Techniques and Applications**
Edited by A. M. Davies
and H. Pettersson

**Radiology of the
Female Pelvic Organs**
Edited by E. K. Lang

**Magnetic Resonance of the Heart
and Great Vessels
Clinical Applications**
Edited by J. Bogaert, A.J. Duerinckx,
and F. E. Rademakers

Modern Head and Neck Imaging
Edited by S. K. Mukherji
and J. A. Castelijns

**Radiological Imaging
of Endocrine Diseases**
Edited by J. N. Bruneton
in collaboration with B. Padovani
and M.-Y. Mourou

Trends in Contrast Media
Edited by H. S. Thomsen,
R. N. Muller, and R. F. Mattrey

Functional MRI
Edited by C. T. W. Moonen
and P. A. Bandettini

Radiology of the Pancreas
2nd Revised Edition
Edited by A. L. Baert
Co-edited by G. Delorme
and L. Van Hoe

Emergency Pediatric Radiology
Edited by H. Carty

Spiral CT of the Abdomen
Edited by F. Terrier,
M. Grossholz, and C. D. Becker

**Liver Malignancies
Diagnostic and Interventional
Radiology**
Edited by C. Bartolozzi
and R. Lencioni

Medical Imaging of the Spleen
Edited by A. M. De Schepper
and F. Vanhoenacker

**Radiology of
Peripheral Vascular Diseases**
Edited by E. Zeitler

Diagnostic Nuclear Medicine
Edited by C. Schiepers

**Radiology of Blunt Trauma
of the Chest**
P. Schnyder and M. Wintermark

**Portal Hypertension
Diagnostic Imaging-Guided
Therapy**
Edited by P. Rossi
Co-edited by P. Ricci and L. Broglia

**Recent Advances in Diagnostic Neu-
roradiology**
Edited by Ph. Demaerel

**Virtual Endoscopy
and Related 3D Techniques**
Edited by P. Rogalla, J. Terwisscha
Van Scheltinga, and B. Hamm

Multislice CT
Edited by M. F. Reiser, M. Takahashi, M. Modic, and R. Bruening

Pediatric Uroradiology
Edited by R. Fotter

**Transfontanellar Doppler Imaging in
Neonates**
A. Couture and C. Veyrac

**Radiology of AIDS
A Practical Approach**
Edited by J.W.A.J. Reeders
and P.C. Goodman

CT of the Peritoneum
Armando Rossi and Giorgio Rossi

Magnetic Resonance Angiography
2nd Revised Edition
Edited by I. P. Arlart,
G. M. Bongratz, and G. Marchal

Pediatric Chest Imaging
Edited by Javier Lucaya
and Janet L. Strife

**Applications of Sonography
in Head and Neck Pathology**
Edited by J. N. Bruneton
in collaboration with C. Raffaelli
and O. Dassonville

Imaging of the Larynx
Edited by R. Hermans

**3D Image Processing
Techniques and
Clinical Applications**
Edited by D. Caramella
and C. Bartolozzi

**Imaging of Orbital and Visual Pathway
Pathology**
Edited by W. S. Müller-Forell

Pediatric ENT Radiology
Edited by S. J. King
and A. E. Boothroyd

**Radiological Imaging
of the Small Intestine**
Edited by N. C. Gourtsoyiannis

**Imaging of the Knee
Techniques and Applications**
Edited by A. M. Davies
and V. N. Cassar-Pullicino

**Perinatal Imaging
From Ultrasound to MR Imaging**
Edited by Fred E. Avni

**Radiological Imaging
of the Neonatal Chest**
Edited by V. Donoghue

**Diagnostic and Interventional
Radiology in Liver Transplantation**
Edited by E. Bücheler, V. Nicolas,
C. E. Broelsch, X. Rogiers,
and G. Krupski

Radiology of Osteoporosis
Edited by S. Grampp

Imaging Pelvic Floor Disorders
Edited by C. I. Bartram
and J. O. L. DeLancey
Associate Editors: S. Halligan,
F. M. Kelvin, and J. Stoker

**Imaging of the Pancreas
Cystic and Rare Tumors**
Edited by C. Procassi
and A. J. Megibow

**High Resolution Sonography
of the Peripheral Nervous System**
Edited by S. Peer and G. Bodner

**Imaging of the Foot and Ankle
Techniques and Applications**
Edited by A. M. Davies,
R. W. Whitehouse,
and J. P. R. Jenkins

Radiology Imaging of the Ureter
Edited by F. Joffre, Ph. Otal, and
M. Soulie

**Imaging of the Shoulder
Techniques and Applications**
Edited by A. M. Davies and J.
Hodler

Radiology of the Petrous Bone
Edited by M. Lemmerling
and S. S. Kollias

Interventional Radiology in Cancer
Edited by A. Adam,
R. F. Dondelinger, and P. R. Mueller

**Duplex and Color Doppler Imaging of
the Venous System**
Edited by G. H. Mostbeck

Multidetector-Row CT of the Thorax
Edited by U. J. Schoepf

**Intracranial Vascular Malformations
and Aneurysms
From Diagnostic Work-Up
to Endovascular Therapy**
Edited by M. Forsting

Functional Imaging of the Chest
Edited by H.-U. Kauczor

**Radiology of the Pharynx
and the Esophagus**
Edited by O. Ekberg

**Radiological Imaging
in Hematological Malignancies**
Edited by A. Guermazi

**Imaging and Intervention in Abdomi-
nal Trauma**
Edited by R. F. Dondelinger

Springer

MEDICAL RADIOLOGY Diagnostic Imaging and Radiation Oncology

Titles in the series already published

Springer